Coverbal Synchrony in
Human-Machine Interaction

Coverbal Synchrony in Human-Machine Interaction

Editors

Matej Rojc

Faculty of Electrical Engineering and Computer Science
University of Maribor, Slovenia

and

Nick Campbell

Stokes Professor, Trinity College Dublin
The University of Dublin, Ireland

CRC Press
Taylor & Francis Group
Boca Raton London New York

CRC Press is an imprint of the
Taylor & Francis Group, an **informa** business
A SCIENCE PUBLISHERS BOOK

CRC Press
Taylor & Francis Group
6000 Broken Sound Parkway NW, Suite 300
Boca Raton, FL 33487-2742

First issued in paperback 2019

ISBN-13: 978-1-4665-9825-6 (hbk)
ISBN-13: 978-0-367-37929-2 (pbk)

Library of Congress Cataloging-in-Publication Data

Coverbal synchrony in human-machine interaction / editors, Matej Rojc, Nick Campbell.
   pages cm
  Includes bibliographical references and index.
 ISBN 978-1-4665-9825-6 (hardback)
1. Human-computer interaction. 2. Speech processing systems. 3. Nonverbal communication. 4. Gesture. 5. Affect (Psychology)--Computer simulation. 6. User interfaces (Computer systems) I. Rojc, Matej, editor of compilation. II. Campbell, Nick, editor of compilation.

  QA76.9.H85C695 2013
  004.01'9--dc23

                                                    2013023190

Visit the Taylor & Francis Web site at
http://www.taylorandfrancis.com

and the CRC Press Web site at
http://www.crcpress.com

Preface

This book provides novel insights into the research, development, and designing of advanced techniques and methods for more natural multimodal human-machine interfaces for use within desktop and pervasive computing environments. The book consists of 15 chapters structured so as to provide an in-depth investigation of novel approaches from both the theoretical and practical perspective. Humans tend to interact using several modalities and communication channels in a highly complex, yet synergetic manner. The key communication concepts are expressed by fusing the modalities and by using two or more modalities simultaneously. Whilst communicating, humans tend to visualize and personalize the information provided. We perform this either to further explain or emphasize the meaning, or to express our emotional attitude or personal view towards the discussed topic. More commonly, the natural way to communicate is to use speech in combination with hand gestures, facial expressions, gaze, and posture, sketches, drawings, and handwriting. Even with the technological advance, we should still expect no-less whilst communicating with computers and machines. The embodied conversational agents (ECAs) and speech-based dialog interfaces represent a concept of how visual, auditory, and other kinds of representative information can be combined into a flexible and more natural flow of information exchange. Such multimodal interaction systems adapt and integrate the contents and their functionalities into dialog that resembles face-to-face conversations. Whether it be human-machine interfaces for driving some application, or different ECA-based human-machine interfaces directly simulating face-to-face conversation, all natural interfaces strive to exploit and use different communication strategies in order to provide additional meaning and personal touch to the content expressed and/or discussed. The synergistic approach when using more than one modality makes the interactions with machines not only flexible, natural, and robust, but also lively and closely imitates interactions with other persons.

This book presents and describes novel concepts for advanced multimodal human-machine interfaces which can be used within different contexts, ranging from simple multimodal web-browsers (e.g. multimodal content reader), to more complex multimodal human-machine interfaces for ambient intelligent environments (e.g. supportive environments for the elderly, agent-guided household environments etc.), and within different computing environments (from pervasive computing to desktop environments). Within the described communication-concepts the authors discuss several communication strategies used to facilitate the different aspects of human-machine interaction. The systems, techniques, and methods covered in the book address several key-issues, like: how to effectively (and 'dynamically') present a dialog topic, how to lead the information exchange and how to 'guide' the conversation, how to listen and respond (affective listeners), how to exchange roles during a conversation, and how to respond (by using facial/body gestures), when the lead-role has been taken or given to the ECA, etc. Viability and credibility established through appropriate responses and by using personal-features and emotional responses is also particularly relevant for highly-natural imitations. The viability and credibility may be achieved by accurately linking the auditory and visual information representation through concepts like: analyzing and annotating relations between facial expressions, emotion, and gestures (e.g. compensation, overriding the meaning, etc.), relations between speech, personality, and gestures (e.g. when human motion is produced, tendency to perform different contexts of motion like e.g. descriptive gestures, enumeration, etc.). Within the context of OCM (Own Communication Management), the book addresses a generation of different body motion types, their tendencies to repeat their dynamics, and different relations between features of human-human communicative dialog (e.g. emotive motion, personality traits related motion, etc.). However, within the context of ICM (Interactive Communication Management), the book also addresses concepts like: active and passive listeners, different motion-related backchannels within different types of feedback, turn-taking, giving/taking speaker roles, accepting speaker roles and responses to situations when the role of speaker was forcefully taken.

To sum up, this book covers abstract behavior-description languages for animating embodied conversational agents, multimodal input processing techniques (speech-based processing, context-based processing, video-based processing/analysis/annotations), multimodal output processing techniques (animating, and driving embodied conversational agents), multimodal machine-learning methods for automatic response generation, behavior and communication management techniques, behavior adaptation to different physical, emotional, and personality traits, designing multimodal frameworks, and evaluation techniques for multimodal systems.

The editors are extremely grateful to all the authors for contributing to this book with their novel and original ideas.

May, 2013

Matej Rojc
Nick Campbell

Contents

List of Contributors

Allwood, Jens: SCCIIL Interdisciplinary Center, University of Gothenburg, Gothenburg, 412 96 Gothenburg, Sweden. E-mail: jens@ling.gu.se

André, Elisabeth: Fakultät für Angewandte Informatik, 86159 Augsburg, Universitätsstr. 6a. E-mail: andre@informatik.uni-augsburg.de

Bergmann, Kirsten: SFB 673, CITEC, Faculty of Technology, Bielefeld University, 33615 Bielefeld, Germany. E-mail: kirsten.bergmann@uni-bielefeld.de

Bertrand, Roxane: Aix-Marseille Université, CNRS, Laboratoire Parole et Langage, Aix-en-Provence, France. E-mail: roxane.bertrand@lpl-aix.fr

Bevacqua, Elisabetta: Lab-STICC—IHSEV, CERV 25, rue Claude Chappe 29280, Plouzané, France. E-mail: bevacqua@enib.fr

Brechmann, André: Special-Lab Non-Invasive Brain Imaging at the Leibniz-Institute for Neurobiology, Brenneckestr. 6, 39118 Magdeburg, Germany. E-mail: andre.brechmann@ifn-magdeburg.de

Brosch, Tobias: Institute of Neural Information Processing, University of Ulm, 89069 Ulm, Germany. E-mail: tobias.brosch@uni-ulm.de

Campbell, Nick: Stokes Professor, Trinity College Dublin, The University of Dublin, Ireland. E-mail: nick@tcd.ie

Esposito, Anna: Viale Ellittico, 34, 81100 Caserta, Italy (Second University of Naples, Department of Psychology) and Via G. Pellegrino 19 84019 - Vietri su Mare (SA), Italy (IIASS). E-mail: iiass.annaesp@tin.it; anna.esposito@unina2.it

Ferré, Gaëlle: Université de Nantes, Laboratoire de Linguistique Nantes, 44300 Nantes, France. E-mail: Gaelle.Ferre@univ-nantes.fr

Frommer, Jörg: Department of Psychosomatic Medicine and Psychotherapy, Faculty of Medicine, Otto-von-Guericke University Magdeburg, Germany, Leipziger Straße 44, 39120 Magdeburg, phone: 0391-67-14200. E-mail: joerg. frommer@med.ovgu.de

Glodek, Michael: Institute of Neural Information Processing , University of Ulm, 89069 Ulm, Germany. E-mail: michael.glodek@uni-ulm.de

Guardiola, Mathilde: Aix-Marseille Université, CNRS, Laboratoire Parole et Langage, 13100 Aix-en-Provence, France. E-mail: mathilde.guardiola@ lpl-aix.fr

Haase, Matthias: Otto-von-Guericke University, Magdeburg, Germany, Leipziger Straße 44, 39120 Magdeburg. E-mail: matthias.haase@med.ovgu. de

Hoffmann, Holger: Medical Psychology, Ulm University, 89075 Ulm, Germany. E-mail: holger.hoffmann@uni-ulm.de

Hrabal, David: Medical Psychology, Ulm University, 89075 Ulm, Germany. E-mail: david.hrabal@uni-ulm.de

Jokinen, Kristiina: University of Helsinki, Institute of Behavioural Sciences, P.O. Box 9 (Siltavuorenpenger 1 A), FI-00014 Finland. Ph. +358-(0)50 3312521, fax +358-(0)9 191 29542. E-mail: kristiina.jokinen@helsinki.fi

Jonsdottir, Gudny Ragna: Icelandic Institute for Intelligent Machines. Menntavegur 1, 101 Reykjavík, Iceland. E-mail: gudny@iiim.is

Kačič, Zdravko: Faculty of Electrical Engineering and Computer Science, University of Maribor, Smetanova ulica 17, 2000 Maribor, Slovenia. E-mail: kacic@uni-mb.si

Kessler, Henrik: University of Ulm, Medical Psychology, Frauensteige 6, 89075 Ulm, Germany, Email: henrik.kessler@uni-ulm.de

Kotzyba, Michael: Otto von Guericke University Magdeburg, Institut für Technische und Betriebliche Informationssysteme (ITI), Universitätsplatz 2, 39106 Magdeburg, Germany. E-mail: michael.kozyba@ovgu.de

Lange, Julia: Otto-von-Guericke University, Magdeburg, Germany, Leipziger Straße 44, 39120 Magdeburg. E-mail: julia.lange@med.ovgu.de

Layher, Georg: Institute of Neural Information Processing, University of Ulm, 89069 Ulm, Germany. E-mail: georg.layher@uni-ulm.de

Limbrecht, Kerstin: Medical Psychology, Ulm University, 89075 Ulm, Germany. E-mail: kerstin.limbrecht@uni-ulm.de

Lingenfelser, Florian: Fakultät für Angewandte Informatik, 86159 Augsburg, Universitätsstr. 6a. E-mail: lingenfelser@informatik.uni-augsburg.de

Mancini, Maurizio: InfoMus Lab, University of Genoa, viale Causa 13, 16145 Genoa, Italy. E-mail: maurizio.mancini@unige.it

Marsella, Stacy: University of Southern California, Institute for Creative Technologies, Los Angeles, CA 90094, USA. E-mail: marsella@ict.usc.edu

Martin, Jean-Claude: LIMSI-CNRS, Rue von Neumann, Batiment 508, 91403 Orsay Cedex, France. E-mail: martin@limsi.fr

Meudt, Sascha: Institute of Neural Information Processing, University of Ulm, 89069 Ulm, Germany. E-mail: sascha.meudt@uni-ulm.de

Mlakar, Izidor: Roboti c.s. d.o.o., Tržaška cesta 23, 2000 Maribor, Slovenia. E-mail: izidor.mlakar@revolutionary-robotics.com

Morency, Louis-Philippe: University of Southern California, Institute for Creative Technologies, Los Angeles, CA 90094, USA. E-mail: morency@ict.usc.edu

Neumann, Heiko: Institute of Neural Information Processing, University of Ulm, 89069 Ulm, Germany. E-mail: Heiko.Neumann@uni-ulm.de

Niewiadomski, Radoslaw: Institut Mines-Telecom, Telecom ParisTech, 37-39 rue Dareau 75014 Paris France. E-mail: niewiado@telecom-paristech.fr

Ohl, Frank: Department of Systems Physiology of Learning, Otto-von-Guericke University Magdeburg D-39118, Germany. E-mail: frank.ohl@lin-magdeburg.de

Palm, Günther: Institute of Neural Information Processing, University of Ulm, 89069 Ulm, Germany. E-mail: guenther.palm@uni-ulm.de

Pelachaud, Catherine: Pelachaud, Catherine: CNRS-LTCI, Telecom Paristech, 75014 Paris, France. E-mail:catherine.pelachaud@telecom-paristech.fr

Piana, Stefano: InfoMus Lab, University of Genoa, viale Causa 13, 16145 Genoa, Italy. E-mail: steto84@infomus.org

Rojc, Matej: Faculty of Electrical Engineering and Computer Science, University of Maribor, Smetanova ulica 17, 2000 Maribor, Slovenia. E-mail: matej.rojc@uni-mb.si

Rösner, Dietmar: Department of Knowledge Processing and Language Engineering, Faculty of Computer Science, Otto-von-Guericke University Magdeburg, Germany, Universitätsplatz 2, 39106 Magdeburg. phone: 0391-67-58314. E-mail: roesner@ovgu.de

Scheck, Andreas: Medical Psychology, Ulm University, 89075 Ulm, Germany. E-mail: andreas.scheck@uni-ulm.de

Schels, Martin: Institute of Neural Information Processing, University of Ulm, 89069 Ulm, Germany. E-mail: martin.schels@uni-ulm.de

Scherer, Stefan: University of Southern California, Institute for Creative Technologies, 12015 Waterfront Dr., Playa Vista, 90094, CA, USA. E-mail: scherer@ict.usc.edu

Schmidt, Miriam: Institute of Neural Information Processing, University of Ulm, 89069 Ulm, Germany. E-mail: miriam.k.schmidt@uni-ulm.de

Schwenker, Friedhelm: Institute of Neural Information Processing, University of Ulm, 89069 Ulm, Germany. E-mail: friedhelm.schwenker@uni-ulm.de

Shapiro, Ari: University of Southern California, Institute for Creative Technologies, Los Angeles, CA 90094, USA. E-mail: shapiro@ict.usc.edu

Thórisson, Kristinn R.: Icelandic Institute for Intelligent Machines and Center for Analysis & Design of Intelligent Agents, Reykjavik University, Menntavegur 1, 101 Reykjavík, Iceland. E-mail: thorisson@ru.is

Traue, Harald C.: Medical Psychology, Ulm University, 89075 Ulm, Germany. E-mail: harald.traue@uni-ulm.de

Tschechne, Stephan: Institute of Neural Information Processing, University of Ulm, 89069 Ulm, Germany. E-mail: stephan.tschechne@uni-ulm.de

Wagner, Johannes: Fakultät für Angewandte Informatik, 86159 Augsburg, Universitätsstr. 6a. E-mail: wagner@informatik.uni-augsburg.de

Walter, Steffen: Medical Psychology, Ulm University, 89075 Ulm, Germany. E-mail: steffen.walter@uni-ulm.de

CHAPTER 1

Speech Technology and Conversational Activity in Human-Machine Interaction

Nick Campbell

1. Introduction

Over the past several years, great strides have been made in both speech synthesis and speech recognition technology, but so far, comparatively little work has been carried out on the higher-level integration of these components to enable them to be used for a more naturally interactive form of conversation between people and machines (or between people by way of machines). This chapter discusses the need for a more proactive form of speech processing, presents some ideas related to the types of speech structure that might need to be modeled for facilitating more spontaneous conversational interactions, and describes some of our recent work in this direction at the Speech Communication Lab of the Centre for Language and Communication Studies at Trinity College Dublin.

Most spoken dialogue interfaces are implicitly asynchronous, as they require a text-based processing step between each utterance. This form of processing enforces a 'ping-pong' type of interaction with explicit turn-taking. The incoming speech is first recognized, or converted into text, then parsed and processed, and the resulting text output is then rendered as speech output by a synthesizer.

The implication of such a processing sequence is that each utterance will be well-formed, relatively complete in itself, and that each 'turn' will provide sufficient lexical content for an intact parse to be obtained. We are not concerned here with details of components that process the propositional content of each utterance to provide a response, but instead with the nature of the interaction per se and examine the so-called 'ill-formedness' of individual utterances in spontaneous conversational speech. Current technology does not perform well when processing normal everyday speech because it often fails to cope with its highly fragmented nature. We will look next at some implications of this for interactive speech processing.

Human spoken interaction in its everyday form is not essentially an asynchronous process, and features considerable fragmentation or overlapping of syntactically ill-formed utterances, along with regular backchannel noises, interruptions and mutual completions of the other partner's utterance as interlocutors interact (Goffman, 1961; Shegloff, 2007). Conversation is structured and balanced much like a dance, with 'listening' being an active and contributive process, i.e. not just passively receiving information but actively collaborating in the production of meaning. Participation in a conversation requires constant mutual monitoring of attention as the themes are conjointly expounded and mutually developed (Chapple, 1939; Kendon, 1990). Speech has evolved to facilitate this mutual structuring which by definition is lacking in written text, which in turn evolved considerably later to perform a significantly different function.

The following example illustrates the difference in style between written text and a spoken form of interaction. The content appears to be highly fragmented, and sentence boundaries are unclear (if the concept of 'sentence' has any meaning at all in spoken sequences). It rambles but in a natural way, easy for us to follow if we think of it as sounds that we are listening to (we can even 'hear' the tone-of-voice change perhaps), but this content would be very difficult for a machine to parse. This sample is from a monologue, from the transcription of a prepared public talk by a top-ranking newspaper journalist, Seymour Hersh, who writes for the New York Times and New Yorker. It is part of his keynote speech to the ACLU in 2004 (from http://www.informationclearinghouse.info/article6492.htm):

"You know, | we all know the story | of how mad they got at General Shinseki, | who I think is going to run for the Senate in Hawaii | and should, | for Inouye's seat, | he's a great general. | The important thing about Shinseki | for me, | and this is just heuristic, | I don't know this, | the important thing about Shinseki is this. | He testifies before the Gulf War | we're going to need a couple hundred thousand troops | and everybody,

| *Wolfowitz and the others —* | *I count Wolfowitz,* | *I lead with him,* | *because he's sort of the,* | *he's the genius in the background,* |*he's the man,* | *very articulate,* | *very persuasive —* | *and so Shinseki testifies* | *we need a couple hundred thousand* | *and everybody's mad at him,* | *it's about two weeks before the war,* | *and it made sense,* | *everybody said,* | *they were mad* | *because he's talking about numbers* | *these guys say you won't need.* | *They're going to go invade Iraq* | *and you know the story,* | *they were going to be greeted with flowers* | *and all that stuff,* | *we all know that story.* | *But it wasn't that.* | *Their complaint with Shinseki was really much more interesting.* | *It was:* | *didn't he get it?* | *Didn't he know what we've been talking about,* | *in the tank with the JCS* | *and the generals —* | *didn't he get it?* | *We could do it with five thousand troops,* | *we have to make these bargains* | *with these crazy Clintonized generals — | I'm talking like Rummy,* | *like Rumsfeld would talk —* | *literally,* | *unfortunately —* | *these soft generals,* | *these Clintonized generals —* | *didn't Shinseki get it?* | *Didn't he understand what we're doing here?* | *We did it in Afghanistan,* | *we're going to do it in Iraq.* | *Some Special Forces,* | *some bombing,* | *we're going to take it over.* | *It's going to be like this.* | *He didn't get it,* | *that was the problem,* | *that's why they had to read him out.* | *He wasn't on the team.* | *And so you have a government that basically has been operating since 9/11 very successfully on the principle that if you're with us you're a genius, if you're against us you're not just somebody [in the] loyal opposition, you're a traitor. They can't deal with you. I'm exaggerating very slightly."* [The bars ('|') have been inserted to mark a chunk boundary.]

This is the talk of a highly educated man addressing an educated audience, but it is very fragmented and quite different in form from what he would write in an article for publication. It reads more like a sequence of almost random thoughts, or small disjointed chunks of meaning, which when strung together begin to create a coherent image. It would take on a very different appearance if printed on a page, yet we argue here that this is *not* ill-formed. It is an excellent example of how information is best transmitted through speech to a present audience. The sequence is optimized to allow feedback after each small chunk (or 'niblet' as we shall propose later) of meaning.

When reading through text on a page, the eye can scan back and forth to reconstruct the author's intended meaning, but with one-dimensional information such as speech, the structure of its content must be shown in real-time through chunking and prosody. The structure of this monologue is optimized for the ear. It reflects the forms used by ordinary people when they speak in everyday conversation, and although it is not ungrammatical per se, it employs a different form of grammar from written text. The speaker is adjusting his phrasing, and his timing, to keep the audience in maximum contact; allowing them to parse and perceive the impact of each small chunk before moving on to the next. He is talking not 'to', but 'with' his audience, repeating chunks where necessary, and constantly testing their comprehension to maintain their attention.

Our task is to provide machines with a similar faculty—to process speech interactively, taking into account the cognition of the listener.

2. Even Dogs and Young Babies Can Do It!

The human is a socially organized animal, and we are unique among animals in spending so much time rearing our young. Our infants are helpless and dependent on a carer for longer than any other animal, but in turn they spend long hours watching and (more often) listening to the people around them, and consequently they learn the norms of human behavioral patterns very early. They become familiar with the patterns of speech sounds and rhythms of spoken interaction from even before birth, as the sounds of the mother's speech are carried into the womb to the hearing infant along with her blood (with its varying adrenaline levels) as she goes about her daily conversational activities (Karmiloff and Karmiloff-Smith, 2001).

The Hungarian ethologist Ádám Miklósi has shown that the feature which most differentiates dogs from wolves, their nearest animal neighbor, is that only the former have really learnt to watch and learn from human behavior. In this way, dogs become companions to the humans who host and care for them, while wolves lack the capacity for such serendipitous coexistence (Miklósi, 2008). The capacity to observe, to interpret behavior and also to empathize is perhaps what underlies the mechanisms of companionship that are fundamental to all social relationships.

A key feature of human communication is that we have learnt to express propositional content alongside social information simultaneously. From earliest times, we have watched our fellow beings and learnt to read information about their cognitive states from their behavior (Dunbar, 1998). We know whether or not they are listening, and paying attention, and make continuous estimates about their levels of comprehension as we speak. We unconsciously structure our speech to facilitate this process.

Our speech processing technologies, however, presently lack any such notion of empathy. They also lack the ability to observe the effects of their actions on others. People consequently feel uneasy with much of present speech technology, and this may be hampering its acceptance in spite of what is now a high level of technical competence in both synthesis and recognition components.

3. Use of 'Social Prosody' in a Conversation

Much of the social or interpersonal information in speech is carried by the prosody and signaled by changes in the intonation, loudness, rhythm, and tone-of-voice of the speaker. It is also carried by the numerous backchannel utterances that intersperse a conversation to show listener feedback. In our JST/ESP corpus of 1500 hours of everyday conversational interaction (www.speech-data.jp), these short nonverbal interjections accounted for more than half of the total number of utterances. These words, like 'ah' and 'um', 'yeah' and 'yeah yeah yeah', are characterized both by phonetic simplicity and prosodic complexity, perhaps serving principally to carry tone of voice information simultaneously signaling both speaker affect and relation to the interlocutor (Campbell and Mokhtari, 2003).

The study of speech prosody has a long history, but much of the science to date has focused on the relations between the intonation of syntactic elements in a sentence—i.e. on linguistic content. More recently, however, the social aspects of spoken interaction have begun to be studied from a prosodic point of view, and it has been shown that prosody functions not only to signal the structure and relationships of morphological, syntactic, and semantic aspects of propositional content but also simultaneously serves as a messenger for affective and cognitive information related to speaker participation status in a discourse and inter-participant relationships (Campbell, 2007).

Traditional studies have been based on read speech. Read speech and broadcast speech stand in contrast to conversational speech in that they function primarily to convey text-based information to an audience that is largely passive. They are one-way processes, as is present-day dialogue technology. In the case of radio broadcasts, for example, no real-time feedback from the audience is even possible and the speaker has no need to take any observable cognitive states of the listener into account when rendering text as speech. The task is simply to render the content intelligibly (content that was originally created as text, and which through its complexity presents a particular prosodic challenge to the broadcaster, who is therefore usually a highly trained performer). In the case of a public lecture, however, the audience may be visible, while passively listening with no right to speak, but an effective presenter will take into account such cues as small head movements and facial expression changes that signal understanding, and adjust the pace and content of the lecture accordingly. In a typical face-to-face conversation, where all participants have equal right to speak, this two-way interaction becomes intense.

Previous work based on the JST/ESP 1500-hour corpus of highly interactive natural speech recorded through head-worn microphones by volunteers in everyday situations has revealed significant changes, particularly in voice quality and tone-of-voice, depending on the nature of the relationship with the interlocutor and the various stages of the interaction. We generalize from this to infer that in daily interactive speech social relationship information is encoded into every utterance as part of the vocal setting and that prosodic variation (which includes voice quality) serves not only to carry linguistic and grammatical information but also to display social and cognitive states, and to signal the changing social relationships between the participants. This is what human listeners are accustomed to processing and what is missing from the stream of information in computer-synthesized speech.

Speakers in an interactive conversation are accustomed to constantly monitor the attentive and cognitive states of the listener (if such a term may still be used for the active co-participant) and to adjust their speech behavior accordingly. In addition to processing the propositional or lexical content of each utterance, the conversation participants also employ a different 'grammar' to process social information carried alongside that linguistic information by the prosody (Campbell, 2008a, 2008b).

4. 'Niblets' and Conversational Fragments

In this section, we look more closely at the issue of fragmentation in interactive speech and introduce the concept of *niblets*, a new term proposed for use in speech processing to describe individual fragments of meaning. We use an example related to *Hi-Fi* audio that illustrates a change in lifestyle and customs that might characterize a feature of present-day conversational speech which poses a challenge to be addressed by future technology.

Whereas the previous (before the web) generation had a passion for *Hi-Fi* audio, often spending large amounts of money to buy the 'best' amplifiers and speakers on which to listen to their favorite record albums, the younger generation now takes its music through lower-quality 'ear-buds', buying or downloading only a single-compressed track at a time. This trend reflects not only fashion changes and advances in technology, but also a move away from 'quality' toward 'convenience'. The present generation are still able to buy CD-quality versions of the compressed music they download, but typically they do not—the marketing model has also changed—and multiple low-price medium-quality purchases, each well below the threshold of pain, have replaced the single high-value high-quality purchase model of the past.

This change in lifestyle has parallels in speech processing, where multiple low-quality utterances might be found more often than well-formed utterances that are closer to text in form. Anyone who has spent time reading through transcriptions of spontaneous conversations will recognize that very few of the chunks of speech form well-formed sentences (or even well-formed phrases) and that this fragmentation is a defining characteristic of spontaneous interactive two-way social communication.

In considering improved technologies for a future generation of spoken dialogue systems, we propose therefore that instead of one high-quality *'speech* → *text* → *speech'* transfer, we might prefer a sequence of multiple low-complexity, low-quality, high frequency 'niblets'. For example, the sequence——*"you ... me ... friends ... okay"*——works well as a functional utterance in the real world despite its obvious fragmentation (consisting of four 'niblets'), and its ungrammaticality. In this example, the word 'friends' might be replaced by the longer phrase *'work well together'* without affecting the niblet count (the processing load). This chunking and collocation of speech fragments closely resembles the broken and interrupted forms that are common in interactive and conversational speech, as illustrated by the Seymour Hersh extract above.

To explain why this type of supposedly ill-formed utterance processing might be beneficial to future dialogue systems, consider for example the practical case of the user-manual that comes with a digital camera purchase, it may contain several hundreds of pages of information, often in several languages, but at any given time the user typically only needs to access one small part of it. If the manual is available online, or provided digitally as part of the camera, and a speech-enabled graphical interface is to provide the relevant information, then (a) the request *"How do I turn off the flash?"* might be rendered more efficiently in niblets by the customer as (b) *"nikon ... s6000 ... flash ... off"*, and the response from the system is a combination of an image, an animated gesture, and a few niblets of its own: (c) *"push here ... top part"* (two niblets) delivered as speech through a multimodal interface.

The key point being made here is that rather than a well-formed, grammatical text-like utterance (a), the fragmented version (b) carries equivalent information in a more efficient (if less elegant) manner, and that the response (c) involving multimodal information is both fast and effective. The recognition load is also simplified.

5. Multimodal Discourse and Dialogue Tracking

Human beings have proven to be remarkably efficient at using every modality available when communicating, and dialogue has naturally evolved as a multimodal activity. It is a logical consequence that speech technology should evolve in the direction of multimodality. Whereas the telephone is evidence that speech alone can function in isolation without any need to see the interlocutor, it is clear that most people do take advantage of the visual channel when talking to each other. People even gesture when talking on the telephone!

Cameras are now ubiquitous, and computer memory and CPU speeds fast enough to enable real-time video and speech processing on a phone, a tablet or a notebook computer, as evidenced by Siri and Google Talk, so we should now begin to model the integration of these several data streams for dialogue processing. We will look next at the types of information that can be relatively simply extracted by a combination of camera and microphone and examine how they might be used to process speech interactivity.

Figure 1 plots six minutes of audiovisual data extracted from a 30-minute conversation between four people sitting round a table casually chatting. The figure shows both camera and microphone data aligned, with speech activity represented as bars of color, overlayed with graphs displaying in this case vertical head movement for each participant (Campbell and Douxchamps, 2007). Each row of the plot shows a 60-second activity, taken from the 5th to the 10th minutes of Day One in the FreeTalk corpus (SSPNet 2010 - FreeTalk).

The plot allows us to visualize the discourse activity without knowing exactly what is being said, which puts us in the position of a machine faced with the task of processing human dialogue. However, it is clear from the plot that much of the activity can be parsed without any knowledge of the actual underlying speech. From such an estimation of the function of each utterance in the overall activity, we can make a better guess at how the dialogue should be processed. For example, laughter or backchannel utterances might be better treated by prosodic or voice quality analysis alone, while longer (propositional?) utterance sequences would need to be processed by speech recognition and semantic analysis. A glance at the figure above shows that these two types are readily distinguishable within the overall structure, as is the dynamics of the discourse.

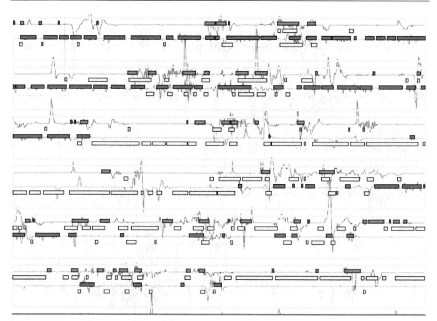

Figure 1. Traces from a multimodal conversation of four participants (color-coded to show speaker id), with the solid horizontal bars indicating speech activity and the lines plotting vertical head movement. Each row represents data from one minute of speech.

(Color image of this figure appears in the color plate section at the end of the book.)

We see that while a single speaker appears to dominate in all but one one-minute slice of the action, the other 'listeners' are also active throughout. For the first two minutes, speaker 1 (voice activity coded by grey bars) dominates, and for the third and fourth minutes, speaker 2 (coded by green) dominates, to be replaced by speaker 3 (coded by yellow) for the remaining two minutes. It is clear, however, that no one in the conversation remains passive during these times. Every row shows overlapping activity, and the majority show all participants offering brief contributions throughout. Row 5 is more fragmented and this particular pattern of joint activity (representing laughter) is frequently observed throughout the corpus (Campbell, 2006). This high level of joint activity from all participants throughout each minute of the conversation reflects the social nature of the interaction, which is not well matched by the strict turn-taking and text-based articulation expected in many current automated spoken dialogue systems. This is multi-party interaction, and two-party dialogue is a special case of this wider norm but similar principles apply: the listener is expected to take an active role in the dialog, not just answering when questioned, but contributing to the discourse throughout. Figure 2 plots data that

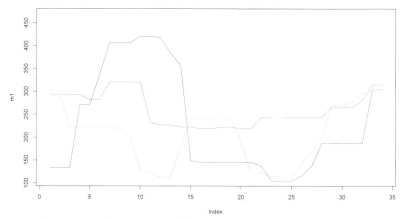

Figure 2. Traces of speech activity for all four participants in Figure 1, showing average head movement for each minute of the 30-minute conversation (units represent pixel shift). Note only a brief low from two members at the start and again between minutes 23 and 25, otherwise the majority are active at all times, with two peaks—the first at around 10 minutes and the second between 15 and 20 minutes.

(Color image of this figure appears in the color plate section at the end of the book.)

confirm a high level of joint speech activity throughout the entire conversation.

Figure 2 plots averaged head movement, for each minute from the camera data illustrated in Figure 1, for each of the four participants across the whole conversation. Distinct peaks of high activity reveal those times when a given speaker is particularly dominant in a conversation, but it is noticeable that with the exception of two brief periods, all participants remain highly active throughout, taking a contributory role in the whole of the conversation. All participants rise to an almost equal level of activity towards the end of the recording when the session comes to a close.

Table 1 shows the correlations between measures of the head activity of the four speakers shown in Figures 1 and 2. The synchrony is confirmed to be highest for participant 3, whereas participant 2

Table 1. Correlations of head activity between all participants (from summed data for each of the 30 minutes). This tabulates the data shown in Figure 2.

Speaker	1	2	3	4
1 (red)	-	0.107	0.335	0.207
2 (blue)	0.107	-	0.117	0.094
3 (green)	0.335	0.117	-	0.353
4 (brown)	0.207	0.094	0.353	-

appears not to be so well synchronized with the others except during the first part of the conversation. Participant 1 is positive throughout, and docks well with the others. (Participant 2 was the author who, as one of the experiment participants, tried to maintain a less dominant role as the dialogue progressed; Participant 1 was a guest.)

6. Multimodal Discourse in the Real World

We carried out an experiment to implement and test the above observations in the form of an advanced dialogue system.

We installed audio and video sensors on a small robot platform. This prototype device used a LEGO NXT (Mindstorms) robot as a mobile platform for a small high-definition webcam and noise-canceling microphone array, with wired and wireless streaming of data to a nearby computer (an Apple Mac-Mini) for processing. The webcam provides the 'eyes' and the microphone serves as 'ears', with a second sound sensor to provide some feedback on background noise levels, in conjunction with an ultrasonic distance sensor to assist in maneuvering the device and to help locate and face the current speaker. The sound sensor provides secondary low-level 'hearing' for the detection of speech activity, and the ultrasonic distance gauge allows for precise positioning of the robot so that audio and webcam capture conditions can be optimized.

A Viola-Jones algorithm was used in a modified implementation of OpenCV for face detection from camera input, and movement was measured in the area of the detected faces as reported in Campbell and Douxchamps (2007). Use was made of fundamental frequency and rms amplitude in a crude speaker-detection and overlapping-speech-detection module. The video information from the face detection proved useful in matching voice activity to identified persons within the areas of vision. The combination of these two types of information appears highly effective, but we have not yet performed formal or quantitative tests to support this claim.

The interaction experiment took the form of a short conversation with visitors who came in freely off the street to view exhibits in the Science Gallery (of which this was one). The 'exhibit' was labelled as illustrating the difficulties of robot speech recognition and the viewer was challenged to test the robot by speaking to it. A speech synthesizer was activated when a person came into view of the robot and initiated a series of predetermined utterances, starting with "hello, hi!". This was repeated after a gap, and most people responded to the second greeting. It helped that they could see the robot's camera had spotted them and drawn a circle around the image of their face on a computer monitor mounted above the display.

As soon as they responded, the robot synthesized another pre-programmed utterance: "What's your name?" followed immediately by "My name's Herme", "H-E-R-M-E". To which a large number of respondents gave their own name in reply. We found that sets of three brief utterances worked best to capture a visitor and lure them into a conversation. In all, the robot persuaded almost 500 visitors to sign a consent form allowing us to make further use of their data (the entire interaction was recorded) while perhaps about twice that number walked away without signing (but who were recorded anyway as part of the corpus of human-machine interaction that we are now working with). Such triples as "Why are you here today", "Really?", "Oh", with suitable gaps, timed according to interlocutor reaction, and "Tell me a joke", "Tell me a funny joke", "Ha ha, he he he", were particularly effective in evoking a response.

The robot spoke in a childlike voice and paid no attention to any responses from the visitor except to trigger the timing of the next utterance in its series. This response time is critical. With a remote human triggering the responses by monitoring the conversation from the lab through Skype, we achieved the majority of sign-offs, but with an automatic system using camera and audio input alone we achieved the most failures. This is to be expected; the intelligence required for even this simple form of dialogue speech processing is enormous, but it turned out that even with such a simple device, we were able to maintain conversations lasting more than three minutes and resulting in a sophisticated response action (the signing of a consent form and reading its identification code to the robot).

The use of judicious niblets was, we suppose, the reason for this success. We triggered a social response in the interlocutor that was perhaps automatic. By engaging them in social chat, we made use of primal patterns of behavior that might have facilitated (in a more sophisticated engine) the transfer of higher-level propositional content. By mimicking the observed patterns of behavior in social chat, we were able to engage with adults coming on off the street for a significant amount of time. Over a three-month period, we obtained an average of about 10 signed consent forms per day.

7. Conclusion and Future Work

This chapter has presented ideas for the improvement of speech processing technology based on observations of human speech performance. It has proposed that the supposed ill-formedness typical of spontaneous interactive speech is not due to 'performance error' as has previously been claimed, but that it is part of a highly evolved system of human social interaction that provides a framework for the exchange of propositional content alongside the signaling of interpersonal details.

The chapter has drawn attention to the function of mutual co-creation of meaning (as expounded by Kendon and others) and shown that a judicious combination of visual and audio information can be used to provide information about the dynamics of a discourse such that deep processing can be constrained to only a relatively small subset of the actual speech that is produced.

The proposal that interactive speech is as much a social process as it is discoursal was tested by the use of a small robot synthesizer that processed response activity (rather than speech content) and was able to maintain a sustained 'conversation' with unpaid volunteer off-the-street participants using a pre-programmed dialogue script.

Being educated people, exposed to the patterns of written text since early childhood, perhaps many of us are blind to the actual nature of the conversational speech that we hear every day, and mentally transform the fragmented input into a well-formed image of what the speaker is 'saying'. Like machines, we do our best to map from the fragmented actual utterance sequence onto a well-formed text equivalent for easier parsing.

The current paradigms of computer-based spoken dialogue systems still remain firmly grounded in text, and require extensive transformations between speech and text for processing. This chapter has proposed that a different system, employing image processing alongside audio information, might be developed, and that the granularity and well-formedness of the speech chunks that it processes might then be closer to that observed in actual human spoken interaction. The term 'niblets' was proposed for the minimal units of meaning that form the basic units for processing, and the importance of prosody in the rendering of each niblet was stressed.

Prosody serves to convey structural information that helps disambiguate propositional content, but it also, and at the same time, provides information about the discoursal and social intentions of the speaker.

There might be a fundamental 'law of nature' that requires interacting humans to first negotiate on the level of companionship and then to broaden the discussion to matters of more general interest.

This empathic nature of conversational activity needs to be included in our future technology.

We have made a start, as shown in Figure 1, simply by observing who is speaking when, and how the other members of the group respond to each utterance in terms of both speech (simply presented as noise) and motion (from observing head movements), and we thereby form an image of the progress of the conversation and of the participant roles of each of the interlocutors. Perhaps this is the best that young babies and dogs can do, as they certainly cannot be expected to follow the text of any conversations they hear, and it is this level of processing that inspires our latest technology developments for dialogue tracking and processing.

This chapter has presented results from work related to multimodal recordings of multiparty conversational interactions, showing that participants typically engage positively throughout a discourse, synchronizing their speech and movements to a very high degree. It has presented the concept of niblets to account for the small and often ungrammatical fragments of speech that are characteristic of such informal conversations, and has argued that there is a strong social component related to empathy in human spoken interactions and that our present technologies for speech processing fail to take this into account. The chapter concluded by describing a small robot that has been developed for the gathering of socially relevant data in a conversational setting.

Acknowledgement

This work is being carried out as part of the FastNet project at Trinity College Dublin (The University of Dublin) with support from Science Foundation Ireland (SFI Stokes Professorship Award 07/SK/I1218).

REFERENCES

Campbell, N. and P. Mokhtari. 2003. Voice quality: The 4th prosodic dimension, in Proceedings of the 15th International Congress of Phonetic Sciences (ICPhS'03), Barcelona, Spain, pp. 2417–2420.

Campbell, Nick. 2006. "A multimedia database of meetings and informal interactions for tracking participant involvement and discourse flow", Proc. LREC 2006, Lisbon.

Campbell, Nick. 2007. "On the Use of Nonverbal Speech Sounds in Human Communication", pp. 117–128 in *Verbal and Nonverbal Communication Behaviors*, LNAI, Vol. 4775.

Campbell, Nick. 2008a. "Individual traits of speaking style and speech rhythm in a spoken discourse", pp. 107–120 in Esposito, A. et al. (eds.). *Verbal and Nonverbal Features of HH and HM Interaction*, Springer-Verlag, Berlin.

Campbell, Nick. 2008b. "Multimodal Processing of Discourse Information: The Effect of Synchrony", ISUC, pp. 12–15, 2008 Second International Symposium on Universal Communication, Osaka, Japan.

Campbell, Nick and Damien Douxchamps. 2007. "Robust real time face tracking for the analysis of human behavior", pp. 1–15, in *Machine Learning and Multimodal Interaction*, Springer's LNCS series, Vol. 4892.

Chapple, Eliot D. 1939. "Quantitative analysis of the interaction of individuals", pp. 58–67 in Proceedings of the National Academy of Science USA. February; 25(2).

Dunbar, Robin. 1998. Grooming, Gossip and the Evolution of Language. Dunbar: Cambridge, Mass. Harvard University Press.

Focus on Actions in Social Talk; Network-enabling Technology an SFI-funded project at TCD. (http://netsoc.tcd.ie/~fastnet/)

Goffman, Erving. 1961. Encounters. Indianapolis: Bobbs-Merrill.

JST/CREST Expressive Speech Processing project, introductory web pages at: http://www.speech-data.jp/

Kendon, Adam. 1990. Conducting Interaction: Patterns of Behaviour in Focused Encounters. Cambridge: Cambridge University Press.

Karmiloff, Kyra and Annette Karmiloff-Smith. 2001. Pathways to Language: From Fetus to Adolescent. Karmiloff: Cambridge, Mass. Harvard University Press.

Miklósi, Ádám. 2008. "Dog Behaviour, Evolution, and Cognition", project website at http://familydogproject.elte.hu/

Schegloff, Emanuel A. 2007. Sequence Organization in Interaction: A Primer in Conversation Analysis, Volume 1, Cambridge: Cambridge University Press.

SSPNet: The Social Signal Processing site is hosting FreeTalk for the download of large files (http://www.sspnet.eu/). This corpus material can also be viewed interactively with subtitles at http://www.speech-ata.jp/taba/nov07/flash/day1_flash_hmy.html. The software for displaying equivalent data in the same way is available for download from the same site.

CHAPTER 2

A Framework for Studying Human Multimodal Communication

Jens Allwood

1. Introduction

This chapter provides an overview of some of the most important functions and processes involved in human face-to-face communication. Special attention is given to multimodal communication. It also gives an overview of the most important factors (e.g. social activity, personality and national-ethnic culture) that influence human communication. The chapter provides definitions for communication, language and culture. It discusses the contents of communication and the dynamics of dialog and presents a model of embodied communication, involving several levels of awareness and intentionality. Finally, there is a discussion of the relation between face-to-face communication and communication technology.

2. A Challenge

Due to greater requirement of communication technology in our lives, more issues of human-human communication are being studied now, than ever before.

Keeping in mind that there is still a lot we do not know about human-human communication, the goal of this chapter is to provide an overview of features in face-to-face communication, which need to be taken into account in constructing both computer-based systems for communication between humans and systems for communication between humans and robots or other artificial agents. In this way, we hope to support the cooperation between computer scientists, engineers, signal processors and communication researchers having a background in the social sciences or linguistics.

3. Nature, Culture, Communication, Cognition and Language

One of the most discussed issues in studying communication concerns the interplay between Nature and culture. What in communication is due to Nature and what is due to culture? Basically, culture is always the result of cultivation of Nature. It is cultivated Nature. Culture, thus, always has a natural foundation, but involves human shaping of naturally given physical, behavioral and cognitive resources. We have natural genetic predispositions for cognition, social bonding, communication and language and through socialization (which is a kind of cultivation) in particular cultures and social communities, we acquire culture-specific, convention-regulated ways of social bonding, thinking, communication and language.

Since our focus in this book is on communication, we start by turning to the question: What is communication? An answer to this question is provided by the following definition: Communication = sharing of information, cognitive content or understanding with varying degrees of awareness and intentionality. For a different definition of communication, Shannon and Weaver (1949) and for a critical discussion of this, Reddy (1979).

Thus, we can say that: A and B communicate if and only if A and B share a cognitive content as a result of A's influencing B's perception, understanding and interpretation and B's influencing A's perception, understanding and interpretation. The influence is mediated through their action and behavior or by the results of their action and behavior, e.g. texts or paintings (Allwood, 2008b.)

It is here important to note that a person can be informative to another person unintentionally, e.g. when the color of their hair or pitch of their voice provides information about their age or sex, or when blushing gives information about their emotional arousal. Similarly, also perception can be unaware; you could, for example, be influenced by another person without noticing it, as when the person's larger pupil size signals interest and this subconsciously is interpreted as a signal of friendliness by you. In this way, both you and your interlocutor could influence each other without being aware of it (Allwood, 2002).

A wide notion of communication will include such cases, i.e. unaware and unintentional sharing of information, while a more narrow notion might, for example, require that communication must always be intentional and/or aware.

Two important characteristics of human beings are that they can be, and mostly are, social and often rational. The primary means for sociality and rationality is communication. We can incorporate a little more of this in our definition of communication in the following way:

Communication = sharing of old or new factual, emotive and conative aspects of cognition through co-activation and co-construction of content, information or understanding, occurring as a part of and means for joint social activities involving degrees of coordination in a way which is often multimodal and interactive. This definition incorporates sociality by stressing coordination and collaboration and rationality through the reference to goal-directed activity.

We should also note that even if face-to-face communication is always multimodal and interactive, there are other forms of communication, which are less multimodal, e.g. telephone conversations or exchange of written information (SMS, e-mail, letters, chat) and forms of communication which are less interactive, like reading a book, watching TV or listening to the radio.

Similarly, communication is not always collaborative and cooperative. We communicate also when we are quarreling or compete. In fact, communication is often essential in carrying out both conflict and competition.

It might here be interesting to compare the use of the term *communication* with use of the term *dialog*, which even if sometimes also used in the general sense we have used *communication*, instead often is used in a more restricted sense for non-competitive and non-conflictual communication: "We want dialog, not conflict". To avoid this more narrow interpretation, we will here use "communication" as the general term.

3.1 Language

In order to facilitate communication, mankind has evolved natural languages as our most important means of collective information processing, enabling coordination, collaboration and cooperation.

Using the definition of communication given above, we can define language as follows:

Language = a social convention-based system of communication for the sharing of complex information, using vocal, gestural or written symbols. For other definitions of language, Bloch and Trager (1942) and Everett (2012).

Among languages, so-called natural languages are especially important and can be defined as means of communication for sharing of complex information between people, using vocal, gestural or written symbols that have developed naturally, i.e. without extensive human planning and construction.

Natural languages, thus, contrast with artificial languages, like computer languages, chemical formulae, mathematical formulae and Morse code, but also with intentionally constructed auxiliary languages, like Volapük, Esperanto, Ido and Klingon.

Natural languages have probably been part of human evolution for at least 200000 years and have in this way acquired physical, biological, psychological and social properties, which are combined in complex systems with systemic properties both on an individual and on a collective level.

In general, when we study language and communication, it is useful to distinguish the "expressions", "contents" and "contexts" of language as three aspects that continuously influence, constrain and reinforce each other.

The "expressions" of language include sound (linguistic sounds), visible behavior (gestures) and artifacts (writing and texts). The three correspond to the three primary expressive modes of language— speech, gesture and writing. While speech and gesture in face-to-face communication have probably evolved together with the development of humans from higher primates, writing is a later cultural development.

All linguistic expressions, whether they are spoken, gestural or written, have a "content", which has cognitive, emotive and conative aspects; the cognitive aspects involve factual information (from everyday topics to more specialized topics), the emotive aspects involve affective-epistemic attitudes and the conative aspects involve intentions and acts of will. The content also includes features related to social identity and personality and features related to "communication management". For more discussion of the content of communication, see Sections 4 and 6.

The expressions with their content are used in different "contexts", that is, settings or situations that have properties that influence both the production and interpretation of the expressions. This means that linguistic expressions are placed in a context that is multimodal and influenced by a particular activity and culture (see Section 4, below).

4. Functions and Processes Involved in Communication

To communicate involves participating in a number of processes that need to be managed so that the actions and behavior in communication can be adapted successfully in order to reach the goals of communication.

We have therefore evolved several mechanisms for communication management (CM). Two of the most important of these are (Allwood, 2008b):

1. Interactive Communication Management (ICM)
2. Own Communication Management (OCM)

Interactive Communication Management involves means for managing the interaction in communication, while Own Communication Management involves ways of managing your own contributions to communication. Both of these types of communication have subsystems.

Interactive Communication Management, for example

- Turn management
- Feedback
- Sequencing

Own Communication Management, for example

- Mechanisms for choice and planning
- Mechanisms for change

Besides mechanisms for managing communication, there are also the features of the message that is managed. We will call this the Main Message (MM) to differentiate it from the auxiliary messages involved in Communication Management.

An overview of the structure and functions of Human Communication is given in Table 1.

The main message is the reason a contribution to communication was made and can contain communicative acts (for instance, statements, questions and requests), referential content and expressed attitudes that are to be shared with the interlocutor. In relation to the aspects of content mentioned above at the end of Section 3, the communicative acts relate to conative (intention and will) aspects, the referential content relates to factual content and the expressed attitudes relate to

Table 1. Structure and functions of human communication.

Main Message (MM)	Communicative acts
	Referential content
	Expressed attitude
Communication management Interactive Communication Management (ICM)	Turn management
	Feedback
	Sequences
Own Communication Management (OCM)	Choice
	Change

the affective-epistemic aspects. Features, connected with social identity and personality, are more complex and can relate to any aspects.

The interactive communication management features help the interlocutors to successively provide new contributions to be shared. Turn management features help them coordinate their contributions and collaborate in construction of joint content. Feedback processes help them communicate successfully, making sure that they have contact, perceive and understand each other's emotional-attitudinal reactions and contributed content. The sequencing features help them adjust communicative acts to each other in a relevant manner, for example, giving answers to questions, answering expressions of gratitude with expressions of generosity, etc.

Finally Own Communication Management (OCM) processes allow interlocutors to keep their turn while planning (Choice function), for example, by using hesitation words or prolonged duration of syllables or gestures. OCM processes also allow speakers to change what they have said or gestured when they feel the need for this, in such a way that their interlocutor(s) can follow what is going on.

5. Multimodal Communication

As we have already noted, face-to-face communication is multimodal. What this means is that more than one of the sensory modalities and more than one of the production modalities in Table 2 are involved. Even if both perception and production can be multimodal, the basis in multimodal communication is multimodal perception, so that in this sense speech in face-to-face communication can be multimodal since we can both hear and see the activity of the speech organs.

This means that content in face-to-face communication is shared through use of multimodal contributions that normally consist of at least vocal verbal elements (with phonology, morphology,

Table 2.　Sensory modalities and production modalities in multimodal communication.

Sensory Modalities	Production Modalities
Sight	Communicative body movements/gestures/writing
Hearing	Voice, speech
Touch	Touch
Smell	Smell
Taste	Taste

lexicon, syntax, semantics and pragmatics) combined with prosody and communicative body movements (Kendon, 2004; Argyle, 1988; Allwood, 2008a).

Multimodality implies multimediality, since the sensory modalities also involve physical media, i.e. optical (sight), acoustic energy (hearing), pressure (touch) as well as molecules affecting taste and smell. In addition, there are other perceptual modalities than the traditional five senses that may be relevant, e.g. modalities for temperature, color, shape, movement, etc.

Studying communication from a multimodal perspective leads to a deeper understanding of many processes connected with communication. Two of these are:

(i)　**Multimodal integration** (sometimes, with a metaphor from physics, also called information fusion). Multimodal integration concerns how we can integrate information from our separate sensory modalities with our memory sources to form a common complex multimodal experience. For instance, in a normal conversation, we integrate what we see (colors, shapes and movements), what we hear, touch and smell with what we epistemically and emotionally experience. Furthermore, this is all integrated with other, already stored information we have in our memory concerning our interlocutor and about what he/she is saying or doing.

(ii)　**Multimodal distribution** (with another metaphor from physics, this is sometimes called information fission). Multimodal distribution concerns how we distribute what we want to communicate or do, using several production modalities. If I want to tell you that I am happy to see you, my message will be distributed into a vocal verbal aspect, a prosodic (intonation, tone of voice) aspect, a gestural aspect (face, head, arms, torso, etc.) and possibly a touch and smell aspect. How exactly these different aspects are related to each other is one of the questions still to be resolved in the study of multimodal communication.

In many contexts, flexibility in the choice of modality is needed. Multimodality gives us this flexibility, and also the possibility of being redundant when this is needed, for example, in a complex noisy environment.

Building on the definition of communication given above and building on the notion of multimodality, we can now give a definition of multimodal communication.

Multimodal communication = co-activation, sharing and co-construction of information simultaneously and sequentially through several modes of perception (and production) (Allwood, 2008a).

In Table 3, we give an overview of how dimensions of production and perception can be related in multimodal communication.

The combination of dimensions can be simultaneous or sequential, occur on varying levels of consciousness and intentionality and involve several communicative orientations (see below).

As we can see from the table, communication can involve many types of communicative expressions over and above the auditory aspects of speech. Table 4 gives an overview (Allwood, 2002).

Most of these expressions can supplement auditory aspects of speech or play an autonomous role in communication.

Table 3. **Multimodal face-to-face communication—Perception and production.**

PERCEPTION	Hearing	Vision	Touch	Smell	Taste
	Understanding + Attitudinal Reactions				
PRODUCTION					
SPEECH:					
Prosody/Phonology	x	x			
Vocabulary	x				
Grammar	x				
GESTURES:					
Facial gestures		x			
Manual gestures		x			
Body movements		x			
Posture		x			
Touch			x		
Smell				x	
Taste					x
Medium	acoustics	optics	physiology	molecules	molecules

Table 4. Communicative expressions over and above auditory aspects of speech.

1.	Facial gestures (nose, eyebrows, cheek, forehead, chin, etc.)
2.	Head movements
3.	Gaze direction, mutual gaze
4.	Pupil size
5.	Lip movements
6.	Hand and arm movements
7.	Leg and foot movements
8.	Body posture
9.	Distance between communicators
10.	Spatial orientation
11.	Clothing and bracelets
12.	Touch
13.	Smell
14.	Taste
15.	Non-linguistic sounds

6. Contents of Communication

As we have seen in Sections 3 and 4, the content of language and communication has many features. We will now consider these features a little more in detail. Perhaps the most important types are information concerning:

(i) Physiological states, like fatigue and hunger

(ii) Character—identity—personality, like haughty, timid, aggressive

(iii) Affective-epistemic attitudes (including emotions), like joy, friendliness, surprise, boredom, interest, etc.

(iv) Factual content, giving information about our beliefs and assumptions concerning facts

(v) Communication management, that is, information about ICM (feedback, turn taking, sequences) and OCM (choice and change).

In many contexts, the verbal part of the auditory, mostly vocal message is the most important. However, for all the types of content, especially the first three types, both prosody and communicative body movements have a major role. Thus, in contexts where information about physiological states, personality or affective-epistemic states is in focus, the importance of prosody together with visible body movements increases. Prosody and visible body movements are also very important for information structure, that is, structuring a message with regard to what is important and needs attention and what can be back-grounded and presupposed. If we compare the list of contents just given for speech with possible contents associated with visible body movements, we find that visible communicative body movements can activate and help share information of all the types mentioned, but, like prosody, they are especially important for all the dimensions of content not concerned with factual information, especially emotions and attitudes. Perhaps this points to a close evolutionary relationship between prosody and gesture.

6.1 *Affective aspects of content*

Communication does not only involve sharing of factual information. It also involves sharing of attitudes and emotions. To some extent, attitudes and emotions are part of what is shared in all types of communication. However, in some types they are perhaps the main focus, like in small talk, quarrels or love making, while in others, like a scientific lecture, they have a more subordinate role. Since they are so pervasive, they play a major role in what we express in communication which means that understanding how affective-epistemic attitudes, like interest, surprise, boredom, uncertainty, friendliness or amusement, are indicated, displayed or signaled (see below) and what reactions, perceptions, understandings and responses they give rise to is essential. Some of the main modes of interaction, like coordination, collaboration, cooperation, competition and conflict, all depend on emotions and attitudes. If we want to understand and facilitate these modes of interaction, we must understand the role of emotions and attitudes in communication.

6.2 *Content in small talk*

Social contact is a basic human need. Human beings need social contact to fully develop. Social contact involves communication. The content shared in communication can be more or less important. In some situations, social contact can be more important than the actual content shared. Contact with other persons becomes the primary motive for communicating and the topics chosen for communication in this kind of situation will tend to be such that, depending on culture and other circumstances, they are seen to be of neutral, general relevance, like the weather, sports, television, politics, economy or family. "Small talk" of this kind has sometimes been called "phatic communion" (cf. Malinowski, 1922), from the Greek (phatos—speech), where the idea is that "phatic communion" is a kind of fellowship, or sharing through speech, where the main function is social contact and emotional togetherness.

7. Communicative Orientation

Every contribution to a dialog has several communicative orientation functions. The four most important types are (Allwood, 2000):

(i)   One or more responsive functions (all contributions except the first)

(ii) One or more expressive functions, mostly involving the expression of emotional or epistemic attitudes

(iii) One or more evocative functions

(iv) One or more referential functions.

The four communicative orientations can be seen in Figure 1 below.

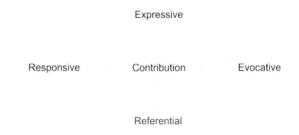

Figure 1. The communicative orientations of a contribution.

Besides these four types of functions, contributions have an "information structure" (see Section 6, above) helping to focus attention on new or noteworthy parts of a contribution and to defocus other aspects, which can be taken for granted to a greater extent.

7.1 Dynamics of dialog

Building on the orientation functions introduced above, we can better understand what drives dialog forward by considering the interplay between the evocative and the reactive/responsive functions of contributions. The double term "reactive/responsive" is used to include both reactions that are automatic and unaware and responses that are more deliberate. The evocative and reactive/responsive functions work like two cogwheels, linking into each other. The evocative functions of Speaker A trigger the reactive/responsive functions of Speaker B who then makes another contribution in which the first part usually is mainly responsive and the second part mainly evocative. Combined with the expectations connected with particular social roles, the two types of functions give rise to the commitments and obligations connected with the roles of main communicator (speaker) and co-communicator (listener) for all participants as they alternate between these two roles. See Table 5.

Table 5.   Dynamics of dialog: Two cog wheels linking into each other.

	Expressive			Expressive	
Reactive/ Responsive	Contribution 1	Evocative	Reactive/ Responsive	Contribution 2	Evocative
	Referential			Referential	
	Communicator Obligations and commitments			Co-communicator Evaluation obligations	

An example of the functions is given below in a short exchange of two contributions between A and B, waiting for the bus on a rainy morning

> A: always raining in Gothenburg
> B: (nodding) mm yeah it is
>   (depressed)

We start by analyzing A's contribution

(i)   *Reactive/responsive*: Since this is the first contribution, there is no previous contribution to respond to. However, one might say that A's contribution is a response to the situation at hand with B present as a potential co-communicator.

(ii) *Expressive*: A is making a predication, which in this particular situation functions as a statement expressing A's belief about the weather (belief is an epistemic attitude).

(iii) *Referential*: A refers implicitly, through the predication, to the meteorological situation.

(iv) *Evocative*: The evocative functions of A's contribution can be described as Contact, Perception, Understanding (CPU) and sharing of belief, i.e. by making the contribution, A is seeking contact with B by attempting to make B perceive, understand and share the belief expressed by A's contribution.

Besides the orientation functions of A's contribution, A is also creating a normative relation with B, which can be described as follows: Being a speaker, A considers B cognitively and ethically, considering questions like: Does A really want to contact B?, Is it ok to contact B?, Will B be able to understand? Over and above these normative requirements, A's contribution also contains an implicit commitment to the belief that A is expressing through the predication in the contribution. This is what in table 5 is called "obligations and commitments" of a communicator.

If we turn to B's contribution, B, in the co-communicator role, first carries out a conscious and less conscious internal appraisal and evaluation of A's contribution and then switches to the communicator role and produces behavioral reactions and responses related to A's contribution. In the co-communicator role, B needs to evaluate his/ her own willingness and ability to react to the evocative functions in A's contribution. Can and does he/she want to continue, perceive, understand and share the belief that A is expressing and how does he/she react to this emotionally and epistemically? This is what in Table 5 is called "the evaluation obligations" of the co-communicator. On the basis of an evaluation/appraisal of this type, B then provides a reaction/response to A's contribution which then are subject to the "obligations and commitments" of a communicator.

> B: (nodding) mm yeah it is (depressed)

The functions of this contribution can be described as follows:

(i)   *Reactive/responsive*: Through the nodding in combination with the feedback words *mhm, yeah*, B expresses contact, perception and understanding (CPU) (I am willing to have contact, perceive and understand).

(ii) *Expressive*: B's responsive functions are also expressive, so his/her contribution also expresses contact, perception and understanding. In this case, this is done with depressed facial gestures and tone of voice. In addition, by the word *yeah* and reformulation *it is* of A's implicit statement, B expresses his agreement and own belief in the meteorological state described by A (shared belief).

(iii) *Evocative*: Like A's contribution (and like most contributions), B's contribution is an attempt to evoke continued contact, perception, understanding and shared awareness (belief) of B's sharing of A's belief. It is, thus, an attempt to evoke a belief about a belief in A, i.e. A should become aware that B shares his belief.

In other words, even small talk with a simple exchange of information about the weather, like the one above, involves what is sometimes called mentalizing and reliance on a so-called theory of mind (Frith and Frith, 2010), allowing a quick build-up of shared beliefs and emotions. This is, to a very great extent, done through use of multimodal feedback mechanisms, which consist of a combination of small words like *m*, *mhm, yeah, ok*, head movements, gaze and facial gestures.

8. An Interactive View of Embodied Communication

In face-to-face communication, we can say that communicators form a more or less integrated dynamic system by establishing communication links on several levels of intentionality and awareness. Figure 2 below presents a simplified model of such a system, involving two communicators (Allwood et al., 2008).

The figure shows two communicators, A and B, who are communicating on several levels of awareness and intentionality. In actuality, awareness and intentionality probably vary continuously from no awareness or intentionality to higher degrees of awareness and intentionality. However, in the model, we distinguish three levels that have somewhat different properties (Allwood, 2008b).

On the level that is least aware and intentional (the indicative level), A is influencing B, without intending to do so, or even being aware of doing it. Similarly, B is being influenced, but is not really aware of this happening.

On this level, subconscious or perhaps better non-conscious reactions and appraisal can take place. Usually, these processes are quick and lead to responsive behavioral reactions that are hard to control. One word for this is "automatic". In general, the model predicts that the more aware a process is, the slower it will be and the easier it will be to control and the less aware it is, the quicker it will be and the more difficult to intentionally control it will be. On the lowest level of awareness, there are many partly overlapping processes that are basic to communication, like co-activation, mirroring, priming, alignment and emotional contagion (Pickering and Garrod, 2004; Arbib, 2002; Tarde, 1903).

On the middle level (the display level), the sender is more aware of what he/she is doing and more in control of his/her behavior. Here, the basic communicative intention (Allwood, 2002, 2008b) is to display or show something to the interlocutor. There are several kinds of display. Often behavior, which has been initiated as indicated and automatic, gradually becomes aware and can then be reinforced by more aware and intentional display.

On the recipient side, the model also predicts degrees of awareness and intentionality, extending from the non-conscious processes already described above to more aware processes involving discrimination and identification of stimuli as separate from each other.

On the third, or most intentional and aware level (the signal level), processes are slower, more aware and deliberate than on the lower levels. The main communicator can now engage in what we will call signaling, which involves not only displaying (showing) something, but also intending that the recipient becomes aware that something is being displayed. Using language normally involves this level of intentionality and awareness, since language is a system for communication that conventionally presupposes "signaling" to a recipient. Linguistic expressions are conventional "signals" or conventional means for displaying that you are displaying (showing that you are showing).

On the recipient side, signaling is related to better perception and understanding. On this level, the recipient moves beyond discrimination and identification to understanding, which in the model is equivalent to connecting perceived input information to stored background information in a meaningful way. Understanding linguistic expressions provides a special case of this process that can be brought out by considering the case of trying to understand a language you don't know. Imagine you hear (or see) the phrase *hao che* from a Chinese person and don't know Chinese. You will probably be able to discriminate and to some extent identify the sounds, but you will not be able to understand, since you cannot connect your perceived input to an already stored background in a meaningful way. However, if you know Chinese, you will be able to do this and understand what is being said. Now compare this example to hearing

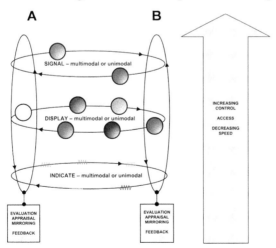

Figure 2. A dynamic system of communication, involving two communicators and three levels of co-activation, which influence each other.

(Color image of this figure appears in the color plate section at the end of the book.)

an English speaker saying *delicious* while eating. In this case, you can probably connect what you perceive to already stored information, enabling you to understand what is being said.

Analyzed this way, understanding linguistic expressions can be seen as a special case of understanding in general, which always involves connecting input information to stored background information in a meaningful way. The depth of understanding is dependent on the extent to which such connections can be made, where more connections mean increased depth of understanding. Understanding a phenomenon means becoming aware of how the phenomenon relates to other phenomena.

In communication, one particularly interesting type of connection, relevant both for production and understanding, relates input information to stored information concerning the social activity that is being pursued. This is done in such a way that the activation involved in perception and understanding involves predictions about what the relevant next stage of the activity could be which means that relevant responses can fairly rapidly be activated and given. In this way, communication relevant to a particular social activity can be driven forward by relevant co-activation on several levels of awareness (Allwood, 2000).

9. Activity, Personality and Culture

9.1 Factors that influence communication—Activity

Actual communication is always multi-causally influenced. Communicators have multiple overlapping roles and purposes that are given by culture, age, gender, occupation and not least activity. They can draw on many communicative resources even in a particular activity, which means that the activity can contain types of communication, which are not strictly speaking part of this activity. A very common example of this is small talk (see above), which can be a part of as diverse social activities as a patient-doctor consultation or a negotiation between politicians.

Thus, there are many factors that influence and are influenced by the way that contributions are produced and understood in dialog. Consider for example the factors influencing a German politician negotiating with a French politician in the context of the European Parliament in Strasbourg. Starting from a general level, we have the influence of human nature and the external physical environment. On a slightly less general level, there is a particular national culture

(German and French, European) and its associated language (French, German or English with possible interferences from German or French) to consider. On the next level, there are social institutions, like organizations (European Parliament) and above all social activities. In this case, the activity is "negotiation". All of the mentioned factors enable us to have certain expectations and make predictions about the behavior of the two politicians.

Social activities, like negotiations, are one of the most profound types of influence on communication (Allwood, 2000). We communicate in very different ways, depending on whether we are talking to colleagues in a project meeting, participate in a lecture, are trying to sell something to a customer, are interviewing someone or are trying to flirt with someone.

All of these activities can be described and analyzed using the following activity parameters:

  (i)   Purpose, e.g. formal meeting
 (ii)   Roles of participants, e.g. chairman, secretary, participant
(iii)   Procedures, media and other instruments
(iv)   Environment (organizational and physical)

However, besides the roles that communicators have in such activities, their communication will also, as already mentioned, be influenced by participant characteristics, such as organizational position, gender and educational background.

But over and above, the mentioned factors, the interaction itself, in many ways, is decisive; as it goes through particular sub-activities or phases, using exchange units as exposed through the particular contributions of the negotiating politicians. In the end, as we have seen above, it is these individual contributions with their more or less aware features in terms of evocative and reactive/responsive functions that drive dialog forward, employing several levels of intentionality and awareness.

Let us now turn to two of the factors that, besides social activity, have an influence on communication, namely personality and national-ethnic culture.

9.2 *Personality*

One of the factors, often thought to influence communication, is personality (identity or character) which can be described as a characteristic set of holistic biographical, psychological and social features regarded as long term and typical of a particular person.

Often personalities are described in terms of emotional attitudes such as warm, cold, generous, stubborn, dominant, shy, aggressive, easy-going, etc., traits that are often assumed to have fairly direct communicative consequences, so that a warm person shows warmth in communication, while a dominant person attempts to dominate.

Theories of personality vary (Hall and Lindzey, 1957) in how permanent or malleable personality features are assumed to be. Some theories see them as always present, based on a genetic disposition or early child development, while others see them as highly fluctuating, based on the type of interaction developed in a particular communicative situation.

9.3 Cultural influence

We have seen above that national-ethnic culture is among the influences on communication. Perhaps this is especially interesting, when two or more persons with different cultural backgrounds communicate and we have what is often called "intercultural communication".

Reflecting on intercultural communication raises the question of what culture is (Kroeber and Kluckhohn, 1952; Geertz, 1973; Allwood, 1985). We can define the culture of a community as their shared patterns of thoughts, behavior, artifacts and traces in the environment, based on, but not determined by, Nature. Thus, the ability to breathe or to walk, although shared in all human communities, is not cultural, since it is directly given by Nature. Culture is based on Nature, but requires humans to create regularities not directly given by Nature. Since such created regularities can differ between communities, they are in many cases relevant for communication. They affect both behavior, activities and the assumptions communicators have about what they can take for granted as shared. Such assumptions are often automatic and can, if not made aware, lead to misperceptions and misunderstandings between interlocutors.

Cultural traits and differences can influence all aspects of communication, that is, production, interpretation, interaction and assumptions about context, for example, assumptions about the proper, or polite, way to carry out various social activities, like greeting, e.g. thanking, introducing yourself, getting to know someone, negotiating, etc.

Our awareness of cultural traits, as well as of cultural differences, often takes the form of more or less stereotypical generalizations concerning what is common or normal on a group level. If seen this way, cultural traits become not deterministic causes valid for every individual. Individual members of a culture may always be exceptional. Assumptions about cultural traits and cultural differences must therefore always be made with caution and checked for validity.

Since cultural differences may have an influence on all aspects of communication, this means that they can have an influence on both main message features (choice of communicative, referential content and expressed attitudes) and features of communication management (turn management, feedback, sequencing, i.e. ICM, choice and change management, i.e. OCM). See above. They may, for example, influence the frequency and the way in which we move our head in order to give positive or negative feedback to our interlocutor or the way in which we hesitate in order to keep the floor.

A very clear example of differences in the main message part of a contribution concerns the content different languages allow us to express in different social activities. What may be easily and directly expressed in one language may perhaps only be awkwardly expressed by paraphrase, if at all, in another language. Usually, differences of this sort reflect the fact that what has been in focus and conceptually developed in one culture and language need not have been of interest and developed in another culture and language.

Cultural differences can concern different types of relations between interlocutors, like power and trust relations. Cultures vary with regard to how much power and what type of power is connected with different social relations, like boss—employee, parent—children or doctor—patient. Likewise, they vary with regard to how much and who you trust. Do you trust your parents, your children, your boss, your employees, your doctor, your patients, the police and your politicians?

Also ethical aspects of communication can be subject to cultural differences. When and to what extent is it acceptable to lie to other people? When and to what extent is it acceptable to hurt other people? Can you lie to or hurt your parents, your children, your spouse, your boss, your employees, your doctor, your patients etc.?

9.4 Culture and activity

Activity parameters like purpose, roles of participants, procedures, media and other instruments as well as environment (organizational and physical) are to some extent different in different cultures: This can play a role for the communicative behavior in the activity and can be seen in features of communicative acts, interpretation and interaction patterns. Thus, cultural differences can be found, for example:

(i)   In role expectations, e.g. concerning power, respect, politeness, and in the manner in which power or politeness is expressed.

(ii) In the sequences opening or closing an activity, e.g. in sequences of greeting, introduction and leave taking.

(iii) In turn management and feedback.

(iv) In what is seen as the purpose of a given activity, especially concerning non-explicit purposes.

(v)   In environmental features like the occurrence of (and attitudes to) cleanliness, dirt, noise and silence. In connection with the environment, it is also interesting to consider natural factors, like the climate or topography that can have an influence on culture and communication.

Investigations of how culture and social activity influence communication should therefore, if possible, be combined. Otherwise there is a clear risk that behavior attributable to an activity difference is attributed to a cultural difference and vice versa.

10. Face-to-face Communication and Communication Technology

What happens when we introduce communicative technology into face-to-face communication? (Allwood and Ahlsén, 2012.) Today, there are many types of communication technology, for example, writing, radio, TV, electronic audio-video communication (Skype, Youtube etc.), email, chat, Facebook and mobile communication devices of many types. Some of these support human-human communication, while others involve communication with a virtual agent (games, tutoring systems), a robot or bot. Many new devices for picking up information have been developed: e.g. sensors for GPS, galvanic skin responses, heart beat and brain activity (for example, Zhang et al., 2006).

Two questions that will be with us for a while and periodically need to be asked again are: What is missing and what are the consequences of new communication and information technology for communication?

As an example of the effects of communication technology, let us consider the extent to which communication involves synchronization in time and space. See Table 6, below.

The table shows how communication technology has enabled us to bypass the constraint of co-presence in time and space, which is a feature of face-to-face communication. But there is a price to pay for this. Some features of face-to-face communication are lost. So far, lack

Table 6.   Communication technology and synchronization in time and space.

Space/Time	Same Time	Different Time
Same Location	1. Face-to face communication	3. Bulletin board
Different Location	2. Phone, Video and audio conferencing, Skype, Chat	4. Writing (letters, email, fax), Voice mail, Blog, Internet, Recording devices

of synchronization has also led to less redundancy, less multimodality, less non-aware sharing, less interactivity and less complexity of certain types. It has, of course, also had positive effects like bridging time and space, i.e. sharing of information across points in time and locations in space. These were a motive for developing the technology in the first place.

Regarding human involvement in communication, communication technology can have two basic functions:

(i)   It can be supplementary, supporting human-human communication. The bridging of time and/or space discussed above are probably the clearest examples of this. Other examples are the online availability of a database or other kinds of information not normally available in face-to-face communication.

(ii) It can replace humans in communication.

Here again, there are two basic cases:

(i)   Replacing one or more humans by a VR agent or a robot. This can be done, for example, for provision of services, like a travel agent, an information officer, an artificial companion, etc., and will involve bridging time and/or space. In every case, we have to ask questions. What properties need to be modeled in the VR agent or robot to provide the service? Do we want features over and above what is necessary for the service? What human features need to be recreated? Here, an interesting problem is the topic we have been discussing in this chapter—what human features are presupposed by language and communication. This is the information we need if we want the artificial agents to communicate like humans.

(ii) Replacing several humans perhaps, perhaps all, by bots who act as electronic representatives, servants or spokespersons for humans. Again, we can ask questions about what properties need to be modeled in the bot-agents to enable them to carry out the tasks for which they are made, especially if the tasks involve language and communication.

11. Conclusions

In this chapter, we have presented an overview of some of the main features of human communication and some of the factors that influence these features in different situations. We have stressed that many of the features work more or less automatically at a low level of awareness and intentionality and that this provides a special challenge for communication technology.

In general, a better understanding of the features of human-human communication gives us a better basis for evaluating which of these features we want to enhance, leave out or emulate through development of communication technology. This, in turn, makes it possible to develop criteria for evaluation of communication technology, something that is becoming more and more necessary, given the steadily increasing amounts of technology that are available.

REFERENCES

Allwood, J. 1985. Intercultural Communication (translation of Tvärkulturell Kommunikation). In J. Allwood (ed.). Tvärkulturell kommunikation. *Papers in Anthropological Linguistics*, 12. Göteborg: Department of Linguistics, University of Gothenburg, 12, Göteborg.

Allwood, J. 2000. Activity Based Pragmatics. 1995. In H. Bunt and B. Black [eds.]. Abduction, Belief and Context in Dialogue: Studies in Computational Pragmatics. John Benjamins, Amsterdam. Also in *Gothenburg Papers in Theoretical Linguistics*, 76, Dept of Linguistics, University of Gothenburg, pp. 47–80.

Allwood, J. 2002. Bodily Communication—Dimensions of Expression and Content. In B. Granström, D. House and I. Karlsson (eds.). *Multimodality in Language and Speech Systems*. Kluwer Academic Publishers, Dordrecht, pp. 7–26.

Allwood, J. 2008a. Multimodal Corpora. In A. Lüdeling and M. Kytö (eds.). Corpus Linguistics. An International Handbook. Mouton de Gruyter, Berlin, pp. 207–225.

Allwood, J. 2008b. A Typology of Embodied Communication. In I. Wachsmuth, M. Lenzen and G. Knoblich (eds.). Embodied Communication in Humans and Machines. Oxford: Oxford University Press.

Allwood, J., K. Grammer, S. Koppand and E. Ahlsén. 2008. A framework for analyzing embodied communicative feedback in multimodal corpora. J-C. Martin (ed.). JLRE (Special Issue on Multimodal Corpora).

Allwood, J. and E. Ahlsén. 2012. Multimodal Communication. In A. R. Mehler, R. Romary and D. Gibbon (eds.). Handbook of Technical Communication. Mouton de Gruyter, Berlin.

Arbib, M.A. 2002. The Mirror System, Imitation, and the Evolution of Language. In C. Nehaniv and K. Dautenhahn (eds.). Imitation in Animals and Artefacts. MIT Press, Cambridge, MA.

Argyle, M. 1988. Bodily Communication. Methuen, London, pp. 229–280.

Bloch, B. and G.L. Trager. 1942. Outline of Linguistic Analysis. Linguistic Society of America, Baltimore.

Everett, D. 2012. Language the Cultural Tool. Profile Books Ltd., London.

Frith, U. and C. Frith. 2010. The social brain: Allowing humans to boldly go where no other species has been. *Philosophical Transactions of the Royal Society* B2010, **365**:165–176.

Geertz, C. 1973. The Interpretation of Culture. Basic Books, New York.

Hall, C. S. and G. Lindzey. 1957. Theories of Personality. John Wiley and Sons, Inc., New York.

Kendon, A. 2004. Gesture: Visible Action as Utterance. Cambridge University Press, Cambridge.

Kroeber, A.L. and C. Kluckhohn. 1952. Culture: A critical review of concepts and definitions. Harvard University Peabody Museum of American Archeology and Ethnology Papers, 47.

Malinowski, B. 1922. Argonauts of the Western Pacific. Routledge and Kegan Paul, London.

Pickering, M. and S. Garrod. 2004. Toward a mechanistic psychology of dialogue. *Behavioral and Brain Sciences,* **27**:169–225.

Reddy, M. J. 1979. The conduit metaphor: A case of frame conflict in our language about language. pp. 284–310. In A. Ortony (ed.). Metaphor and Thought. Cambridge University Press, Cambridge.

Shannon, C.E . and W. Weaver. 1949. A Mathematical Model of Communication. University of Illinois Press, Urbana, IL.

Tarde, G. 1903. The Laws of Imitation. Henry Holt, New York.

Zhang, Z., D. Potamianos, M. Liu and T. S. Huang. 2006. Robust multi-view multi-camera face detection inside smart rooms using spatio-temporal dynamic programming. In: The Proceedings of the 7th International Conference on Face and Gesture Recognition (FGR), Southampton, United Kingdom, pp. 407–412.

CHAPTER 3

Giving Computers Personality? Personality in Computers is in the Eye of the User

Jörg Frommer, Dietmar Rösner, Julia Lange and Matthias Haase

1. Giving Computers Personality— What is the Current Approach?

A review of the literature shows that most researchers who attempt to assign personality to computers narrow their focus towards only a small number of personality aspects. Often interactive characters or embodied conversational agents are analyzed with regard to the impact of their introverted- vs. extravertedness. For example, it was investigated as to whether users preferred characteristics that are similar or dissimilar to theirs in terms of introversion vs. extroversion. Nass and Moon (2000) examined the effects of introverted and extroverted signals as presented both textually and visually, and the paralinguistic features of human or computer-generated voices. Other studies have examined those computer systems which present submissive and dominant characteristics, and the preference of users for similar or dissimilar characteristics to their own (Nass et al., 1995). In addition, the impact of introvert or extrovert styles' assistance during a decision problem has been researched, together with its impact on any trust in the advice (Hess et al., 2009). McKeown et al.

(2012) examined the influences of four different characters within a simulated computer-based "Sensitive Artificial Listener" (SAL). Each character had a mix of different characteristics regarding various personality traits. It was determined that the subjects' reactions to the simulated characters were clearly distinguishable. These studies demonstrated that the influence of a computer system's presented personality, or more specifically particular personality traits, plays an important role. Nass and Moon (2000, p. 91) summarized that: "In computer science literature, 'personality' has traditionally been one of the 'holy grails' of artificial intelligence".

But what exactly do we mean by 'personality'? Personality refers to the unique psychosocial characteristics of a subject that affect numerous consistent traits within different situations (Zimbardo and Gerrig, 2004). In personality psychology and psychological research, quite a variety of different personality theories and models describing distinct personality traits have been developed and discussed. In the *psychodynamic theory*, importance is attached to unconscious experience and behavior with reference to drives, maturation, and early childhood experiences. In the *phenomenological and philosophical approaches*, the observations of self and others using classification systems are connected with the classifications of experience and behavior. In the *cognitive theory*, there is neither objective nor independent reality, but rather a subjective, individually experienced, and interpreted world (John et al., 1991; McAdams, 1995). In *behavioral and inter-actionist approaches*, individual learning differences develop because of people's life-experience histories (Fisseni, 2003). In the *factor-analytical theories* of personality, the individual is viewed as a unique structure which can be best understood and described through traits (Guilford, 1959). Traits are distinct and temporally constant characteristics that describe specific aspects of a person's behavior. This theory is based on dimensions which are determined using a statistical method of factor analysis. Eysenck (Eysenck and Levy, 1972) attempted to reduce the complexity of personality to two key dimensions (introversion vs. extroversion and stability vs. instability). Within the factor-analytical theory, Goldberg (1990) described the "Big Five" model. Later research added evidence of the "Big Five" factors of human personality (Ozer and Benet-Martinez, 2006). Nowadays the "Big Five" model is the most influential personality model. This model assumes that behavior in situations (state) is influenced by steady characteristics (traits). The "Big Five" factors are: Extroversion (high: externally focused, active, outgoing—low: internally focused, shy), Agreeableness (high: cooperative, willing, conforming—low: less willing, argumentative, non-conformist), Conscientiousness (high: good impulse control, stable, reliable—low: poor impulse control, unstable, unreliable), Neuroticism (high: tendency to worry, unpredictable, emotionally unstable—low: calm, fairly predictable, rational), Openness (high: explores, tries new things, accepts new values—low: rigid, accepts new values slowly) (Goldberg, 1990). All in all, there is more than one definition of personality. This makes it difficult for researchers to agree on one personality model or theory. For this reason, they use the one which is the most useful for their particular research questions.

2. Users Do Ascribe Personality— A Theoretical Perspective

As already pointed out, it is apparent that current studies only deal with the topic by asking "Which personality should a system have?" In our opinion, this should be complemented by asking the question "Which personality does a user ascribe towards the system?"; whether a system is experienced as to be helpful, supportive, and overlooking depends on how the user himself/herself experiences it (see Figure 1).

It has become apparent that users do not ascribe personality to computers according to the models or theories of personality mentioned

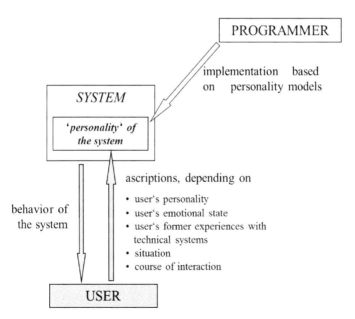

Figure 1. Extended model for "giving computers personality"; current model in white, extension in grey.

above. In fact, they may also ascribe a variety of mental states amongst these various personality traits and features. But where does this idea that users ascribe mental states towards technical systems come from? In literature, there are numerous examples of people's implicit tendencies to treat technical systems like entities with intentions and to form close and emotional relations with them: Weizenbaum (1996) reported that persons interacted with ELIZA—a relatively simple program which simulates a Rogerian psychotherapist—by ascribing background knowledge, insights, and logical reasoning to it. Further, the owners of a Tamagochi ascribed feelings of longing to this virtual pet when they left it alone for a time and even expressed guilt in response (Wilks, 2005). The female owner of an Aibo (a robotic dog) reports that she prefers the robot to a living dog, because it will not betray her or annoy her by dying suddenly (Turkle, 2006), and the speaking doll Primo Puel has been found to reduce loneliness and feelings of insecurity in an elderly Japanese woman. People apply a variety of social rules and expectations from human-human communication to computers, e.g., they overemphasize human social categories (e.g., by applying gender stereotypes to computers). For an overview on these studies, Nass and Moon (2000), who explain these phenomena in relation to Langer's (1989) concept of "mindless behavior".

2.1 Dennett's intentional stance as a possible explanation of ascriptions by users

The works by Daniel C. Dennett (1971, 1987) offer a philosophical framework of such ascriptions by users. He explains three stances which people can adopt towards both biological and non-biological systems (like computers).

1. *Physical stance:* In the physical stance, explanations and predictions are based upon the real physical state of the system, as well as upon knowledge of the laws of nature. In order to predict, for example, the moves of a chess computer at this particular stance, you would have to follow the effects of the input energies on their way through the computer. In addition, you would have to know all the critical variables within the physical condition of the computer, which are very complex and would need enormous calculations. This stance is commonly adopted towards 'simple' systems, especially for predicting malfunctions. As an example, you can predict that water in a pot will start boiling if the pot is placed on a hob long enough; moreover, you can predict that the water will not start boiling if the plug of the hob is not placed in

the socket.

2. *Design/functional stance:* In the functional stance, explanations and predictions are based upon knowledge or presumptions of the system's construction plan and its functional components. In this stance, predictions are based on the assumption that the chess computer has a certain design (is programmed in a certain way) and will act as it is designed (programmed) to do under certain circumstances. So you will be able to predict the countermoves of the computer for each of your moves by knowing the programmed tactics of the program. Predictions will be accurate on the condition that the computer functions without any bugs. In this stance, you ignore the actual physical condition or inner stance of the computer. This stance is generally adopted to predict the behavior of mechanical objects. For example, you will be able to predict when an alarm clock will ring by exploring the exterior, without needing to know if it works by battery or solar energy, or if it is made of brass wheels or silicon chips.

3. *Intentional stance:* Explanations and predictions during this stance are based upon the presumption that the system acts rationally: it holds certain information, pursues certain goals, and chooses a behavioral pattern based on that particular background which is rational and appropriate to the situation. This means that humans ascribe intentional states towards the system such as hopes, worries, perceptions, emotions, etc. Therefore, the system becomes an intentional system. It undergoes an anthropomorphization process within its user. In this stance, you would explain and predict the moves of the chess computer, for example, by its desire to win. This stance is adopted if a system is organized, seems to be optimally constructed and the predictions of its behavior in physical or design stances are not feasible because of its complexity. For example, if you see a person (a complex biological system) waving on the other side of the street, you can hardly explain this behavior on a physical or design stance but only on an intentional stance, as his/her wish to greet someone.

2.2 Examples of other theories dealing with the ascription of mental states

There are theoretical concepts, mainly in psychology, which deal with the ascriptions of thoughts, emotions, aims, personality traits, and so on. Amongst these the concept of mentalization is very popular. It is akin to research of theory of mind (Fonagy et al., 2002) and is currently used in neuroscience and psychoanalysis. The ability to mentalize is defined as the ability to understand and interpret one's own behavior and the behavior of other people through ascribing mental states like thought, emotions, aims or wishes. This is a more application-oriented concept than Dennett's explanations and is used quite often for empirical research (Fonagy et al., 2002).

Apart from the implicit nature of humans ascribing mental states towards technical systems, the examples above show how humans develop even close, socio-emotional relations with technical systems. This can be described theoretically by using the psychological concept of emotional attachment (Bowlby, 1958). Following this concept, every person has the inherent need to establish intense close emotional relations with others. We now see the first efforts in transmitting this concept into relations towards technical systems. For example in Turner and Turner's (2012) research, no differences could be found in the emotional attachment to non-digital artifacts when compared to digital ones like mobile phones.

3. Users' Ascriptions in HCI: An Empirical WoZ Approach

3.1 Need for empirical investigations

We will follow the theory of intentional stance because it enables the expansion of considerations dealing explicitly with the ascription of mental states from humans to technical systems. Systematic empirical studies for establishing the concept of intentional stance adoption towards technical systems currently focus mainly on human-robot interaction (HRI). Thus, numerous, especially motion oriented, features of robots have been identified that increase the probability of users adopting this stance, e.g. rationality, goal-directedness, self-propelled motion, equifinality, spatial contingencies (Terada and Ito, 2008), and attention towards objects and humans (reactive movements) (Terada et al., 2007), as well as unexpected behavior (e.g. deception) from the robot (Terada and Ito, 2010; Short et al., 2010).

There are various methods used today for examining intentional stance. For example, neuroimaging (PET or fMRI) is used to identify those brain areas active when people ascribe intentions towards biological or non-biological systems (Gallagher and Frith, 2003). In HRI, self-report questionnaires are used, where subjects report their subjective perceptions of animations, usually using predefined answering options. These instruments are discussed according to methodological limits: Subjects report their ascriptions when prompted, but may not make them when instructed not to do so (Mar and Macrae, 2007).

If the objective is to systematically and empirically examine the way in which users ascribe personality towards computers and how this changes in reaction to different conditions of an interaction, there are four points that are, in our opinion, particularly relevant:

- developing a suitable experiment that is standardized to a large extent, to guarantee the comparability of user-specific results, which by design allows many opportunities for user ascriptions based on an experimental scenario, thus guaranteeing high involvement of the subjects, and is realistic and challenging;
- a sample which makes it possible to check the effects of age, sex, and education;
- suitable methods for identifying users' ascriptions during the interaction, which supplement studies based on self-report methods or neuroimaging;
- a design containing system interventions, which makes it possible to examine possible changes in users' ascriptions as a result of these interventions.

3.2 LAST MINUTE: A WoZ research project

This research project aims at a systematic empirical investigation of users when ascribing in HCI, and at possible changes in these as a result of an affect-oriented system intervention. Therefore, we tried to meet the requirements explained above for a suitable study design. Our study was based on a WoZ experiment wherein a subject interacts with a simulated computer system that is operated by a hidden experimenter (operator) (Gibbon et al., 1997). The first part of our results deals with the linguistic aspects of user utterances during this experiment, thus allowing for conclusions on how subjects experienced the system, and for further discussion of the indirect conclusions from user ascriptions. Next, changes in linguistic features are highlighted following the system's intervention. We finish with the results from analyzing the semi-structured interviews conducted as a subsample after the experiment. These interviews attempted to examine the subjective experience of the entire interaction, as well as the identification of the ascriptions which occurred during the interactions; furthermore, this was also investigated in regards to the intervention. The central result is the uncertainty experienced regarding the expectations of the simulated system, in particular regarding the ascriptions towards it and, as follows, the communication with it. These are discussed in connection with the development of mistrust and uncanny valley effects.

4. Description of the WoZ Experiment

4.1 Design issues

Our WoZ scenario was designed in such a way that many aspects that are relevant during the mundane situations of planning, re-planning, and strategy change (e.g. conflicting goals, time pressure, etc.) would be experienced by the subjects (Rösner et al., 2012a). The overall structure of the experiment was organized in modules as follows (Frommer et al., 2012):

- Greetings and Purpose,
- Prevention of Negative Courses of Dialogs (sub-module Initial Dialog, sub-module LAST MINUTE (see Figure 2)),
- Module Saying Goodbye.

The Initial Dialog served as a personalization phase, followed by the problem-solving at the 'LAST MINUTE' sub-module. These sub-modules served quite different purposes and were further sub-structured in a different manner (for more details, cf. Rösner et al., 2012a).

Personalization sub-module

Throughout the whole personalization phase the dominant mode of interaction was system initiative only, i.e. the system asked a question or provided a prompt. In other words, this sub-module was a series of dialog turns or 'adjacency pairs' (Jurafsky and Martin, 2008) that were made up of a system question or prompt followed by the user's answer or reaction. In some sense, this sub-module thus more resembled investigative questioning than a symmetric dialog.

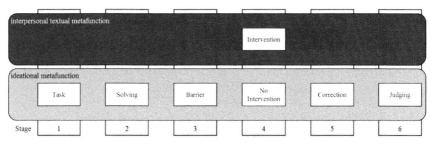

Figure 2. LAST MINUTE: stages of the experiment (The names for the levels of communication refer to Halliday's model of 'systemic functional grammar' (Halliday and Christian, 2004).)

The LAST MINUTE sub-module

Selection: During the bulk of 'LAST MINUTE', the subject was expected to pack a suitcase for a two-week holiday trip by choosing items from an online catalogue with twelve different categories, presented in a fixed order. In a simplified view, we thus had an iterative structure made up from twelve repetitions of structurally similar sub-dialogs of selections from a single category. For each category the available options were given as menu on the subject's screen.

Normal packing sub-dialog: In a normal packing sub-dialog, we essentially had a series of adjacent pairs made up of a user request for a number of items (more precisely, a user request for a number of instances from an item type) from the current selection menu (e.g. 'ten t-shirts') followed by a confirmation by the system (e.g. 'ten t-shirts have been added').

Barriers: The normal course of a sequence of repetitive sub-dialogs with choices from a total of twelve categories was modified for all subjects at specific time-points. These modifications or barriers were: Firstly, after the sixth category, the current contents of the suitcase were listed verbally (listing barrier), secondly, during the eighth category, the system for the first time refused to pack selected items because the airline's weight limit had been reached for the suitcase (weight limit barrier), thirdly, at the end of the tenth category, the system informed the user that there would now be more detailed information available about the target location Waiuku (Waiuku barrier).

Additional barriers could occur depending on the course of the dialog. These are typically caused by user errors or limitations of the system or a combination of both.

Intervention: After the Waiuku barrier, a randomly selected subset of half of the subjects received an empathic intervention that was designed according to psychotherapy principles (Elliot et al., 2011). The intervention created room for reflection and allowed subjects to verbalize any emotions they may have experienced.

4.2 Aspects of the experiments

Involvement

Our subjects were highly engaged during the experiments and all the subjects completed the WoZ sessions of approx. 30 minutes each. The subjects were involved in solving the planning task and experienced the need for a strategy change after the information about the weather conditions at the target location. In other words, the suggestive presentation in the initial legend of a presumable summer holiday of fourteen days had worked well and, after being informed that Waiuku is in the southern hemisphere, most subjects tried hard to readjust the contents of their suitcases. This was clearly expressed in the answers of those subjects with interventions as well in the answers to the final questions by all the subjects.

The role of the computer voice

We had specifically chosen a computer voice for the experiment that sounded more technical and machine-like than the state-of-the-art text-to-speech (TTS) systems. This decision was deliberate and was intended to support the illusion of interacting with a technical system. As the reactions of the subjects showed, this rationale proved to be successful. We could not find any indications that the subjects doubted that they were interacting with a technical device. Those subjects who referred to the quality of the computer voice during the interviews consistently emphasized voice quality as being highly relevant for their subjective impressions of the interaction.

Challenges for the subjects

During their initial briefing, the subjects were informed that all interactions would be based on speech only, and that neither a keyboard nor a mouse would be available to them. Since the briefing did not comprise any detailed information about the natural language processing and the problem-solving capabilities or limitations of the system, the subjects were more or less forced to actively explore these aspects during the course of the interaction. The challenge for the subjects was twofold: They had to find out how (i.e. with which actions) they could solve problems that they encounter during interaction, and also to discover which linguistic means were available for them to instruct the system to perform the necessary actions. In other words, in order to be successful, they had to build a model of the capacities and limitations of the system based on their experiences of successful and unsuccessful interactions. The user's model of the system would, of course, strongly influence the user's behavior and the subsequent course of the interaction.

4.3 Sample

The generated LAST MINUTE corpus was based on 135 experiments. Two experiments were aborted due to technical problems; three experiments were excluded because the subjects did not complete all the necessary questionnaires. Thus, the corpus consisted of data sets from 130 subjects. Seventy of them were between 18 and 28 years old ('young'; mean = 23.2; median = 23.0; std = 2.9; 35 males, 35 females), and 60 of them were 60 years old or older ('elderly'; mean = 68.1; median = 67.0; std = 4.8; the oldest subject was 81 years old; 29 males, 31 females). Within the young group, 44 subjects had a high school diploma, and 26 had none. Within the elderly group, 35 subjects had a high school diploma or a university degree and 25 subjects had none. In addition to the main experiment, 73 of the subjects, balanced in age, educational level, sex, and group (control group vs. experimental group), underwent a semi-structured interview (Lange and Frommer, 2011).

5. Analyses on Linguistic Structures, Rapport and Politeness

Linguistic analyses of the LAST MINUTE transcripts were an essential prerequisite for an in-depth investigation of the dialog and problem-solving behavior of the subjects during the WoZ experiments (Rösner et al., 2012b). A long-term goal is the correlation of findings from these analyses with socio-demographic and psychometric data from the questionnaires.

5.1 Motivation

The subjects in our WoZ experiments were only told that they would interact in spoken language with a new type of system that would be personalized, and thus would therefore ask some questions and pose some tasks. They did not receive explicit instructions about the linguistic constructions and interaction patterns that were possible or impossible to use during the interaction. How do people with differing technical backgrounds interact in such a situation? In the following, we report on the three areas investigated:

- What were the preferred linguistic structures employed by the subjects?
- Did the users adapt to the language of the system, i.e. did they create rapport?
- Did the subjects express politeness during their utterances?

The answers to these questions were relevant for interpreting from the linguistic evidence as to how the subjects may have experienced the system that they talked to and that talked to them during the experiments.

Linguistic structures

First inspections of transcripts quickly revealed that there were many variations in lexicalization but only a small number of linguistic constructs that the subjects employed during the packing and unpacking sub-dialogs of the LAST MINUTE experiments. For issuing packing (or unpacking) commands, these structural options were used:

- full sentences with a variety of verbs or verb phrases and variations in constituent ordering,
- elliptical structures without verbs in a (properly inflected) form like <number> <item(s)>,
- 'telegrammatic structures' in a (mostly uninflected) form with an inverted order of head and modifier like <item> <number>.

What was the distribution of these types of linguistic constructs? During a quantitative analysis of the 'LAST MINUTE' phase, the absolute and relative numbers for the usage of these constructs had already been calculated from the full set of transcripts.

Rapport

The next area explored was the question of rapport: Did the users mirror the language of the system, e.g. on a lexical or syntactic level? For example, this system uses the general and somewhat non-colloquial term 'hinzufügen' (engl. to add) in its feedback for selection operations. Similarly the system always uses 'wurde entfernt' (engl. approx. 'was removed') when confirming unpacking requests. Did the users mirror this usage?

Politeness

In all its utterances, the system uses the polite German 'Sie' (polite, formal German version of 'you') when addressing the user. In requests the system employs the politeness particle 'bitte' (engl. 'please'). How polite were the users during their utterances?

5.2 Methods and results

The general approach for all three questions was to perform quantitative analyses with appropriate search patterns within the complete set of *N*

= 130 transcripts. In those cases where the occurrences of the tokens searched for may be ambiguous, the respective findings were manually inspected and qualitatively interpreted within their context.

Linguistic structures

The packing/unpacking phase of the N = 130 transcripts can be summarized as follows: We had a total of 8,622 user utterances. When we performed part of speech (POS) tagging (with a slightly modified version of STTS[1]) and then counted the varying POS tag patterns, we found 2,041 different patterns for the 8,622 utterances. The distribution was strongly skewed (cf. Table 1): A small number of (regular) POS patterns did cover a significant fraction of utterances.

Table 1.   Most frequent POS patterns for elliptical structures.

Class	Sem	POS pattern	No. of occurrences
E	P	ART NN	1,020
E	P	CARD NN	657
E	P	NN	537
E	C	ADJ NN	355
E	C	ADJ, ADV	349
E	C	NN	148

When classifying the POS tag sequences, we distinguished four categories: full sentences and sentence-like structures (S; with an obligatory verb), elliptical constructs without a verb (E), telegrammatic constructs (T, cf. above) and meaningful pauses, i.e. user utterances that more or less consisted of interjections only (DP). In descending order of occurrences we had the following counts:

- 5,069 user utterances or 58.79% (realized with 223 patterns) were classified as E,
- 807 user utterances or 9.36% (realized with 135 patterns) as S,
- 551 user utterances or 6.39% (realized with 21 patterns) as T, and finally
- 178 user utterances or 2.06% (realized with 8 patterns) as DP.

At the time of writing, 2,017 utterances that were realized by 1,654 different patterns could not be uniquely classified. In many cases, this was due to the typical phenomena of spontaneous spoken language, e.g. repairs, restarts, and the use of interjections.

---

[1]   www.ims.uni-stuttgart.de/projekte/corplex/TagSets/stts-table.html

Rapport

Within the total of $N = 130$ transcripts, only in $N_1 = 25$ transcripts was there at least one occurrence of a form of the verb 'hinzufügen' (engl. 'to add') that could be found in the user utterances during the packing/unpacking phase. For these $N_1 = 25$ transcripts, we had a range from 1 to maximally 8 occurrences with mean: 2.36, std: 1.85, and median: 2.0. Within $N_2 = 68$ transcripts, at least one occurrence of a form of the verb 'entfernen' (engl. 'to remove') could be found in user utterances during the packing/unpacking phase. For these $N_2 = 68$ transcripts, we had a range from 1 to maximally 13 occurrences, with mean: 4.22, std: 3.34, and median: 3.0. In the intersection of both groups, i.e. at least one occurrence each of a form of the verbs 'entfernen' and 'hinzufügen', we had $N_3 = 20$ transcripts. For these $N_3 = 20$ transcripts, we had a range from 2 to maximally 19 combined occurrences with mean: 8.85, std: 4.64, and median: 8.0.

Politeness

Politeness when addressing the system was demonstrated in a variety of ways. We first took a detailed look at the usage or avoidance of personal pronouns and then we analyzed the usage of politeness particles.

Pronouns: The following counts were all (unless otherwise noted) taken from the packing/unpacking phase of the transcripts: Within a total of $N = 130$ transcripts only in $N_1 = 21$ transcripts at least one occurrence of 'Sie' as a formal personal pronoun was used when addressing the system. Only within $N_2 = 4$ transcripts was the informal 'du' (or one of its inflected forms) used to address the system (other uses of 'du' were within idiomatic versions of swear words like 'ach du lieber gott!', engl 'oh god!'). Within $N_3 = 18$ transcripts the subjects employed the plural personal pronoun 'wir' (engl 'we'). Some occurrences of 'wir' during off-talk could be seen as more or less fixed phrasal usages (like 20101115beh 'ach das schaffen wir locker', engl. '. . .we will make this with ease' or 20110401adh 'wo waren wir', engl. 'where have we been'), but when used in commands (packing, unpacking, . . .) then this pronoun could be given an inclusive collective reading as referring to both subject and system as a joint group. Please note: The pronoun 'wir' thus allows users to avoid explicitly approaching the system. Two examples of this latter usage:

20110307bss: ja dann nehm wir eine jacke raus
[engl.: yeah then we take a jacket off]
20110315agw: dann streichen wir ein hemd
[engl.: then we cancel a shirt]

In summary: How the users approached the system differed significantly. Most subjects avoided any personal pronouns when addressing the system, some employed the German 'Sie' (formal German version of 'you') and only very seldom was the informal German 'du' used.

Politeness particles

From the total of $N = 130$ subjects $N_1 = 67$ used one of the politeness particles 'bitte' or 'danke' at least once within the packing/unpacking phase. The maximum number of uses is 34, with a mean of 7.57, std of 7.89, and median of 4.0. If we do neglect those subjects at and below the median as only occasional users of these particles, we do obtain $N_2 = 32$ subjects that used these particles much more frequently. Intersecting the group of subjects with at least one occurrence of 'Sie' (cf. above) with the users of the politeness particles 'bitte' or 'danke,' resulted in a subgroup of $N_3 = 19$ subjects. Their combined numbers of occurrences were ranging from 2 to 32 with mean: 8.60, std: 8.38, and median: 4.00. In other words, most users of 'Sie' were also users of the politeness particles.

5.3 Discussion and remarks

Linguistic structures

The use of elliptical structures is a typical aspect of efficient communication in naturally occurring dialogs (Jurafsky and Martin, 2008). Thus, the dominance of elliptical structures within the users' contributions in the LAST MINUTE corpus could be seen as a clear indication that most subjects had experienced their dialogs with the system in a way that licensed their natural dialog behavior. The empirical analysis of the structures regarding user utterances was fed into the implementation of an experimental system as well, thus allowing to replace the wizard with an automated system based on the Nuance commercial speech recognizer.

Rapport

Given the figures about the users' employing specific verbs used by the system, we had to conclude that lexical rapport—i.e. users mirroring the lexical items of the system—was the exception rather than the rule. Nevertheless it would seem worthwhile to explore whether—and how—the subgroup of subjects that did so, is differing from those subjects that did not. This finely grained analysis is now on the agenda.

Politeness

Politeness is one of a number of indicators of the way subjects experience the system. The difference between system and users in using personal pronouns and politeness particles was another example for the finding that most users do not try to build rapport with the system on the level of lexical choices. As with other subgroups of subjects, the following questions have to be further investigated: Were there differences in the overall dialog success or failure between the 'normal' and the 'polite' users? Were there correlations between user politeness, socio-demographic data, and those personality traits measured using psychometric questionnaires?

6. Changes in Speech Behavior System Controlled Intervention

6.1 Motivation

In order to communicate meaningfully, certain necessary conditions for language must be met, amongst which is the assumption that the object of the communication is conscious (Stroud, 1999). Intentionality can be examined, like all states of consciousness, through introspection, but is also reflected in the language produced by a speaker. Therefore, the analysis of language features can provide additional information about the relations between the subject, a target object (which can be with or without consciousness), and the context within which the communication occurs. However, it is unclear which particular language features correspond to whatever is ascribed. In this section, attention turns towards two linguistic phenomena: informal vs. formal language styles and user-initiated voice overlaps. Informal language, defined only in that it is not the fully formal form of the given language, can be understood in the broadest sense as dialectal linguistic behavior. Linke (1996) described dialects as coded alienations of linguistic form. People who are able to speak both dialectal and formal language show a rudimentary form of multilingualism. The switching between different language forms is called code-switching (Denkler, 2007). This is understood to be an expression of increased emotion or increased arousal (Dewaele, 2010; Grosjean, 2001). At this point, it should be emphasized that this does not necessarily mean that code-switching is always an indicator of increased emotion.

Overlapping utterances in interactions can be seen as an expression of a high involvement style (Tannen, 1984). Empirical studies have shown that overlaps are an expression of involvement (Murata, 1994; Li et al., 2005) and can therefore be considered a valid interpretive feature. Our research focused on a psychotherapeutically oriented intervention, and its influence on the language behavior described above.

- What effect does a system-initiated psychotherapeutically oriented intervention have on the emotional state of a user?
- Can ego-involvement be increased by the implementation of an intervention?

6.2 Methods and results

The transcribed interactions before and after the intervention phase were examined, along with the dialectal speech and the overlaps induced by the subjects. First, the transcripts were analyzed for the verbosities of the individual subjects. The verbosities of the subjects within those phases before and after the barrier did not differ significantly ($F = 1.48$, $p = .226$) between the experimental ($M_{pre} = 127.4$; $M_{post} = 141.9$), and the control group ($M_{pre} = 136.8$; $M_{post} = 147.6$). This was advantageous for further analysis, because the intervention had no effect on the verbosities.

Dialectal speech behavior

Code-switching between formal and dialectal speaking styles was only expected for those people using dialectal language. Therefore, only those subjects that produced at least one example of dialectal linguistic behavior were included in the analysis ($N = 100$). First, the number of dialectal forms of language was divided by the total number of each subjects' verbosity to calculate a percentage value. This enables a between subjects comparison of the dialectal speech behavior. During the corrective phase (stage 5, see Figure 2), immediately after the subjects received either the empathic intervention or no intervention, statistically significant differences were determined between the experimental ($M = 1.4$) and control groups ($M = 0.7$). The subjects of the experimental group produced significantly less dialectal speech, whereas before the intervention phase the groups did not differ from each other.

Overlaps

Again, the number of overlaps was scaled according to verbosity, in order to create a percentage value. A distinct variance was found between the determined overlap of the experimental and control groups. The subjects of the experimental and control group not significantly differ before the intervention (phase 2) and after the correction phase (phase 5). In the experimental group, however, a clear increase in overlapping was observed for the correction phase (Figure 3).[2]

6.3 Discussion and remarks

The preliminary results obtained show that a system initiated intervention can influence the state of the user.

Dialectal speech behavior

The reduced expression of dialectal linguistic behavior after the intervention within the experimental group suggested a low arousal regarding the emotional state (Grosjean, 2001; Dewaele, 2010).

Overlaps

Furthermore, statistically insignificant differences between the experimental and control groups were found in terms of subject-initiated overlaps of speech. Descriptively, a clear difference in the results between the second and the fifth phases are shown, which directly followed the intervention. This indicates a higher ego-involvement of subjects within the experimental group.

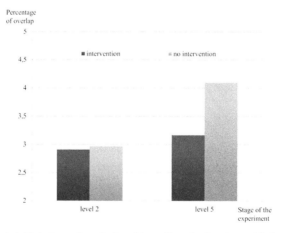

Figure 3. Subjects initiated overlaps before (stage 2) and after (stage 5) the intervention/no intervention.

---

[2] The results of the analyses carried out were regarded as preliminary, since the scope of the automatic detection of overlaps and informal language (dialectal linguistic behavior) is still limited, but improving steadily.

7. Users' Subjective Experiences and Ascribing During HCI

7.1 Motivation

The aforementioned philosophical and psychological theories and concepts, as well as the examples from empirical studies, suggest that the development of ascribing, like aims, motives, even personality characteristics, to computers does occur in HCI. In terms of Dennett's argumentation, it means: Because computers are perceived by ordinary users as complex, but optimally constructed and with highly complex physical processes occurring inside, it is impossible to explain or predict their behavior within physical or functional stances. Thus, to our minds, it is probable that users adopt the intentional stance; the computer's behavior is explained by ascribing mental states, and therefore they experience the computer as an intelligent, humanoid counterpart (they anthropomorphize it). We presume that positive and negative ascribing towards computers affect the experience and form of the interaction with the system, in such a way that a systematic empirical investigation of users' ascriptions is worthwhile. Therefore, we ask ourselves the following questions:

- How do users experience the interaction with the simulated system and what ascriptions do they make?
- Which effects arise from the subjective experience and the ascribing of the form and development of the interaction?
- How do users experience a system-initiated affect-oriented intervention within a critical dialog situation during HCI?

7.2 Methods and results

We conducted semi-structured interviews using a subsample of subjects after the WoZ experiments (Lange and Frommer, 2011). Such interviews enable people to reflect and explain their thoughts, beliefs, feelings, and processes in a free, non-restricted way. Besides the other advantages of this method, the individual inhibition threshold can be reduced, e.g. if someone has the feeling of being naive when ascribing humanlike characteristics towards a technical system. Each of our interviews starts by encouraging spontaneous narration about subjective experiences during the experiment. This was prompted by the interviewer: "You have just done an experiment. Please try to put yourself back in the experiment and tell me how you did during the experiment. Please tell me in detail what you thought and experienced!" Other topics were examined according to the interview guidelines, e.g. former feelings and experiences with technical devices.

We noticed that the initial narration of the subjects comprised a lot of ascriptions towards the system, so during the first step we concentrated on those parts of the interview within our analysis. During the second step, we focused the interview sequence on the experience of the affect-oriented intervention using 'Interpersonal Process Recall' (IPR). IPR is a special interview situation, which is established in psychotherapy research (Elliott, 1986; Frommer and Lange, 2010). It facilitates the identification of the subtle, hidden aspects of an interaction, and stimulates more accurate reports of experiences: The record of the interaction is promptly shown to the subject and he/she should think back through the situation and describe what he/she felt and experienced. The analysis of the initial narrative interview sequence, as well as the sequence referring to the intervention, was done using established qualitative social research methods, specifically systematic content analysis and interpretative phenomenological analysis, which allow for the condensing of the material, and extraction of the main topics (Krippendorff, 2004; Smith and Osborn, 2003).

In the following paragraphs, we concentrate on some of the main results from our interview analysis: first, regarding uncertainty when ascribing towards the system during the whole interaction, and second, regarding their experiences when ascribing during the system's intervention.

The first part of the analysis revealed that the users were relatively uncertain about what they can expect from the system, what characteristics, aims, motives, and so on, they should ascribe towards it and how they should optimally communicate with it. This uncertainty can be recognized on three levels: first, actualization of presumptions, ascriptions and memories; second, the behavior and actions of the counterpart; third, the strategies when in contact with the counterpart/reactions of the user. In the following, there are examples for each of these uncertainty levels introduced and illustrated by subjects' utterances in the interviews.

Actualization of presumptions, ascriptions and memories

Users experience "uncertainty of expectation caused by the phenomena of hybridization". That is, the system is experienced as a kind of hybrid, which is neither a clear technical interaction partner (like a computer or machine), nor a clear human interaction partner ("after all you felt like you were talking to a human, but when you still remember that it's a computer voice it makes you afraid that it heard pretty much everything"—CT[3]). Features like the sound of the "voice", lack of reciprocity, and asymmetry during communication lead to the impression of interacting with a computer or machine (or even an object). On the other hand, features like humanlike speech-recognition and the ascription of the ability to exude confidence evoke the impression of interacting with a conscious being. This impression of hybridization causes uncertainty and irritation in the user about what he or she can expect from the counterpart, and demands for tolerance of ambiguity and the development of adequate and representative expectations. This is accompanied by uncomfortable uncertainty in regards to a possible abuse of confidence by the hybrid, and feelings of foreignness occur because of a communication pattern that is unusual for human-human communication ((reference: self introduction) "above all I didn't even get a reaction on anything (.) it takes note of it but doesn't react (.) maybe a conversation or even an interjected question would have been indicative but it didn't happen (.) you have to speak in turns with each other (.) that's unfamiliar"—FW).

Behavior and actions of the counterpart

Users tend to differ when experiencing the system's question for user information as "loss of anonymity vs. development of relatedness". It appears that questions from the system are accepted (even favored) although users evaluate these as inadequate, as the purpose of the questions remains unclear and so the discourse is difficult to make sense of ("no no it was good that I had to tell the computer how I feel and how old I am (.) well shoe-size weight and stuff is nonsense but apart from that asking about hobbies was ok for me too (.) I liked that because I thought that gives a fuller picture (.) you knew he wanted to know that in order to see what she does and how is she (.) maybe even what her IQ is (.) well you don't do that, but I mean it is still important that a computer knows that (.) because I think it can rate me better"—MW). Such questions are accompanied by the impression of being tested, uncomfortable uncertainty and skepticism in regard to the subsequent use of given data, an alternation between the feeling of being questioned (loss of anonymity) and the feeling of being personally identified (development of relatedness) and the assumption of the system's adaptation to the user ("well now in hindsight I'm asking myself why it did that (.) when it was asking me I thought we were playing some kind of game or it will show me a movie about soccer or I don't know something about my hobbies triggers something

---

[3] CT, FW, MW, UK: refers to subject codes; users' utterances were translated from German to English.

specific to me and when I see it in hindsight I don't know what it needed it for (.) so it was doing it just to learn something about me (.) I guess I don't know why, but yeah"—CT). Surprisingly, the subjects presented here consistently provide the requested information and thereby bow to the system.

Strategies in contact with the counterpart/reactions of the user

Users "work for adaptation and personalization" of the system. They do not seem to experience the system as adaptive, but instead they try to adapt to the system themselves ("when I was asked to introduce myself (.) somehow I was already thinking (.) introducing ok what can be interesting for a computer program and then I told it my age what I do where I live and then I was waiting for what else it wanted to hear (.) and then I was asked to say where I live where I work and something about my family (.) so at first it rattled off everything and I tried to keep that all in mind in order to work everything off so to speak"—UK). They anticipate the system's aims and abilities and attempt to find adequate strategies for his or her responses ("and also being asked about these experiences that I was especially happy about and those experiences I found especially annoying (.) at first I was rummaging around for something to tell this machine that it might understand, so to say"—UK). This adaptation and personalization to the system takes cognitive effort, which causes the stress subjects tolerated according to the objectives of establishing an interaction with the system, to gratify its "requirements" and to get past feelings of foreignness ("well it was a little bit like talking to someone who's not German and just started to learn the language (.) it's really like, yeah almost like an accent so to speak (.) you have to get used to it first so after having heard the first words and adapting yourself to talking to someone and then again so to speak needing twenty thirty seconds to understand the one talking"—UK).

In the second part of our interview study, we conducted a detailed analysis on how users experience the affect-oriented system intervention given within a critical dialog situation and what characteristics, aims, etc. they ascribe towards the system in this specific situation. Five main themes arose (Wahl et al. (in press): First, emotional support versus intention to mislead; second, the opportunity to reflect as support and demand for self-criticism; third, wish for assuming responsibility and relief; fourth, suspicion and skepticism concerning privacy; fifth, unexpected system features. In the following, each theme is explained in detail and is illustrated by the utterances of subjects during the interviews.

Emotional support versus intent to mislead

Some of the subjects felt understood and taken care of ("And then there are those queries asking if everything is still alright and that makes you feel safer"—UK), whereas others felt they were not being taken serious ("Actually it doesn't matter what I say, if I say it is this way or that way, the computer isn't interested anyway"—KK) and intentionally deceived ("Yeah I felt punked a bit"—MB).

The opportunity to reflect as support and de mand for self-criticism

Subjects took the chance to reflect upon their affect, and the system was perceived as supportive ("In that moment it was definitely a relief that it asked me about that"—CK), but at the same time they felt the system placed a demand on them "to admit the mistakes you've made and introspect a little bit" (BS).

Wish for assuming responsibility and relief

Being in a critical situation, the subjects wanted the computer to assume responsibility ("There is this tendency, every now and then that you want to blame the computer for something"—UK). They felt that this would be a relief when dealing with that situation ("If it would have tried to vindicate itself, I would have known what was going on, now I don't know why it did this to me"—TB).

Suspicion and skepticism concerning privacy

Subjects felt uncomfortable with disclosing personal information ("It wants to know something about your feelings, but you always have this kind of protective wall around yourself and you don't want to open up about yourself, especially not to a computer"—MB) and mistrust about how the system would use it ("You don't know how it wants to use the information and so some kind of insecurity and mistrust appears"—UK).

Unexpected system features

Subjects underestimated what the computer system was capable of and were surprised ("You always think that a computer is a predetermined program, but when it reacts that individually it's kind of strange, impressive or great (…) you don't expect that from a computer"—BP). As a result they adapted their communication style to a level they thought the system was able to process ("Well it's a medium you don't expect that much from, you always try to keep things simple"—UK).

Users experienced the intervention individually. Their reports varied from feelings of emotional support or having the opportunity for self-reflection based on the system's recognition of the affectivity that occurred during the critical situation, to the ascriptions of intentional deception or the demand for self-criticism. Surprisingly, similar patterns of uncertainty and attempts to understand and adapt to the system, which occur throughout the interaction, and arise during the intervention when the system interacts in a way that could be perceived as empathetic, the unexpected perception of which can lead to yet more uncertainty, and further attempts to adapt. Ascribing empathy towards the system leads to the ascriptions of higher cognitive functions such as the ability to abuse confidence. In addition, the demands of users towards the system arise, e.g. the demand to assume responsibility for the development of a critical situation, to disclose its aims, and to reinforce the behavior of the user. All in all, individual differences when experiencing and accepting the intervention have become apparent and necessitate the design of user-type specific contents and strategies of system interventions in order to enable individualized support.

7.3 *Discussion*

Our analysis was surprising in the number and variety of users' who ascribed during their interactions with the system. The phenomenon of hybridization proved to be important; the resulting uncertainty about the nature of the counterpart (human or machine?) could be related to diverse further ascriptions and behaviors of the users. These show that a higher anthropomorphization of the computer system is not necessarily accompanied by higher acceptance or trust in the system—a phenomenon known as the "uncanny valley effect" (Pollick, 2010).

Trust and trustworthiness in HCI have been investigated frequently (Steghöfer et al., 2010), but mainly in connection with aspects of safety, privacy, reliability, and effectiveness (Hasselbring and Reussner, 2006; Sutcliffe, 2006). According to our findings, trust and mistrust seem to have resulted from users' ascribing abilities. It is a question of feelings, which have a regulating function during interpersonal relations. In a broader sense, trust can be defined as "confident expectations of positive outcomes from an intimate partner" (Holmes and Rempel, 1989, p. 188); more specifically it is described as follows "according to the dyadic (interpersonal) perspective, trust is a psychological state or orientation of an actor (the trusting) towards a specific partner (the trusted) with whom the actor is in some way interdependent (that is, the trusting actor needs the trusted's cooperation to attain valued outcomes or resources)" (Simpson, 2007). Most studies try to increase trust by designing systems in a more anthropomorphic way, but encounter contrary effects in doing so. For example, Nowak (2004) found that the agent with the highest level of anthropomorphizm in the study was reported as being the least socially attractive, reliable and credible, whereas the agent with the lowest level of anthropomorphizm was reported as being highly socially attractive, reliable and credible. The uncanny valley effect, as reflected in these findings, describes the complicated relation between realism/anthropomorphizm and users' acceptance of artificial figures like avatars, interface agents, etc. (Pollick, 2010): The assumption that an artificial figure will be more acceptable the more realistically/anthropomorphically it is designed, has to be rejected. In fact, acceptance does not linearly increase with any increase in anthropomorphizm, but at a certain point of 'very realistic but not perfect', it abruptly decreases into unacceptability, and then increases again when the representation of life is indistinguishable from reality, such as in films. The impressions of being uncanny and confused as experienced by our subjects seemed to have been increased further by the system's disregard of the need for distance and privacy (loss of anonymity). This was followed by mistrust regarding a possible abuse of confidence. The ascription of intentional deception further increased this feeling. If you follow the consequences further, the users' co-operation up to the point of reacting, in terms of a complex reaction as a defense against experiencing outer and inner constraints (Miron and Brehm, 2006), seems probable, e.g. for example the ascription of intentions to pressurize, compel or demand subservience. Several studies have shown that reactance during HCI and human-robot interaction is an observable phenomenon (Lui et al., 2008). It seems that reactance occurs more readily if (animated) agents/avatars are involved. For example, there was the highest level of reactance (operationalized as feelings of anger and negative cognitions) when presented with information by an animated robot, the moving mouth of which was accompanied by text shown within a speech-bubble, in contrast to the same information using a non-animated image of the robot, or as text alone without an agent (Roubroeks et al., 2011). When users ascribe more, and therefore the anthropomorphization of the system increases, impressions of uncanniness (uncanny valley) and thereby mistrust, seemed more probable, which could have led to reactance. The users seemed to be trying to deal with such feelings of foreignness, skepticism, etc. by adapting to the identified abilities and aims of the system. The effect of user characteristics known from personality psychology (age, sex, education, technophile, stress handling, interpersonal problems, and personality traits according to the Big Five model) on the maintenance and development of trust and co-operation within HCI, is a topic of ongoing research (Saariluoma and Oulasvirta, 2010; Saleem et al., 2011; Schweer, 2008).

However, aside from negative ascribing, numerous positive outcomes were ascribed showing that the user formed the impression that the system wanted to connect with him/her on a personal level by asking for private information and experiences, wanting to develop a relation with the user, wanting to adapt to the user's individuality. It was particularly during the system's intervention that positive ascribing arose, such as the ascriptions of understanding or supportiveness towards the system. These experiences enhanced the users' continued co-operation and the development of trust, even under conditions of feasible anthropomorphization.

8. Summary and Conclusion

When discussing spoken dialog systems, Edlund et al. (2008) have suggested and convincingly argued for distinguishing between different metaphors that users may use during interactions with such a system. Do users treat the system more like a tool, i.e. do they choose the interface metaphor, or do they prefer the human metaphor by accepting the system as an interlocutor and behave more like they would during human-human dialogs? In the light of the metaphor discussion, we can summarize and re-interpret our results from Analyses on Linguistic Structures, Rapport and Politeness as follows: those subjects whose linguistic behavior provided a strong indication for the dominance of one of these metaphors were in the minority within our sample. This holds for the minority group of those that preferred technically sounding "telegrammatic structures" (cf. above) and thus obviously preferred the interface metaphor on the one hand, as well at the other extreme, for the group of those that heavily employed interpersonal signals such as formal pronouns and politeness particles, thus indicating a human metaphor at work. Although further investigations are necessary, the majority of our subjects—see as well the discussion of hybrids in Chapter 7—seem to have worked with "a metaphor that lies between a human and a machine—the android metaphor" (Edlund et al., 2008).

The results of the linguistic analyses regarding changes in speech behavior due to the system's intervention revealed that, based on the mean scores, the intervention had a twofold effect on the users: firstly, it made them more involved in the task and secondly, the users were less emotional. These effects are traceable to the characteristics of the intervention: on the one hand, it addressed the difficulty of the actual situation and the emotional state of the user and, on the other hand, it communicated on a more interpersonal level. In providing the intervention, the system created a relation offer to the user through empathy, which he or she could accept or deny. We can therefore assume that a user will only accept the system's relation offer if it is in keeping with his/her of mental states to the system. An intervention which is inappropriate to the situation or the relation the user has with the system will be unacceptable.

The quantitative analysis of speech behavior showed a tendency of the users' to ascribe human characteristics towards the system. In order to respond individually to users, more information *is needed* about users' individual experiences when ascribing. The analyses of the interviews gave insights into the ranges and variances of these experiences. It became apparent that these results sometimes differed from the quantitative analysis.

The ascriptions identified during our interview analysis, as well as their effect on users' behavior, have led to numerous resulting theoretical considerations according to their relevance: On the one hand, the ascriptions can increase users' co-operation and trust in the system, if they ascribe positive personality traits, motives, aims, attitudes, etc. (for example willingness to personally identify the user). On the other hand, negative ascriptions can lead to a reduction of co-operation, accompanied by feelings of mistrust and skepticism (like ascribing the possibility of abuse of confidence to the system), which could lead to reactive user behavior.

All in all, our results seem to fortify our assumption, that beyond the programmer, who implements personality within computers, it is the user who is "giving" personality to the computer by ascribing different mental states like motives, aims, feelings, as well as personality traits to it. Further investigations are needed, especially regarding relations between the linguistic features of users' utterances, as well as users' experiences when ascribing during HCI. Moreover, investigating the impact of different system interventions on these relations, which should be individualized for users or user-types, seems worthwhile for maintaining and increasing users' co-operation in HCI.

Acknowledgements

The presented study is performed in the framework of the Transregional Collaborative Research Centre SFB/TRR 62 "A Companion-Technology for Cognitive Technical Systems" funded by the German Research

Foundation (DFG). The responsibility for the content of this paper lies with the authors.

REFERENCES

Bowlby, J. 1958. The nature of the child's tie to his mother. *International Journal of Psycho-Analysis*, **39**:350–373.

Denkler, M. 2007. Code-switching in Gesprächen münsterländischer Dialektsprecher. Zur Sprachvariation beim konversationellen Erzählen. *Zeitschrift für Dialektologie und Linguistik*, **74(2+3)**:164–196.

Dennett, D.C. 1971. Intentional systems. *The Journal of Philosophy*, **68**:87–106.

Dennett, D.C. 1987. *The Intentional Stance*. Cambridge, UK: The MIT Press.

Dewaele, J.M. 2010. Emotions in Multiple Languages. Basingstoke, England: Palgrave Macmillan.

Edlund, J., J. Gustafson, M. Heldner and A. Hjalmarsson. 2008. Towards human-like spoken dialogue systems. *Speech Communication*, **68**:87–106.

Elliot, R., A.C. Bohart, J.C. Watson and L.S. Greenberg. 2011. Empathy. *Psychotherapy Research*, **48(1)**:43–49.

Elliott, R. 1986. Interpersonal Process Recall (IPR) as a psychotherapy process research method. In L. Greenberg and W. Pinsof (eds.), *The Psychotherapeutic Process* (pp. 503–527). Guilford, USA: The Psychotherapeutic Process.

Eysenck, H.J. and A. Levy. 1972. Conditioning, introversion-extraversion, and the strength of the nervous system. In V.D. Gray and J.A. Nebylitsyn (eds.), *Biological Bases of Individual Behavior*. New York, USA: Academic Press.

Fisseni, H. 2003. Persönlichkeitspsychologie. Ein Theorienüberblick. 5th Edition. Göttingen, Germany: Hogrefe.

Fonagy, P., G. Gergely, E.L. Jurist and M. Target. 2002. Affect Regulation, Mentalization, and the Development of the Self. New York, USA: Other Press.

Frommer, J. and J. Lange. 2010. Psychotherapieforschung. In G. Mey and K. Mruck (eds.), *Qualitative Forschung in der Psychologie* (pp. 776–782). Wiesbaden, Germany: Verlag für Sozialwissenschaften.

Frommer, J., D. Rösner, M. Haase, J. Lange, R. Friesen and M. Otto. 2012. Teilprojekt A3: Früherkennung und Verhinderung von negativen Dialogverläufen. Operatormanual für das Wizard of Oz-Experiment. Lengerich, Germany: Pabst Science Publishers.

Gallagher, H.L. and C.D. Frith. 2003. Functional imaging of 'theory of mind'. *Trends in Cognitive Sciences*, **7(2)**:77–83.

Gibbon, D., R. Moore and R. Winski. 1997. Handbook of Standards and Resources for Spoken Language Systems. Berlin, Germany: Mouton de Gruyter.

Goldberg, L.R. 1990. An alternative "description of personality": The Big-Five factor structure. *Journal of Personality and Social Psychology*, **59**:1216–1229.

Grosjean, F. 2001. The bilingual's language modes. In J. Nicol (ed.), One Mind, Two Languages: Bilingual Language Processing (pp. 1–22). Oxford, England: Blackwell.

Guilford, J.P. 1959. *Personality.* New York, USA: McGraw-Hill.

Halliday, M.A. and M.I. Christian. 2004. An Introduction to Functional Grammar. 3rd Edition. London, UK: Hodder Arnold.

Hasselbring, W. and R. Reussner. 2006. Toward trustworthy software systems. *Computer*, **39(4)**:91–92.

Hess, T., M. Fuller and D. Campbell. 2009. Designing Interfaces with Social Presence: Using Vividness and Extraversion to Create Social Recommendation Agents. *Journal of the Association for Information Systems*, **10(12)**:889–919.

Holmes, J. and J.K. Rempel. 1989. Trust in close relationships. In C. Hendrick (ed.), Close Relationships (pp. 187–220). Newbury Park, USA: Sage.

John, O., Hampson, S. E. and L.R. Goldberg. 1991. Is there a basic level of personality description? *Journal of Personality and Social Psychology*, **60**:348–361.

Jurafsky, D. and J.H. Martin. 2008. Speech and Language Processing: An Introduction to Natural Language Processing, Computational Linguistics, and Speech Recognition. 2nd Edition. New Jersey, USA: Prentice Hall.

Krippendorff, K. 2004. Content Analysis: An Introduction to Its Methodology. 2nd Edition. Thousand Oaks, USA: Sage.

Lange, J. and J. Frommer. 2011. Subjektives Erleben und intentionale Einstellung in Interviews zur Nutzer-Companion-Interaktion. Lecture Notes in Informatics, volume 192, p. 240. Berlin, Germany: GI-Jahrestagung.

Langer, E.J. 1989. Mindfulness. Reading, USA: Addison-Wesley.

Li, H., Y.O. Yum, R. Yates, L. Aguilera, Y. Mao and Y. Zheng. 2005. Interruption and involvement in discourse: Can intercultural interlocutors be trained? *Journal of Intercultural Communication Research*, **34(4)**:233–254.

Linke, A., M. Nussbaumer and P. Portmann-Tselikas. 1996. Studienbuch Linguistik. Tübingen, Germany: Niemeyer.

Lui, S., S. Helfenstein and A. Wahlstedt. 2008. Social psychology of persuasion applied to human-agent interaction. An interdisciplinary journal on humans in ICT environments. *Interdisciplinary Journal on Humans in ICT Environments*, **4(2)**:123–143.

Mar, R.A. and C.N. Macrae. 2007. Triggering the intentional stance. In Novartis Foundation, Empathy and Fairness (pp. 111–133). London, UK: John Wiley & Sons, Ltd.

McAdams, D. 1995. What do we know when we know a person? *Journal of Personality*, **63**:365–396.

McKeown, G., M. Valstar, R. Cowie, M. Pantic and M. Schröder. 2012. The SEMAINE database: Annotated multimodal records of emotionally coloured conversations between a person and a limited agent. *IEEE Transactions on Affective Computing*, **3(1)**:5–17.

Miron, A.M. and J.W. Brehm. 2006. Reactance Theory—40 years later. *Zeitschrift für Sozialpsychologie*, **37**:9–18.

Murata, K. 1994. Intrusive or co-operative? A cross-cultural study of interruption. *Journal of Pragmatics*, **21(4)**:385–400.

Nass, C. and Y. Moon. 2000. Machines and Mindlessness: Social Response to Computers. *Journal of Social Issues*, **1(56)**:81–103.

Nass, C., Y. Moon, B. Fogg, B. Reeves and D.C. Dryer. 1995. Can computer personalities be human personalities? *International Journal of Human-Computer Studies*, **43(2)**:223–239.

Nowak, K.L. 2004. The influence of anthropomorphism and agency on social judgment in virtual environments. *Journal of Computer Mediated Communication*, **9(2)**:00.

Ozer, D.J. and V. Benet-Martinez. 2006. Personality and the prediction of consequential outcomes. *Annual Review of Psychology*, **57**:401–421.

Pollick, F.E. 2010. In search of the uncanny valley. *Social Informatics and Telecommunications Engineering*, **40(4)**:69–78.

Rösner, D., J. Frommer, R. Friesen, M. Haase, J. Lange and M. Otto. 2012a. LAST MINUTE: A multimodal corpus of speech-based user-companion interactions. In LREC 2012 Conference Abstracts (pp. 96–104). Istanbul, Turkey.

Rösner, D., M. Kunze, M. Otto and J. Frommer. 2012b. Linguistic analyses of the LAST MINUTE corpus. Proceedings of KONVENS 2012 (pp. 145–154). Vienna, Austria.

Roubroeks, M., J. Ham and C. Midden. 2011. When artificial social agents try to persuade people: The role of social agency on the occurrence of psychological reactance. *International Journal of Social Robotics*, **3(2)**:155–165.

Saariluoma, P. and A. Oulasvirta. 2010. User psychology: Re-assessing the boundaries of a discipline. *Psychology*, **1(5)**:317–328.

Saleem, H., A. Beaudry and A.M. Croteau. 2011. Antecedents of computer self-efficacy: A study of the role of personality traits and gender. *Computers in Human Behaviour*, **27(5)**:1922–1936.

Schweer, M. 2008. Vertrauen und soziales Handeln–Eine differential psychologische Perspektive. In E. Jammal (ed.). Vertrauen im interkulturellen Kontext (pp. 13–26). Wiesbaden, Germany: VS. Verlag für Sozialwissenschaften, Springer VS.

Short, E., J. Hart, M. Vu and B. Scassellati. 2010. No fair!! An Interaction with a Cheating Robot. In Proceedings of the 5th ACM/IEEE International Conference on Human-robot Interaction (pp. 219–226). Osaka, Japan.

Simpson, J. 2007. Psychological foundations of trust. *Current Directions in Psychological Science*, **16**:264–268.

Smith, J. and M. Osborn. 2003. Interpretative phenomenological analysis. In J. Smith (ed.), Qualitative Psychology (pp. 53–80). London, UK: Sage.

Steghöfer, J., R. Kiefhaber, K. Leichtenstern, Y. Bernard, L. Klejnowski, W. Reif, T. Ungerer, E. André, J. Hähner and C. Müller-Schloer. 2010. Trustworthy Organic Computing Systems Challenges and Perspectives. 7th International Conference, ATC 2010 (pp. 62–72). Xi'an, China: Proceedings.

Stroud, B. 1999. The Goal of Transcendental Arguments. In R. Stern (ed.), Transcendental Arguments: Problems and Prospects (pp. 155–172). Oxford, UK, Clarendon Press.

Sutcliffe, A. 2006. Trust: From Cognition to Conceptual Models and Design. 18th International Conference, CAISE 2006 (pp. 3–17). Luxembourg: Springer, Berlin/Heidelberg, Germany.

Tannen, D. 1984. Conversational Style: Analyzing Talk among Friends. Norwood, USA: Ablex Publ.

Terada, K. and A. Ito. 2008. Effects of Attention on Intention and Goal Attribution toward Robot's Behavior. 7th International Workshop on Social Intelligence Design (SID 08). San Diego, USA: San Juan, Puerto Rico.

Terada, K. and A. Ito. 2010. Can a Robot Deceive Humans? Proceedings of the 5th ACM/IEEE International Conference on Human Robot Interaction (HRI'10) (pp. 36–37). Osaka, Japan.

Terada, K., T. Shamoto, H. Mei and A. Ito. 2007. Reactive Movements of Non-humanoid Robots Cause Intention Attribution in Humans. The Proceedings of the IEEE/RSJ International Conference on Intelligent Robots and Systems (IROS 2007) (pp. 3715–3720). San Diego, USA.

Turkle, S. 2006. Diary. *London Review of Books*, **28(8):**36–37.

Turner, P. and S. Turner. 2012. Emotional and aesthetic attachment to digital artefacts. Cognition, Technology and Work, pp. 1–12.

Wahl, M., J. Lange, D. Rösner and J. Frommer. (in press). Subjektives Erleben einer affektorientierten Intervention in kritischen Dialogsituationen der HCI. *The Proceedings of the Workshop Kognitive Systeme.* Duisburg, Germany.

Weizenbaum, J. 1996. ELIZA—A computer program for the study of natural language communication between man and machine. *Communications of the ACM*, **9(1):**36–45.

Wilks, Y. 2005. Artificial companions. *Interdisciplinary Science Reviews*, **30(2):**145–152.

Zimbardo, P.G. and R.J. Gerrig. 2004. *Psychologie*. München, Germany: Pearson Studium.

CHAPTER 4

Multi-Modal Classifier-Fusion for the Recognition of Emotions

Martin Schels, Michael Glodek, Sascha Meudt, Stefan Scherer, Miriam Schmidt, Georg Layher, Stephan Tschechne, Tobias Brosch, David Hrabal, Steffen Walter, Harald C. Traue, Günther Palm, Heiko Neumann and Friedhelm Schwenker

1. Introduction and Motivation

Research activities in the field of human-computer interaction increasingly addressed the aspect of integrating features that characterize different types of emotional *intelligence*. Human emotions are expressed through different modalities such as speech, facial expressions, hand or body gestures, and therefore the classification of human emotions should be considered as a multi-modal pattern recognition problem. In recent time, a multitude of approaches have been proposed to enhance the training and recognition of multiple classifier systems (MCSs) utilizing multiple modalities to classify human emotional states. The work summarizes the progress of investigating such systems and presents aspects of the problem namely fusion architectures and training of statistical classifiers based on only marginal informative features. Furthermore, it describes how the effects of missing values, e.g. due to missing sensor data or classifiers exploiting reject options in order to circumvent false classifications, can be mitigated. Another aspect is the usage of partially supervised learning, either to support annotation or to improve classifiers. Parts of these aspects are then exemplified using two recent examples of emotion recognition, showing a successful realization of MCSs in emotion recognition.

Research in affective computing has made many achievements in the last years. Emotions begin to play an increasingly important role in the field of human-computer interaction, allowing the user to interact with the system more efficiently (Picard, 2003) and in a more natural way (Sebe et al., 2007). Such a system must be able to recognize the users' emotional state, which can be done by analyzing the facial expression (Ekman and Friesen, 1978), taking the body posture and the gestures into account (Scherer et al., 2012) and by investigating the paralinguistic information hidden in the speech (Schuller et al., 2003; Oudeyer, 2003). Furthermore, biophysiological channels can provide valuable information to conclude to the affective state (Cannon, 1927; Schachter, 1964).

However, the emotions investigated so far were in general acted and the larger part of research was focused on a single modality, albeit the problem of emotion recognition is inherently multi-modal. Obviously, the entire emotional state of an individual is expressed and can be observed in different modalities, e.g. through facial expressions, speech, prosody, body movement, hand gestures as well as more internal signals such as heart rate, skin conductance, respiration, electroencephalography (EEG) or electromyogram (EMG). Recent developments aim at transferring the insights obtained from single modalities and acted emotions to more natural settings using multiple modalities (Caridakis et al., 2007; Scherer et al., 2012; Zeng et al., 2009; Chen and Huang, 2000). The uncontrolled recording of non-acted data and the manifold of modalities make emotion recognition a challenging task: subjects are less restricted in their behavior, emotions occur more rarely and the emotional ground truth is difficult to determine, because human observers also tend to disagree about emotions.

In this chapter, MCSs for the classification of multi-modal features are investigated, the numerical evaluation of the proposed emotion recognition systems is carried out on the data sets of the 1st AVEC challenge (Schuller et al., 2011) and a data set recorded in a Wizard-of-Oz scenario (Walter et al., 2011). Combining multi-modal classifiers is a promising approach to improve the overall classifier performance (Schels and Schwenker, 2010). Such a team of classifiers should be accurate and diverse (Kuncheva, 2004). While the requirement to the classifiers to be as accurate as possible is obvious, diversity roughly means that classifiers should not agree on misclassified data. In our studies, various modalities and feature views have been utilized on the data to achieve such a set of diverse and accurate classifiers.

The rest of this chapter is organized as follows: In Section 2, we will present the latest approaches to improve the recognition of emotion. Section 3 describes real-world data collections for affective computing. Furthermore, adequate features are described together with a numerical evaluation. Finally, Section 4 concludes.

2. Multi-modal Classification Architectures and Information Fusion for Emotion Recognition

2.1 Learning from multiple sources

For many benchmark data collections in the field of machine learning, it is sufficient to process one type of feature that is extracted from a single representation of the data (e.g. visual digit recognition). However, often in many real-world applications, different independent sensors are available (e.g. microphone and camera) and it is necessary to combine these channels to obtain a good recognition performance and to achieve a robust architecture against sensor failure.

To create a classifier system, which is able to handle different sources of information, three widely used approaches have been proposed and evaluated in the literature, namely *early fusion, mid-level fusion* and *late fusion* (Dietrich et al., 2003). Using early fusion, the information is combined on the earliest level by concatenating the individual features to a higher dimensional vector, as depicted on the left-hand side of Figure 1. The converse strategy is to combine the independent streams as late as possible, which is called late fusion or multiple classifier system (MCS), see the right-hand side of Figure 1. The third approach, which recently gains more attention, is known as mid-level fusion (Scherer et al., 2012; Eyben et al., 2012; Glodek et al., 2012; Dietrich et al., 2003) and combines the channels in an intermediate abstraction level, as for example conducted in a combined hidden layer of an artificial neural network. The corresponding classifier architecture is shown in the middle of Figure 1.

The selection of an optimal architecture is strongly related to the respective problem. An important clue for choosing the appropriate architecture could be drawn by judging the dependency and

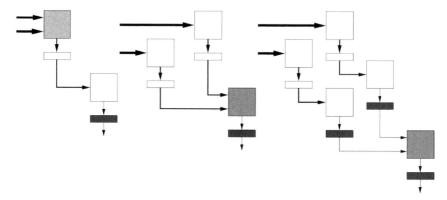

Figure 1. Schematic depiction of different classifier architectures: early fusion, mid-level fusion and late fusion (left to right).

information-gain of the features, and the complexity of the classifier function. Concatenating features of different sources is advantageous because the classification task may become separable. However, extending the dimensionality also implicates to run into the so-called *curse of dimensionality* (Bishop, 2006). Furthermore, in the application of emotion recognition, early fusion is not intuitive as the individual sources are likely to have different sampling rates.

Further, it is often necessary to compensate failing sensors that may occur for example when subjects move away from the camera or when physiological sensors lose contact to the subject's skin. Hence, it is intuitive to combine the individual features as late as possible in an abstract representation.

The mid-level fusion is a good compromise between the two extremes. Figure 2 shows a layered classifier architecture for recognizing long-term user categories. According to the key concept, the patterns are always classified based on the output of the proceeding layer such that the temporal granularity likewise the level of abstractness constantly increases. According to the theory, the architecture is able to recognize classes which are not directly observable (e.g. the affective state) based on the available evidences (Glodek et al., 2011; Scherer et al., 2012).

MCSs are widely used in the machine learning community (Kuncheva, 2004). The performance of an MCS not only depends on the accuracies of the individual classifiers, but also on the diversity of the classifiers, which roughly means that classifiers should not agree on the set of misclassified data. MCSs are highly efficient pattern recognizers that have been studied by various numerical experiments and mathematical analysis, and lead to numerous practical applications

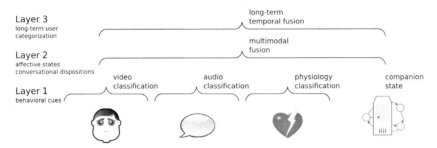

Figure 2. Layered (mid-level) classification architecture to recognize dispositions in human-companion interaction. The level of abstraction increases in each layer to obtain high-level symbolic information.

such as activity recognition (Glodek et al., 2012), EEG analysis (Schels et al., 2011) and classification of bio-acoustic signals (Dietrich et al., 2003) to mention just a few examples. There are different techniques in the literature to attain diverse ensemble classifiers. The individual classifiers can, for example, be trained on different subsets of the training data (Breiman, 1996). Another way is to conduct multiple training runs on the data using different base models or different configurations of a model (model averaging). Furthermore, different subsets of the available feature space (so-called feature views) are often used to construct individual classifiers, which are then treated as independent data streams.

In order to formally reflect that the accuracies of individual classifiers in real-world scenarios and especially in non-acted affective computing are generally low, it is useful to implement mechanisms to increase the robustness and to assess the quality of a decision for a sample. While the robustness can be achieved using the aforementioned ensembles of classifiers, the self-assessment of classifiers can be obtained by defining an appropriate uncertainty measure. Common ways to establish uncertainty measures are to use probabilistic or fuzzy classifiers, or to use the degree of agreement of an ensemble of classifiers, i.e. the more the individual classifiers agree on a specific value or label, the more confident this decision can be seen as. When combining multiple decisions, the uncertainty can be used as a weight in the fusion (Glodek et al., 2012).

Especially in real-world scenarios, it has been proven to be successful to stabilize weak decisions by integrating individual results over time (Glodek et al., 2012). Hereby, the confidence of the classifier can also help to assess weak decisions. This integration could also slow down the sample rates to match the sample rates of the sensory channels.

Human emotions occur in many variations and are often not directly accessible even for human experts when annotating affective corpora. Hence, a severe issue in affective computing is that the labeling procedure is inevitably expensive and time consuming. It would be desirable to incorporate unlabeled data in the overall classification process. This can be done either to improve a statistical learning process or to support a human expert in an interactive labeling process (Meudt et al., 2012). In order to integrate unlabeled data in a supervised machine learning procedure, two different partially supervised learning approaches have been applied, namely *semi-supervised learning* and *active learning*. *Semi-supervised learning* refers to group of methods that attempt to take advantage of unlabeled data for supervised learning (*semi-supervised classification*) or to incorporate prior information such as class labels, pair-wise constraints or cluster membership (*semi-supervised clustering*). *Active learning* or *selective sampling* (Settles, 2009) refers to methods where the learning algorithm has control on the data selection, e.g. it can select the most important/informative examples from a pool of unlabeled examples, then a human expert is asked for the correct data label. The aim is to reduce annotation costs. In our application—the recognition of human emotions in human-computer interaction—we focus more on active learning (Schwenker and Trentin, 2012; Abdel Hady and Schwenker, 2010). An iterative labeling process is displayed in Figure 3, where a machine classifier proposes labels for different areas in a recording for an expert to acknowledge. Based on this, new propositions can be made by the system.

In affective computing, it is not likely that it is necessary to make a decision for every given data sample extracted from a short time analysis. Additionally data samples are delivered relatively often compared to the expected lengths of the observed categories. Hence, it is intuitive to use techniques of sample rejection, i.e. deciding (yes or no) whether a certain confidence level has been achieved or not. Various attempts have been made to introduce confidence-based rejection criteria. Commonly thresholds-based heuristics are used on probabilistic classifier outputs utilizing a distinct uncertainty calculus, for instance doubt and conflict values computed through Dempster's rule of combination in the very well-known Dempster-Shafer theory of evidence (Thiel et al., 2005). Fusion architectures which are making use of reject options not only have to deal with missing signals and different sample rates but also with missing decisions due to rejection.

For these reasons and also as mentioned above, a classification architecture that is designed for a real-world application has to

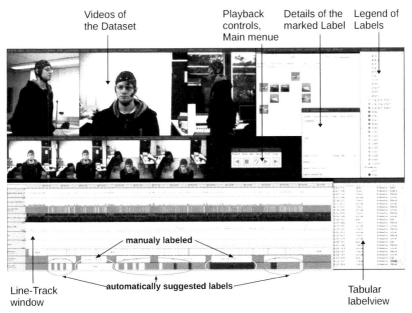

Figure 3. Depiction of the individual channels in an affectively colored human-computer interaction. Furthermore, a semi-automatic annotation process is depicted: The labels encircled in green are set by the human annotator and the system proposes the red labels additionally.

(Color image of this figure appears in the color plate section at the end of the book.)

be robust against missing data. This can be achieved by temporal smoothing of the results as seen in Figure 4. There, the blue short lines represent decisions for the audio channel in word granularity, the orange dots represent decisions for video frames and the short green lines are decisions based on physiological signals such as skin conductance. When a line is drawn in a lighter color, the sample is rejected due to a low respective confidence. But still a decision for every time step is returned by exploiting the hypothesis that the lateral differences over time are low. Further, it is possible to stack multiple layers of classifiers in order to assess more complex categories based on simpler observations. When using, for example, statistical models that can incorporate time series, this architecture can reflect high level concepts that are not directly observable in the data.

Based on this, we propose a classification architecture, as depicted in Figure 5, for the recognition of affective states in human-computer interaction. Here, every individual channel is classified separately with the usage of an uncertainty measure. Based on this, a sample reject mechanism is applied in order to prevent false classifications. The subsequent temporal integration is used not only to further improve

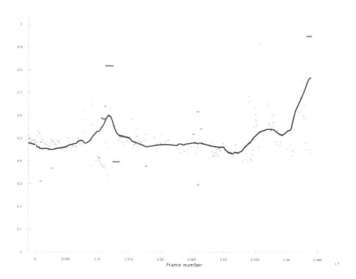

Figure 4. Example for the classification of arousal in human-computer interaction. The strong colors represent decisions that were used in the fusion (solid line), whereas the lighter dots represent rejected samples.

(Color image of this figure appears in the color plate section at the end of the book.)

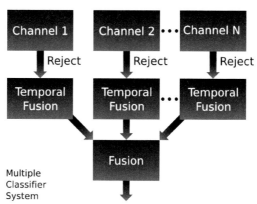

Figure 5. Multiple classifier architecture, which is making use of the reject option. The classification result of each channel has to pass a rejection step, in which decisions with low confidences are filtered out. The outcome is temporally fused and combined.

classification but also to reconstruct missing values. Finally a classifier combination is conducted.

2.2 Base classifiers

In our architectures, linear classifiers, artificial neural networks (e.g. multilayer perceptrons (MLPs)) and support vector machines (SVMs) were used as base classifiers. Linear classifiers are advantageous for noisy data and to avoid over-fitting, because the decisions are based on a rather simple function, namely the linear combination of the features. We obtained the mapping by computing the Moore-Penrose pseudo-inverse function. MLPs are based on a superposition of multiple functions (e.g. linear or sigmoid functions), which are represented by the neurons in the hidden layer (Haykin, 1999). As a result, the complexity of the MLP can be conveniently adjusted by varying the number of hidden neurons.

The SVM is a supervised learning method following the maximum margin paradigm. The classical implementation of the SVM is a typical representative of a kernel method, and therefore the so-called kernel trick can be applied. The kernel trick conducts a mapping to a new feature space that allows classification using non-linear hyper-planes. Within our study, we used the Gaussian radial basis function (RBF) kernel, which transforms the input data into the Hilbert space of infinite dimensions and is calibrated by a width parameter. However, due to noise or incorrect annotations, it is convenient to have a non-rigid hyper-plane, being less sensitive to outliers in the training. Therefore, an extension to the SVM introduces a so-called slack term that tolerates the amount of misclassified data using the control parameter. A probabilistic classification output can be obtained using the method proposed in Platt (1999). Detailed information of these algorithms can be found for instance in Bishop (2006).

Furthermore, Markov models such as the hidden Markov model (HMM) have proven to be a suited method for emotion algorithms (Glodek et al., 2011). The HMM is a stochastic model, applied for temporal/sequential pattern recognition, e.g. speech recognition and recognition of gestures. It is composed of two random processes, a Markov chain with hidden states for the transitions through the states and a second random process modeling the observations. The transition probabilities and also the emission probabilities for the outputs are estimated using the Baum-Welch algorithm. Given the parameters of an HMM and an observed output sequence, the most likely state sequence can be computed (Viterbi algorithm) and the posteriori probability for the observation sequence can be estimated (forward algorithm). This probability can be utilized to classify sequences, by choosing the most likely model (Rabiner, 1989).

3. Experiments

In the following section, several of the aspects presented in the previous sections are evaluated using non-acted emotional data sets.

3.1 Data collection

We will put focus on two different data sets, namely the EmoRec II collected at the Ulm University and the AVEC 2011 dataset.

3.1.1 EmoRec II

A simulation of a natural verbal human-computer interaction was implemented as a Wizard-of-Oz (WOZ) experiment (Kelley, 1984). The WOZ experiment allows the simulation of computer or system properties in a manner such that subjects have the impression that they are having a completely natural verbal interaction with a computer-based memory trainer. The design of the memory trainer followed the principle of the popular game "Concentration". The variation of the system behavior, in response to the subjects, was implemented via natural spoken language, with parts of the subject's reactions taken automatically into account.

In order to induce target emotions during the experiment, we considered the following affective factors that are implemented as natural language dialog:

- Delaying the response of a command
- Non-execution of the command
- Simulate incorrect speech recognition
- Offer of technical assistance
- Lack of technical assistance
- Propose to quit the game ahead of time
- Positive feedback

The procedure of emotion induction is structured in differentiated experimental sequences (ESs) in which the user is passed through VAD octants (valence: positive, negative, neutral; arousal: low, high, neutral; dominance: low control, high control, neutral) by the investigator (compare Figure 6). Audio, video and physiological data (namely electromyography, skin conductance level) were recorded.

Within this study, we focus on the recognition of the emotional octants in ES-4 and ES-6. The database comprises eight subjects with an average age of 63.5 years.

3.1.2 AVEC 2011 Data Collection

The second data collection used in this study has been provided within the Audio/Visual Emotion Challenge (AVEC) 2011 of the ACII 2011 workshop. Overall three sub-challenges were proposed: an audio

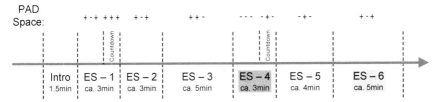

Figure 6. Experimental sequences the subject is guided through for the recordings.

challenge on word-level, a video challenge on frame-level and an audiovisual also on video frame-level.

The data was recorded in a human-computer interaction scenario in which the subjects were instructed to interact with an affectively colored artificial agent. Audio and video material was collected from 13 different subjects in overall 63 recordings. The recorded data was labeled in four affective dimensions: *arousal, expectancy, power* and *valence*. The annotations of the raters have been averaged for each dimension, resulting in a real value for each time step. Subsequently, the labels are binarized using a threshold equal to the grand mean of each dimension. Two to eight raters annotated every recording. Along with the sensor data and the annotations, a word-by-word transcription of the spoken language was provided which partitions the dialog into conversational turns. For the evaluation of the challenge, only arousal was taken into account as classification of the other dimensions yielded poor results.[1] ((Schuller et al., 2011; McKeown et al., 2010) for a detailed description of the data set.)

3.2 Features

In the following, the proposed method for the classification from multiple sources is described. We begin with the description of the individual modalities and extracted features.

3.2.1 Audio Features

From the audio signal, the following features have been applied:

- The *fundamental frequency* values are extracted using the f_0 tracker available in the ESPS/*waves*+[2] software package. Besides the track, the *energy* and the *linear predictive coding* (LPC) of the plain wave signal is extracted (Hermansky, 1990). All three

---

[1] http://sspnet.eu/avec2011/

[2] http://www.speech.kth.se/software/

features are concatenated to a 10-dimensional early fusion feature vector.

- The *Mel frequency cepstral coefficient* (MFCC) representation is inspired by the biological known perceptual variations in the human auditory system. The perception is modeled using a filter bank with filters linearly spaced in lower frequencies and logarithmically in higher frequencies in order to capture the phonetically important characteristics of speech. The MFCCs are extracted as described in Rabiner and Juang (1993).

- The *perceptual linear predictive* (PLP) analysis is based on two perceptually and biologically motivated concepts, namely the critical bands, and the equal loudness curves. Frequencies below 1 kHz need higher sound pressure levels than the reference, and sounds between 2 and 5 kHz need less pressure, following the human perception. The critical band filtering is analogous to the MFCC triangular filtering, apart from the fact that the filters are equally spaced in the Bark scale (not the Mel scale) and the shape of the filters is not triangular, but rather trapezoidal. After the critical band analysis and equal loudness conversion, the subsequent steps required for the relative spectral (RASTA) processing extension follow the implementation recommendations in Zheng et al. (2001). After transforming the spectrum to the logarithmic domain and the application of RASTA filtering, the signal is transformed back using the exponential function.

3.2.2 Video Features

We investigated a biologically inspired model architecture to study the performance of form and motion feature processing for emotion classification from facial expressions. The model architecture builds upon the functional segregation of form and motion processing in primate visual cortex. Initial processing is organized along two mainly independent pathways, each specialized for the processing of form as well as motion information, respectively.

We have directly utilized the two independent data streams for visual analysis of facial features in (Glodek et al., 2011) and already achieved robust results for automatic estimation of emotional user states from video only and audio-visual data. Here, we extended the basic architecture by further subdividing the motion-processing channel. We argue that different types of spatio-temporal information are available in the motion representation, which can be utilized for robust analysis of facial data. On the *global scale* the overall, external, motion of the face is indicative for pose changes and non-verbal communication signals, e.g., head movements during nodding or selective shifts of attention through pose changes. On the other hand, on the local scale, the internal facial motions are indicative of fine-grained changes in the facial expression and emotional exposition. Examples are, e.g., eye blinks, smiles or mouth openings. We reasoned that the segregation of this information should be helpful to further improve the analysis of emotion data and, thus, process the visual input stream along *three* independent pathways. In order to make use of more detailed task-related information, we propose here an extended model architecture which aims at first segregating form and motion, as briefly outlined above, and further subdivides the motion stream into separate representations of global and local motion, respectively. An overview of the outline of the architecture is presented in Figure 7. Motion and form features are processed along two separate pathways, composed of alternating layers of filtering (S) and non-linear pooling (C) stages. In layer S1, different scale representations of the input image are convolved with 2D Gabor filters of different orientations (form path) and a spatio-temporal correlation detector is used to build a discrete velocity space representation (motion path). The initial motion representation is then further subdivided to build separate representations of global and local facial motion. Global motion is approximated by the best-fit affine motion. To achieve this, the facial region is detected by searching for horizontally oriented barcode like structures within a Gabor-filtered input image (Dakin and Watt, 2009) which is refined into facial regions-of-interests around eyes, nose and mouth. These regions are excluded from the successive random sampling process used for the estimation of the affine transformation parameters representing the global flow (affine flow).The residual or local flow is then calculated by subtracting the affine flow from the unmodified flow to provide the input representation for extracting local motion responses. All three streams, or channels, are then further processed in parallel by hierarchical stages of alternating S- and C-filtering steps. Layer C1 cells pool the activities of S1 cells of the same orientation (direction) over a small local neighborhood and two neighboring scales and speeds, respectively. The layer S2 is created by a simple template matching of patches of C1 activities against a number of prototype patches. These prototypes are randomly selected during the learning stage (for details, see (Mutch and Lowe, 2008)). In the final layer C2, the S2 prototype responses are again pooled over a limited neighborhood and combined into a single feature vector which serves as input to the successive classification stage.

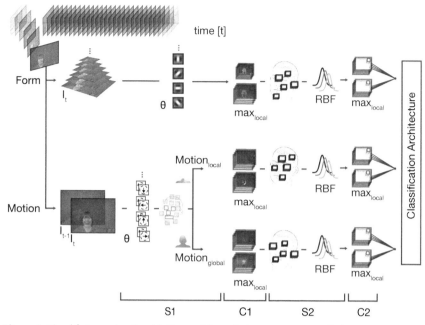

Figure 7. *Visual feature extraction.* Motion and form features are processed along two separate pathways, one form- and one motion pathway. The initial motion representation is further subdivided to build separate representations of global and local facial motion. All three streams, or channels, are further processed in parallel by hierarchical stages of alternating S- and C-filtering stepsand finally combined into a single feature vector which serves as input to the successive classification stage.

(Color image of this figure appears in the color plate section at the end of the book.)

Figure 8 demonstrates the capability of the approach to analyze differential features in non-verbal communication represented in segregated channels of visual motion information. In the case shown, the subject moves the head to point out disagreement or even disgust. This expressive communicative feature is encoded in the *global affine flow pattern* showing head motion to the right (color coded in accordance to the color map in Figure 8, right). The *local motion* activity overall depicts a brief moment in which the person opens her eyes (upward motion of eye lids (color code)) and also the chin region moves left-downwards while closing the mouth. Both motion features are now available to feed forward into the emotion classifier network for analyzing the motion related non-verbal communication behavior. Notice that in the residual flow pattern overall motion is reduced and solely local motion that is caused by facial expression remains (in the shown example caused by eye-, mouth- and cheek-movement).

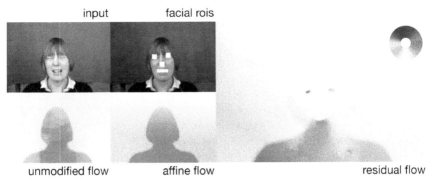

input   facial rois

unmodified flow   affine flow   residual flow

Figure 8. *Segregation of motion into separate channels of global and local motion. Four small images, in reading order:* A still image from the input sequence (input), localized eye-nose-mouth regions (facial regions of interest), optical flow calculated between two successive frames (unmodified flow) and estimation of the affine flow transformation parameters representing the global flow (affine flow).*Right image:* The residual, or local, flow is calculated by subtracting the affine flow from the unmodified flow, which reduces the overall motion energy. Residual flow caused by facial expressionremains (in theexample caused by eye-, mouth- and cheek-movement).

(Color image of this figure appears in the color plate section at the end of the book.)

3.2.3 Physiological Features

The physiological signals were acquired using a NEXUS-32 polygraph, a flexible 32 channel monitoring system. Three physiological channels were recorded: the electromyogram (EMG) of the corrugator supercilii and zygomaticus major, and the skin conductance (SCL).

To measure the SCL, two electrodes of the sensor are positioned on the index finger and the ring finger. Since the sweat glands are innervated exclusively sympathetically, i.e. without influence of the parasympathetic nervous system, the electrodermal activity is considered a good indicator of the "inner" tension of a person. This aspect can be reproduced particularly impressively by the observation of a rapid increase in skin conductance within one to three seconds due to a simple stress stimulus (e.g. deep breathing, emotional excitement or mental activity).

Electrical muscle activity is also an indicator of general psycho-physiological arousal, as increased muscle tone is associated with increasing activity of the sympathetic nervous system, while a decrease in somatomotor activity is associated with predominantly parasympathetic arousal. We used two channel EMGs for corrugator and zygomaticus muscles. EMG responses over facial muscle regions like corrugator supercilii, which draws the brow downward and medial ward to form a frown, and zygomaticus major, which elevates the corner of the mouth superiorly and posteriorly to produce a smile, can effectively discriminate valance (pleasure) and intensity of emotional states.

In general, a slow low- or band-pass filter is applied together with a linear piecewise detrending of the time series at a 10-s basis. From the subject's respiration, the following features (Boiten et al., 1994) are computed (low-pass filtered at 0.15 Hz): mean and standard deviation of the first derivatives (10-s time window), breathing volume, mean and standard deviation of breathe intervals and Poincaré plots (30-s time window each). The EMG signals were used to compute the following features (band-pass filtered at 20–120 Hz, piecewise linear detrend): mean of first and second derivatives (5-s time window) (Picard, 2003) and power spectrum density estimation (15-s time window) (Welch, 1967). The following features are extracted from the skin conductance (SCL) (low-pass filtered at 0.2 Hz): mean and standard deviation of first and second derivative (5-s time window).

3.3 Statistical evaluation

In this section, the statistical evaluation of the architecture is described for the mentioned data. All reported results originate from leave one subject out experiments.

3.3.1 EmoRec II

Classification of Spoken Utterances

The MFCC features have been calculated using 40-ms windows and were averaged to form 200-ms blocks such that all features have a uniform alignment. The three available features were separately classified using a SVM with an RBF kernel function and a probabilistic output function. For the individual features, accuracies from 52.5% to 57.9% were accomplished. Furthermore, these results were combined using the average of the confidence values and a temporal fusion of 10 s was conducted. This resulted in an accuracy of 55.4% (compare Table 1).

Classification of Facial Expressions

We classified the facial expressions using a multivariate Gaussian. In order to render a stable classification, a bagging procedure was conducted and a reject option was implemented: hereby 99% of the test frames were rejected with respect to the confidence of the classifier. An accuracy of 54.5% was achieved and only 52.3% without reject option (see Table 1).

Table 1. Accuracy of each unimodal classifier. Results in percent with standard deviation.

Feature	Accuracy	Accuracy (reject)
MFCC 40 ms	57.1% (2.4%)	n/a
MFCC 200 ms	57.9% (2.1%)	n/a
ModSpec	52.5% (2.2%)	n/a
Fusion of Audio channel	55.4% (4.6%)	n/a
EMG (5 s)	56.5% (10.7%)	69.5% (14.0%)
EMG (20 s)	50.2% (11.0%)	52.4% (24.6%)
SCL (5 s)	52.8% (4.5%)	44.1% (3.4%)
Respiration (20 s)	52.6% (8.4%)	44.6% (17.0%)
Fusion of physiology channel	55.6% (14.5%)	59.7% (13.0%)
Facial Expression	52.3% (4.4%)	54.5% (4.5%)

Classification of Biophysiological Signals

The described features were partitioned into different sets defined by the feature type and window size. This results in four different feature sets. Each of the sets was classified by a perceptron using bagging. Furthermore, a reject option of 80% is used. The setting results in accuracies ranging from 69.5% to 50.2%. Without reject option, the accuracies were lower.

These intermediate results were combined based on a new superordinate time window of 60 s with an offset of 30 s: first the confidences of the individual channels are averaged followed by the combination of the channels, resulting in accuracies of 59.7% (55.6% without reject; compare Table 1).

Multi-modal Combination

The intermediate results of the unimodal classifiers are combined to obtain the final class membership. The final time granularity was set to 60 s to evaluate all combinations of modalities. If no decision can be made (e.g. due to rejection of samples), the overall window is rejected. The weights for the classifiers have been chosen equally for all classifier combinations. An additional study, putting a weighting with focus on audio and physiology according to the respective performance, was conducted. The results are shown in Table 2 and vary around an accuracy of 60%. The highest accuracy is achieved using solely audio and video. Generally, the standard deviations are high. When all available sources are combined, the mean accuracy slightly drops, but on the other hand, the standard deviation decreases. Further details can be found in Schels et al. (2012).

Table 2.   Accuracies of every multi-modal combination. Results in percent with standard deviation.

Combination	Accuracy
Audio (1) + Video (1)	62.0% (15%)
Video (1) + Physiology (1)	59.7% (13.4%)
Audio (1) + Physiology (1)	60.2% (9.3%)
Video (1) + Audio (1) + Physiology (1)	60.8% (9.1%)
Video (1) + Audio (2) + Physiology (3)	61.5% (12.6%)

3.3.2 AVEC 2011 Data Collection

For each label dimension and for each audio feature, a bag of hidden Markov models (HMMs) have been trained (Breimann, 1996; Rabiner, 1989). The hidden states and the number of mixture components of the HMM have been optimized using a parameter search, resulting in the selection of three hidden states and two mixture components in the Gaussian mixture model (GMM) having full covariance matrices.

The evaluation of the optimization process further inferred that some features appear to be inappropriate to detect certain labels. It turned out that only the label *arousal* can draw information from all features, *expectancy* and *power* performed better using only the energy, fundamental frequency and the MFCC features. The label *valance* favored only the MFCC features. For each label, the log-likelihoods of every HMM, trained on the features, are summed. To obtain more robust models, we decided to additionally use five times as many models per class and summed the outcome as well.

Furthermore, the assumption was made that the labels are changing only slowly over time. We therefore conducted the classification on turn basis by collecting the detections within one turn and multiplied the likelihoods to obtain more robust detections. A schema visualizing the applied fusion architecture is shown in Figure 9. The results of this approach are reported in Table 3.

Within the video challenge, the n-SVM was employed as base classifier (Schölkopf et al., 2000). The implementation was taken from the well-known LibSVM repository. We concatenated 300 form and 300 motion features and used them to train a n-SVM using a linear kernel and probabilistic outputs according to Platt (1999). Due to memory constraints, only 10.000 randomly drawn samples were used.

Again a parameter search was applied to obtain suitable parameters, resulting in setting $n = 0.3$ for *arousal* and *power* and $n = 0.7$ for *expectancy* and *valence*. Based on the results of all label dimensions,

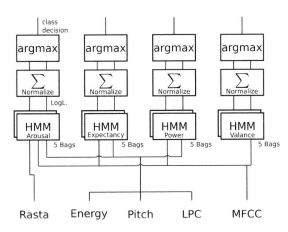

Figure 9. Architecture of the audio classifier. For each label, a bag of HMMs have been trained on selected feature sets.

Table 3. Classification results of the AVEC 2011 development data set. The weighted accuracy (WA) corresponds to the correctly detected samples divided by the total number of samples. The unweighted accuracy (UA) is given by the averaged recall of the two classes.

	Arousal		**Expectancy**		**Power**		**Valence**	
	WA	UA	WA	UA	WA	UA	WA	UA
Audio	66.9	67.5	62.9	58.5	63.2	58.4	65.7	63.3
Visual	58.2	53.5	53.5	53.2	53.7	53.8	53.2	49.8
Audio/Visual	69.3	70.6	61.7	60.1	61.3	59.1	68.8	66.4

an intermediate fusion was conducted using an MLP to obtain the final prediction. A schema illustrating the architecture used is shown in Figure 10. The results are reported in Table 3.

Considering the audiovisual challenge, we used the same approach for each modality as described in the earlier sections but omitted the last layer in which the class decision was performed. The probabilistic outputs of the video stream are collected using averaging and multiplication with a subsequent normalization such that the decisions are on word level. The HMM log-likelihoods of the label dimensions are transformed and normalized such that they are ranging between zero and one.

By concatenating the results of all label dimensions, a new 12-dimensional feature vector is obtained. The new features are then used to train an intermediate fusion layer based on an MLP.

Like in the audio challenge, the final decision is done on a turn basis by collecting the outputs within one turn and fusing them using

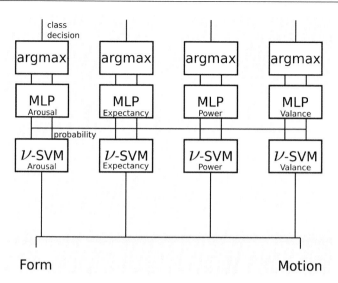

Figure 10. For the video-based classification, form and motion features are concatenated and used to train *n*-SVM for each label dimension. The outputs of the classifiers are used to train an intermediate fusion layer realized by MLPs.

multiplication. Figure 11 shows the audiovisual classifier system, while the results are given in Table 3.

4. Conclusion and Future Work

Classifying the emotion is generally a difficult task when leaping from overacted data to realistic human-computer interaction. In this study, the problem was investigated by combining different modalities. The result of the evaluation shows that the usage of different modalities

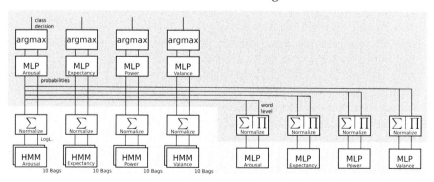

Figure 11. Overall architecture of the audiovisual classifier system, the outputs of all modalities are integrated on word level and used to train a multilayer neural network for each label dimension.

can reduce the testing error. On the other hand, the variances of the classification are relatively high.

Rejecting samples when classifying such kind of data turns out to be a sound approach leading to more robust results, especially when the distribution of the classes in the data is heavily overlapping. In future work, it could be promising to implement an iterative classifier training procedure, where the training data can be rejected.

The results presented in Tables 1 to 3 are preliminary and must be further evaluated in several directions:

1. Feature extraction techniques as described in the previous sections have been successfully applied to the recognition of Ekman's six basic emotions for benchmark data sets consisting of acted emotional data. In these data sets, emotions shown by the actors are usually over-expressed and different from the emotional states that can be observed in the AVEC data set.

2. The classifier architecture is based on the so-called late fusion paradigm. This is a widely used fusion scheme that can be implemented easily just by integrating results of the pre-trained classifier ensemble by fixed or trainable fusion mappings, but more complex spatio-temporal patterns on an intermediate feature level cannot be modeled by such decision level fusion scheme.

3. The emotional states of the AVEC2011 data set are encoded by crisp binary labels, but human annotators have usually problems to assign a confident crisp label to an emotional scene (e.g. single spoken word or a few video frames) or disagree, and thus dealing with fuzzy labels or labels together with a confidence value during annotation and classifier training phase could improve the overall recognition performance.

Acknowledgments

This chapter is based on work done within the Transregional Collaborative Research Centre SFB/TRR 62 "Companion-Technology for Cognitive Technical Systems" funded by the German Research Foundation (DFG). Tobias Brosch and Miriam Schmidt are supported by scholarships of the graduate school Mathematical Analysis of Evolution, Information and Complexity of the University of Ulm.

REFERENCES

Abdel Hady, M. and F. Schwenker. 2010. Combining committee-based semi-supervised learning and active learning. *Journal of Computer Science and Technology,* **25(4)**:681–698.

Argyle, M. 1988. Bodily Communication. Routledge, London (GB).

Bayerl, P. and H. Neumann. 2007. A fast biologically inspired algorithm for recurrent motion estimation. *IEEE Trans. on Pattern Analysis and Machine Intelligence,* **29(2)**:246–260.

Bishop, C. 2006. Pattern Recognition and Machine Learning. Springer, New York (NY).

Black, M. and Y. Yacoob. 1997. Recognizing facial expressions in image sequences using local parameterized models of image motion. *International Journal of Computer Vision,* **25**:23–48.

Blake, A., B. Bascle, M. Isard, and J. MacCormick. 1998. Statistical models of visual shape and motion. *Phil. Trans. Royal Society of London, Series A,* **386**:1283–1302.

Boiten, F., N. Frijda and C. Wientjes. 1994. Emotions and respiratory patterns: Review and critical analysis. *International Journal of Psychophysiology,* **17(2)**:103–128.

Breiman, L. 1996. Bagging predictors. *Journal of Machine Learning,* **24(2)**:123–140.

Cannon, W. 1927. The James-Lange theory of emotions: A critical examination and an alternative theory. *The American Journal of Psychology,* **39(1/4)**:106–124.

Caridakis, G., G. Castellano, L. Kessous, A. Raouzaiou, L. Malatesta, S. Asteriadis and K. Karpouzis. 2007. Multimodal emotion recognition from expressive faces, body gestures and speech. In Artificial Intelligence and Innovations, from Theory to Applications, Vol. 247, pp. 375–388. Springer.

Chen, L. and T. Huang. 2000. Emotional expressions in audiovisual human computer interaction. IEEE International Conference on Multimedia and Expo (ICME), New York, **1**:423–426.

Cootes, T. and C. Taylor. 1992. Active chape models—"smart snakes". Proceedings of the British Machine Vision Conference (BMVC'92), pp. 266–275.

Cootes, T., G. Edwards and C. Taylor. 2001. Acttive appearance models. *IEEE Trans. on Pattern Analysis and Machine Intelligence,* **23(6)**:681–685.

Dakin, S. and R. Watt. 2009. Biological "bar codes" in human faces. *Journal of Vision,* **9(4)(2)**:1–10.

Dietrich, C., G. Palm and F. Schwenker. 2003. Decision templates for the classification of bioacoustic time series. *Journal of Information Fusion,* **4(2)**:101–109.

Dietrich, C., F. Schwenker and G. Palm. 2001. Classification of time series utilizing temporal and decision fusion. Workshop on Multiple Classifier Systems. LNCS 2096, pp. 378–387. Springer.

Ekman, P. and W. Friesen. 1978. Facial Action Coding System: Investigator's Guide. Consulting Psychologists Press, Palo Alto (CA).

Eyben, F., M. Wöllmer and B. Schuller, 2012. A multitask approach to continuous five-dimensional affect sensing in natural speech. *ACM Transactions on Interactive Intelligent Systems (TiiS)*, **2(1)**:6.

Fischler, M. and R. Bolles. 1981. Random sample consensus: A paradigm for model fitting with applications to image analysis and automated cartography. *Comm. ACM*, **24(6)**:381–395.

Glodek, M., L. Bigalke, M. Schels and F. Schwenker. 2011. Incorporating Uncertainty in a Layered HMM architecture for Human Activity Recognition. Proceedings of the Joint Workshop on Human Gesture and Behavior Understanding (J-HGBU), pp. 33–34, ACM.

Glodek, M., G. Layher, F. Schwenker and G. Palm. 2012. Recognizing Human Activities using a Layered Markov Architecture. Proceeding of the International Conference on Artificial Neural Networks (ICANN). LNCS 7552, pp. 677–684. Springer.

Glodek, M., M. Schels, G. Palm and F. Schwenker. 2012. Multi-Modal Fusion Based on Classification Using Rejection Option and Markov Fusion Network. Proceedings of the International Conference on Pattern Recognition ICPR, pp. 1084–1087, IEEE.

Glodek, M., M. Schels, G. Palm and F. Schwenker. 2012. Multiple Classifier Combination Using Reject Options and Markov Fusion Networks. Proceedings of the International Conference on Multimodal Interaction, pp. 465–472, ACM.

Glodek, M., S. Scherer, F. Schwenker. 2011. Conditioned Hidden Markov Model Fusion for Multimodal Classification. Proceedings of the 12th European Conference on Speech Communication and Technology (Interspeech'11), pp. 2269–2272. ISCA.

Glodek, M., S. Tschechne, G. Layher, M. Schels, T. Brosch, S. Scherer, M. Kächele, M. Schmidt, H. Neumann, G. Palm and F. Schwenker. 2011. Multiple Classifier Systems for the Classification of Audio-Visual Emotional States. Proceedings of the International Audio/Visual Emotion Challenge (AVEC) and Workshop. LNCS 6975, pp. 359–368, Springer.

Haykin, S. 1999. Neural Networks: A Comprehensive Foundation. Prentice Hall, East Lansing (MI).

Hermansky, H. 1990. Perceptual linear predictive (PLP) analysis of speech. *Journal of the Acoustical Society of America*, **87(4)**:1738–1752.

Isard, M. and A. Blake. 1998. CONDENSATION—Conditional density propagation for visual ttracking. *International Journal of Computer Vision*, **29**:5–28.

Kelley, J. 1984. An iterative design methodology for user-friendly natural language office information applications. *ACM Transactions on Information Systems*, **2(1)**:26–41.

Kuncheva, L.I. 2004. Combining Pattern Classifiers: Methods and Algorithms. Wiley & Sons, Hoboken (NJ).

Layher, G., S. Tschechne, S. Scherer, T. Brosch, C. Curio and H. Neumann. 2011. Social Signal Processing in Companion Systems—Challenges Ahead. Proceedings of the Workshop Companion-Systeme und Mensch-Companion Interaktion (41st Jahrestagung der Gesellschaft für Informatik).

McKeown, G., M. Valstar, R. Cowie and M. Pantic. 2010. The SEMAINE corpus of emotionally coloured character interactions. In Proceedings of the IEEE International Conference Multimedia & Expo (ICME), pp. 1079–1084.

Meudt, S., L. Bigalke and F. Schwenker. 2012. ATLAS—An Annotation Tool for HCI Data Utilizing Machine Learning Methods. Proceedings of the 1st International Conference on Affective and Pleasurable Design (APD'12) [Jointly with the 4th International Conference on Applied Human Factors and Ergonomics (AHFE'12)], pp. 5347–5352, CRC Press.

Mutch, J. and D. Lowe. 2008. Object class recognition and localization using sparse features with limited receptive fields. *International Journal of Computer Vision,* **80(1):**45–57.

Oudeyer, P. 2003. The production and recognition of emotions in speech: Features and algorithms. *International Journal of Human-Computer Studies,* **59(1–2):**157–183.

Picard, R. 2003. Affective computing: Challenges. *International Journal of Human-Computer Studies,* **59(1):**55–64.

Platt, J. 1999. Probabilistic outputs for support vector machines and comparisons to regularized likelihood methods. Advances in Large Margin Classifiers, pp. 61–74.

Poggio, T., U. Knoblich and J. Mutch. 2010. CNS: A GPU-based framework for simulating cortically-organized networks. *MIT-CSAIL-TR-2010-013/ CBCL-286.*

Rabiner, L. 1989. A tutorial on hidden markov models and selected applications in speech recognition. *IEEE,* **77(2):**257–286.

Rabiner, L. and B. Juang. 1993. Fundamentals of speech recognition. Prentice-Hall Signal Processing Series, East Lansing (MI).

Rolls, E. 1994. Brain mechanisms for invariant visual recognition and learning. *Behavioural Processes,* **33(1–2):**113–138.

Schölkopf, B., A. Smola, R. Williamson and P. Bartlett. 2000. New support vector algorithms. *Neural Computation,* **12(5):**1207–1245.

Schachter, S. 1964. The interaction of cognitive and physiological determinants of emotional state. In L. Berkowitz (ed.), Advances in Experimental Social Psychology (Vol. 1, pp. 49–80). Academic Press.

Schels, M. and F. Schwenker. 2010. A multiple classifier system approach for facial expressions in image sequences utilizing gmm supervectors. Proceedings of the 20th International Conference on Pattern Recognition (ICPR'10), pp. 4251–4254. IEEE.

Schels, M., M. Glodek, S. Meudt, M. Schmidt, D. Hrabal, R. Böck, S. Walter, F. Schwenker. 2012. Multi-Modal Classifier-Fusion for the Classification of Emotional States in WOZ Scenarios. Proceedings of the 1st International Conference on Affective and Pleasurable Design, pp. 5337–5346. CRC Press.

Schels, M., S. Scherer, M. Glodek, H.A. Kestler, G. Palm and F. Schwenker. 2011. On the discovery of events in EEG data utilizing information fusion. *Journal on Multimodal User Interfaces*, **6(3-4):**117–141.

Scherer, S., M. Glodek, G. Layher, M. Schels, M. Schmidt, T. Brosch, S. Tschechne, F. Schwenker, H. Neumann, G. Palm. 2012. A generic framework for the inference of user states in human computer interaction: How patterns of low level communicational cues support complex affective states. *Journal on Multimodal User Interfaces*, **2(1)**:4:1–4:31.

Scherer, S., M. Glodek, F. Schwenker, N. Campbell and G. Palm. 2012. Spotting laughter in natural multiparty conversations: a comparison of automatic online and offline approaches using audiovisual data. *ACM Transactions on Interactive Intelligent Systems,* **2(1)**:1–31.

Schuller, B., G. Rigoll and M. Lang. 2003. Hidden markov model-based speech emotion recognition. IEEE International Conference on Acoustics, Speech, and Signal Processing (ICASSP), **2**:401–404.

Schuller, B., M. Valstar, F. Eyben, G. McKeown, R. Cowie and M. Pantic. 2011. The First International Audio/Visual Emotion Challenge and Workshop (AVEC 2011).

Schwenker, F. and E. Trentin. 2012. Proceedings of the First IAPR TC3 Workshop on Partially Supervised Learning, LNCS 7081. Springer.

Sebe, N., M.S. Lew, Y. Sun, I. Cohen, T. Gevers and T.S. Huang. 2007. Authentic facial expression analysis. *Image Vision Comput.,* **25(12)**:1856–1863.

Serre, T. and M.A. Giese. 2011. Elements for a neural theory of the processing of dynamic faces. In C. Curio, H. Bülthoff and M.A. Giese (eds.), Dynamic Faces. Insights from Experiments and Computation, pp. 187–210. MIT Press.

Settles, B. 2009. Active Learning Literature Survey. Department of Computer Sciences, University of Wisconsin-Madison.

Thiel, C., F. Schwenker and G. Palm. 2005. Using Dempster-Shafer Theory in MCF Systems to Reject Samples. Proceedings of the 6th International Workshop on Multiple Classifier System, LNCS 3541, pp. 118–127. Springer.

Walter, S., S. Scherer, M. Schels, M. Glodek, D. Hrabal, M. Schmidt, R. Böck, K. Limbrecht, H.C. Traue and F. Schwenker. 2011. Multimodal emotion classification in naturalistic user behavior. In Human-Computer Interaction, Part III, HCII 2011, J. Jacko (ed.), LNCS 6763, pp. 603–611. Springer, Berlin (DE).

Welch, P. 1967. The use of fast Fourier transform for the estimation of power spectra: A method based on time averaging over short, modified periodograms. *IEEE Trans. Audio and Electroacoustics,* **15(2)**:70–73.

Zeng, Z., M. Pantic, G. Roisman and T. Huang. 2009. A survey of affect recognition methods: Audio, visual, and spontaneous expressions. *IEEE Trans. on Pattern Analysis and Machine Intelligence,* **31(1)**:39–58.

Zheng, F., G. Zhang and Z. Song. 2001. Comparison of different implementations of MFCC. *Journal of Computer Science and Technology,* **16(6)**:582–589.

CHAPTER 5

A Framework for Emotions and Dispositions in Man-Companion Interaction

Harald C. Traue, Frank Ohl, André Brechmann,
Friedhelm Schwenker, Henrik Kessler, Kerstin Limbrecht,
Holger Hoffmann, Stefan Scherer, Michael Kotzyba,
Andreas Scheck and Steffen Walter

This research was supported by grants from the Transregional Collaborative Research Center SFB/TRR 62 Companion Technology for Cognitive Technical Systems funded by the German Research Foundation (DFG). Translation of the German version by Ute von Wietersheim.

> *Pleasant company alone makes this life tolerable.*   *Spanish.*

1. What are Companions as Cognitive-Technical Systems?

Digital companions are embodied conversational agents (ECA). They communicate in natural spoken language and realize advanced and natural man-machine interactions. It is the main goal of such companions to provide not only functionality but also empathetic responding to the user's needs. In terms of etymology, the English term 'companion'[1]

---

[1] Webster's New World Dictionary (1970) defines companion this way:
A person associated with another person
A person employed to live or travel with another person and
A member of the lowest rank order of knighthood

means *fellow, mate, friend,* or *partner*. It originates from late Latin and literally means *companies,* i.e. *with bread*, an individual close to us, which is able to give us something (in this case: bread). It is unclear whether it is a translation from the German word *gahlaiba*, which mutated in the German language to *hlaib* and finally *Leib* (Eng.: loaf). Companions are individuals close to us *with bread* that replaced the old English word *gefera*, the travel companion, which was derived at from *faran*, and in German to the word *fahren* (drive) and finally to the word *Gefährte* (which translates into companion, closing the circle between traveling together and being able to supply with food). What is also noteworthy is a cross connection to Arabic, because the word *Faran* is a male name, which means baker.

The etymological origin is relevant because current research not only focuses on technical realization problems, but the nature of possible relationships between humans and digital companions is under debate. This was especially expressed in the compilation of contributions titled "Close Engagements with Artificial Companions," which was published by Yorick Wilks in 2010 as a result of a comprehensive seminar at the Oxford Internet Institute. This compilation investigates the topic of cognitive-technical intelligence as a constantly available, selfless and helpful "software agent" from different perspectives as a future vision more intensively and with more facets than ever before. Aside from presenting already existing prototypes, work platforms, and application areas, this discussion gave a lot of room to ethical, philosophical, social and psychological issues because all experts taking part in the discussion believe that in just a few years companion technologies will have enormous communication capabilities. These communication capabilities will allow for human-companion interactions in many areas of life such as at work, in daily life, with regard to health maintenance, mobility, and social networking through highly selective information flows, which exceed the capabilities of currently available assistance systems, humanoid robots, or entertainment technology. This vision emphasizes the need for a practical theory on companion features that must take into account the psychological and social capabilities of cognitive-technical companions with respect to their human users. Companion technologies will make various sources of information available (for example, from the internet) for the interaction between humans and technical systems. Human-companion interactions will not be identical with human-human interactions, but they will probably be very similar. This similarity is maybe not due to the humanoid design of the companion, but the structural similarities of communication as well as information transfer/processing. When asked to compare and contrast positive and negative experiences with technical devices (human-machine interaction, HMI) or with other human beings (human-human interaction, HHI), and whether the emotional content of such experiences can be analyzed, there are significant similarities, but also a few differences because, with regard to HMI, feelings such as shame would be very rare, but do play a role in interpersonal relationships. In addition, negative emotions showed more variety in HHI (Walter et al., 2013b).

Table 1.   Fundamental differences between humans and cognitive-technical companions.

Characteristics	Human Companion	Cognitive-Technical Companion
Determination	Unsure	Determined
Materiality	Organismic, not deterministic	Technical and algorithmic
Availability	Depending on will and activation	At will
Autonomy, personality, and awareness	Yes	No
Emotionality	Subjectively experiencing, socially expressive and emphatic, embodiment	Sensory recognizant and expressive (Avatars)
Communication and Ability to talk	Potentially very comprehensive, multi-modal, natural speech	Very limited, multi-modal
Needs, motivation	Varied, psychobiological	Technical energy supply
Sensitivity	Mental and physical	Device-related

Table 1 lists some of the fundamental differences, which particularly refer to technical and biological characteristics. From the view of an interacting user, however, not all of the differences are relevant. What are most important are the corporeality of a human and its ability to communicate naturally, and the ability of the cognitive-technical system to interpret the natural speech of the users in a semantically correct manner. If the user can use natural speech to interact with the companion, albeit in a limited manner, the character differences become less important and the relationship that a human user develops with an object will depend on the emphatic capabilities or a technological system in responding to human emotions (valence, moods, and discrete emotions) and dispositions (motives, action tendencies, and personality). The user does not approach a new technical system with companion characteristics as an unknown entity, but will transfer his/her "inner world" of earlier effective experiences to the new situation, according to Kernberg (1992). From the self-psychology perspective (Kohut, 1987), it must furthermore be assumed that the symbolic and emotional inner representation of objects fulfills a function, i.e. to maintain and improve the functionality of the individual (for example, as an enhancement). Turtle (2010, p. 5) illustrates this assumption with the help of a student, who commented that she would love to replace her "real" boyfriend for a social robot if the robot were nice to her: "I need the feeling of civility in the house and I don't want to be alone … If the robot could provide a civil environment, I would be happy to help produce the illusion that there is somebody really with me."

A good companion makes a heaven out of hell.   German.

2. From the Assistance to the Companion System: A Qualitative Leap!

Nobody would currently expect a navigation system to be able to respond to the frustrated undertones we use to respond to the repeated instruction, *"When possible, make a U-turn"* when traffic on the opposite side of the highway has come to a standstill. A human passenger would not be forgiven for saying that because he is aware of the situation and the fact that it is simply not possible to do a U-turn. Furthermore, we would expect that navigation system reacts with empathy to the driver's emotional response caused by the repeated insistence *"When possible, make a U-turn"*. The continuous and senseless repetition of the instruction can cause anger in this case. Even more, a lower frustration tolerance as a personal characteristic can intensify this. A traffic jam can also create emotionality. Consequently, there are three possibilities: 1. The emotionality is the result of the interaction between a human and a companion, 2. The emotionality is the result of an external situation or 3. The emotionality is the result of both.

What would turn such an inadequate assistance system into a companion system? It would have to say to the driver, for example: *"Why are you not turning?"* and when the answer is: *"I can't. There is a traffic jam"* it would have to respond by saying: *"Okay, let me try to find another route, stay in this lane."* Such a companion would, at that moment, be more communicative and competent than the driver, if it has information about the traffic jam, and it would be empathetic. The automotive industry is on its way there. It presented at the CeBIT 2011 the "connected car", an upgraded electric Smart, which allows the driver to activate many functions using voice commands. Furthermore, it makes use of external data sources and services over the Internet. For example, the driver can order movie tickets. Ford has presented its embedded *Sync* system, which, in case of an accident, automatically generates an emergency call with location information and informs the driver of the vehicle involved in the accident that help is on the way (Asendorpf, 2011).

Biundo and Wendemuth (2010, also refer to the research request by SFB-TRR 62[2]) describe companion systems as cognitive-technical systems, whose functionality is completely adapted to the individuality of its user. Companion systems are personalized in respect to the user's abilities, preferences, requirements and current needs, and reflect the user's situation and emotional state. They are always available, cooperative and trustworthy, and interact with their users as competent and cooperative service partners. The functionality named in the work definition is not further explained in detail, but it lists, as a central assistance function of a companion system, planning and decision-making systems with which the user and system are equally confronted. The assistance function is a decision-making support function which provides the user with options and the respective reasons that the user can then accept or reject. Companion technologies are therefore not only intended to be an improved interface. Furthermore, they should make the functionalities of the technical systems individually available, but also make new, complex application domains feasible. The following are often cited as applications or domains:

- Assistant for technical devices
- Household and telecommunication devices
- Entertainment electronics
- Ticket dispensing machines
- Medical assistance systems
- Telemedicine
- Organizational assistants
- Health prevention
- Support systems for patients in rehabilitation clinics
- Support systems for individuals with limited cognitive abilities and much more.

In all these application domains, planning and decision-making processes play an important role. In some examples such as telemedicine or organizational assistance, they have top priority with regard to the explicit functionality (Wendemuth and Biundo, 2012). Immediate emotions have an impact on information processing, planning and decision making (Loewenstein and Lerner, 2003). Emotions are therefore considered not only as a subjectively experienced emotional

---

[2] www.sfb-trr-62.de

by-product, but as an action-managing affect factor of companion functionality. Sloman (2010) argues that the complexity of companion features as such should be discussed, but so should the quality and complexity of the requirements. His starting point is the presentation of an illustration of possible interaction flows (world knowledge), with which a companion can compare the respectively current behavior and therefore knows *"what they have done, what they could do, what they should not do, why they should not do it, what the consequences of actions will be, what further options could arise if a possible action were performed, how all this relates to what another individual could or should do, and can also communicate some of this to other individuals"* (p. 180). This would create some sort of situational awareness of the companion, not necessarily a consciousness (that would relate to a subjective experience). At this point, it is apparent that the goals that companion technologies have set for themselves can only be achieved with patient scientific work, because the euphoria about the thinkable and desirable system features of companions should not lead to the wrong conclusion that the necessary formal descriptions of these system features have yet been solved and if, in a rudimentary fashion, only relates to very special, mostly simple cases. In this context, Sloman (2010) initially views two very narrowly defined "target functions" that will become relevant for companions in the near future, and that can be assigned to companion technologies, i.e. "engaging function," which mostly refer to the quality of the interactions that entertain, draw attention, are fun or are just interesting and "enabling functions," which support users with regard to their goals, motives and intentions. The latter can help to solve various everyday problems, since they provide information, teach, organize the user's social and physical environment and enable the user to participate in society. Sub-functions of these target functions are currently offered by existing companion-like assistance systems, which are therefore very helpful as an inspiration for our visions of future technologies.

Beverly Park Woolf from the Department of Computer Science at the University of Massachusetts leaves no doubt: *"If computers are to interact naturally with humans, they must express social competencies and recognize human emotion."* Using the example of tutorial companion systems (CS), she shows that the sensory capturing of dispositions such as boredom, interest, and frustration by companions makes the tutoring functions of learning supports significantly more effective, increases motivation, and reduces adverse emotional states such as frustration, anger, or fear (Woolf, 2010, p. 5). The companions that Woolf refers to as social tutors capture emotional and dispositional responses during the learning process by measuring posture, movements, grip strength, physiological agitation, and facial movements through sensors in the chair, the monitor, the mouse, and the skin. A relatively simple model is used, in which the values of four parameters can be allocated to the following four dispositions (here referred to as emotions) with a precision ranging from 78% to 87.5%: boredom, flow, interest, and frustration. Empirical tests have shown that by using emotionally adequate responses (50 variations of support and encouragement, to keep going and trying hard), students work significantly longer on frustrating and difficult tasks and that their stress level is decreased. The tutorial systems not only respond verbally, but are also able to show an emotional response to the users in the form of avatars, or support the interaction with an interested facial expression as well as positive gestures. The empirical research has shown significant gender differences: Female test subjects benefited more from the emotion-based responses of the digital tutor than male test subjects.

Simply because of the fact that language is the most important form of communication for human beings shows that companion systems should also be able to communicate verbally. Nass and Brave (2005), however, impressively documented which possible consequences must be considered when technical systems are equipped with verbal interaction capabilities. Verbal technical systems could potentially be associated with certain social competencies or personality traits that could have a conscious or unconscious influence on the user's interaction with a technical system. If the expectations are not fulfilled, the system might be less accepted and not trusted. Notwithstanding these risks, however, verbal communication also offers the unique opportunity to increase trust and acceptance not only by providing information in an effective and natural manner, but also by using voice modulation (prosody) capabilities. Humans are probably more aware of what is being said in a conversation, but how something is said is often just as important, especially in social interactions. In the tutorial area, the prosody tool is used to positively influence the learning situation. Prosody can, for example, support the motivating character of comments and help in learning situations to increase perseverance and avoid errors. Consequently, this tool also seems suitable for interactions with technical systems to increase effectiveness. While there have been numerous studies in recent decades about the production and processing of prosody in human-human interactions (Frick, 1985; Scherer, 2003b; Baum and Pell, 1999; Friederici and Alter, 2004), this topic has not yet been investigated intensively with regard to the human-machine interaction area. Until now, the recognition and classification of human prosody by the technical system have been the main topic of the research that has been done in this regard (Cowie and Cornelius, 2003; Schuller et al., 2003). The question of how the use of prosody by a technical system can positively influence human users is still fairly unanswered. A recently published study was able to show that praising and reproaching prosody instead of neutral comments from a system (such as *"correct"*, *"incorrect"*, *"yes"*, *"correct"*, *"no"*, *"incorrect"*), in response to answers selected by test individuals in a learning situation, can lead to significantly higher rates of learning (Wolff et al., 2011). In addition, this study showed that the use of a synthetically generated version of the spoken comments (compared to prosodically neutral, naturally spoken comments) led to a significantly worse learning performance. This was all the more remarkable because the information about the accuracy of the test person's answer was also contained in the synthetically generated response and therefore the decisive information to solve the task. These findings emphasize the risks of using verbal communication, especially by companion systems, because the scope of such systems' communication capabilities must be so complex that a naturally spoken vocabulary would not be practical in contrast with a mere navigation system and its limited vocabulary.

He who has a companion has a master. French.

3. Relationship between Humans and Companions

Why should a human being surround himself with a companion that is more than just a passive assistance system because it reacts emphatically to his or her emotions, needs, and motives? Two perspectives are important here: First, users want to reach a goal, solve a problem, or improve their capabilities. Taking a mountain hiker as an example, the companion could be used to reach a certain objective or avoid bad weather. The interaction with a companion supports users with its specific problem-solving competencies and ability to carry on a dialogue. The companion system takes into account the user's emotional and motivational state as well as his or her cognitive abilities to solve a problem. Therefore, if the fact that bad weather is approaching creates concern, the companion would urge the user to avoid the weather and if that is not possible, it would help prepare for surviving bad weather in the open. In order to do so, the companion must have specific capabilities and must be able to dialogue. In this case, the emotions, motives and intentions are created by the context in which the need for using the system originated. Second, usage is the result of a need for contact, entertainment and activity. To be able to do so, the companion system also requires specific capabilities, but is not activated in the interest of solving a problem. A mountain hiker could also use the companion for entertainment or relaxation purposes when taking a break. Emotionality and motivation change in this case in the dialogue with the companion system itself and not necessarily in the achievement of goals.

If companion systems are able to perform certain technical functions, independently perceive their physical and social environment with the help of their cognitive capabilities, map this information in internal models, and draw conclusions from this information and embed it in internal plans and objectives so that they can subsequently communicate with their users, for example to align human intentions in certain situations with factual requirements, then companion systems are capable of doing things that otherwise only humans can do in an interaction: they can be a friend giving advice, a guide, a therapist, a coach, an expert, or a teacher. It is also feasible that the companion can be used to support an inner dialogue. Humans often use such inner dialogues to look for support in ambivalent or critical situations, to substitute something, to explore possibilities, to bond, to improve themselves, to gain insights, or to self-guide themselves. In such a function, the companion system could use the user's voice (Puchalska-Wasyl, 2007).

To the extent to which the companion's functions relate to its empathy and adaptability and to the extent its communication behavior is geared toward the user's individuality, users will develop a relationship, feelings and a bond with the companion, in which there is an I, a YOU, and a social environment. Users will also form a model of the companion that reflexively includes assumptions about the user model in the companion. Users have preconceived notions about the companion's characteristics and will continue to dynamically develop these over the course of the interaction (also refer to Figure 1).

The quality of such a relationship between human and companion depends on different factors: Prior experience (priming as described by James et al. (2000)) and attribution of the companion's behavior by the user (Bierhoff, 2011) and the projection and transfer of the user's conscious and unconscious wishes and expectations to a given companion system. These factors describe cognitive filters that influence the processing of information in the HCI, but are not created per se in the context of the HCI. They describe earlier experiences, expectations and personal characteristics of the users, if these are relevant for the HCI. Turtle (2010) describes this facilitated projection process, when not just the function but also the design is human or animal-like (even without any special cognitive functions): When

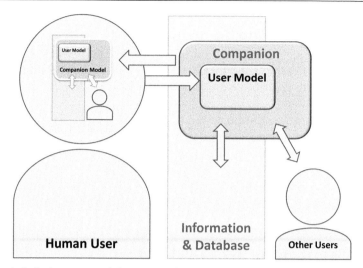

Figure 1. Reflexive nature of the user model in the companion system and the working model of the companion's user.

robots make eye contact, recognize faces, or mirror human gestures, they push our Darwinian buttons, exhibiting the kinds of behavior people associate with sentience, intentions, and emotions. Once people see robots as creatures, people feel a desire to nurture them. With this feeling comes the fantasy of reciprocation: as we begin to care for robots, we want them to care about us…. Eleven-year-old Fara reacts to a play session with Cog, a humanoid robot at MIT by stating "it's like something that's part of you, you know, something you love, kind of like another person, like a baby." (p. 4)

Keep company with good men and good men you'll learn to be.   Chinese.

4. Process Component Model of Moods, Core Affect, Emotions, and Dispositions

4.1. Emotional and dispositional behavior components

The user's mental states that are relevant for companion technologies are summarized under the term emotions and dispositions. These refer to the totality of moods, emotions, motives, action tendencies, and personality. The special companion feature, i.e. the ability to empathically recognize mental states and be able to adjust its own technical functions to the human user, is intended to increase the acceptance of technical systems for certain functions that are useful to humans, and thus make them continuously available to the human user via an interaction cycle. This is to prevent reactance as the result of insufficient empathy.[3]

Emotions and dispositions[4] comprise all psychobiological states that, at varying degrees and complexities, influence the dialogue between a human user and the companion as well as its functional use: Newness and valence, core affect, discrete emotions, moods, motives, action tendencies, and personality. Table 2 hypothetically describes several criteria for the different emotions and dispositions. The time dynamics for the situation assessment of valence (positive vs. negative) is very fast, probably in the range of 200 ms. By contrast, a user's personality only changes very gradually or not at all over months or years. The influence on behavior management probably acts according to a U-function, because the fast assessment of the valence has a strong influence on behavior as well as personality. Moods and action tendencies have less influence within this hypothetical model. The feedback strength

Table 2. Hypothetic illustration of emotions and dispositions as well as characteristic developments for dynamics, the strength of behavior management, feedback strength, and complexity

Characteristics Emotions and Dispositions	Dynamics	Impact on Behavior Control	Feedback Strength	Complexity and Operationalization
Newness and Valence	ms–s	Very high	Very high	Bi-modal and simple
Core Affect and Discrete Emotions	s–min	High	Very high	Average and difficult
Moods	min–days	Average	High	Average and difficult
Action Tendencies	min–hours	Average	High	High and difficult
Motives	hours–months	High	Average	High and simple
Variable and Stable Personality Traits	months–years	High	Low	High and simple

[3] However, it is quite possible to imagine situations in which a user expects decisive interventions from the companion technology, or, for example, when the companion technology is expected to avert danger. The prerequisite is an initiated dialogue between human and companion, during which the realization of companion characteristics is agreed upon (Bryson, 2010).

[4] Disposition (v. lat.: disposition = distribution, allocation, structure, listing, plan). In psychology, the term is used within the meaning of a readiness for (usually pathological) reactions, but is mostly used as a common expression. According to the general definition, which suits the term disposition as it is used within the context of man-companion relationships is "the organized totality of the individual's psychophysiological tendencies to react in a certain way" Chaplin, J.P. (1968, 1975) Dictionary of Psychology. New York: Dell Publishing Co.

describes the immediate effect of a disposition in interaction sequences. A necessary condition for immediate impact is the dynamic of emotions and dispositions, since feedback is only possible if there is change. The strength of the feedback and the impact on behavior control are correlated. Due to the complexity of emotional responding, some emotions are difficult to measure. Personality is also complex, but can easily be measured with self report scales and there is no need to capture personality over time during the interaction. That is also the case for the motivational structure. Action tendencies, discrete emotions and moods must be dynamically captured. So far, that has not been sufficiently achieved. Valence is bi-modal or a one-dimensional value and dynamically recordable.

Emotions and dispositions are the result of the processing of information about emotion-relevant stimuli and their unconscious and conscious cognitive assessments. Dispositions comprise the willingness to respond to emotional and non-emotional stimuli. Emotions and dispositions have an influence on each other. The general consensus is that emotions are composed of several components (Frijda, 1988; Scherer, 2001; Traue and Kessler, 2003; Traue et al., 2005; Frijda, 2007):

- Subjective experience (feeling, mostly semantically codable)
- Cognitive assessment of inner and/or outer stimuli (appraisal)
- Expressiveness of facial movements, gestures, and the body as a whole
- Psychobiological, neuronal, and endocrine activation
- Cognitive drafting of action tendencies and actions

The sensory groups that capture the respective behavior of an individual are also structured according to these components. Each of the components has its own chronological dynamic. This dynamic and the pattern of the sensory parameters result in a clear allocation of emotional processes in humans. Emotions are subjective experiences that, in different situations, are perceived similarly by different individuals. Emotions can also be understood as flexible adjustments between an individual reaction and situation, which lead to action tendencies and facilitate intra-individual and inter-individual interaction regulations (Traue and Kessler, 2003).

With regard to the description of emotions, a distinction can be made between the structural and functional views. The structural perspective describes the inner relationship of emotional components for the temporal processing. For the objective of differentiating between emotions with pattern recognition processes, the structuralism position is particularly suitable (Witherington and Crichton, 2007). From the functional perspective, the emotion components serve different goals. For example, facial expressions serve to communicate the emotions in the social environment, cognitions serve to evaluate stimuli and serve to plan behavioral activity, while physiological reactions, among others, regulate the energy budget, and finally subjective experience serves the conscious awareness of emotions. Emotions are behavioral units, whose components belong to each other from a structural perspective, that develop through an interaction between emotional stimuli and the individual over time and that lead to a process: "Whereas the functionalist approach focuses principally on the nature of emotion, the dynamic systems approach focuses principally on the nature of emotional development on the process by which emotions emerge in real-time contexts and undergo change across developmental time" (Witherington and Critchton, 2007, p. 629).

Moods are potentially long-term emotional states that may affect the quality of the individual experience, but are less intensive. Discrete emotions emerge from mood states. Frijda (1988) sees in the blocking of actions and the triggering of action tendencies a main component of emotions: "Individuals experience the urge to come closer or to turn away, to start screaming, or to sing and move; some just want to withdraw and do nothing, to no longer have any interest or to lose control" (1988, p. 351).

Certain emotions can be allocated to the initiation of action tendencies: Positive emotions activate people to approach other people and objects. In the form of desire, it is a strong, contact-promoting emotion. In contrast, fear triggers avoidance, but also the need for protection or help. Anger leads an individual to turn to someone because it creates the mental energy for coping with or even eliminating the issue. Sadness is an approach to loss. This emotion serves the (imagined) existence of a lost object. Contempt is the socially expressive avoidance, but also fear leads to avoidance. Surprise creates attention and interest. More complex and secondary discrete emotions, such as embarrassment or shame, lead to less clear action tendencies because they usually depend on complex appraisals. The structural similarity of secondary emotions such as guilt, shame, embarrassment, pride, self-confidence, honor or jealously consists in their dependence on the relationship between the ideal and the real self and therefore on complex, conscious, cognitive appraisals.

Emotional components of an emotional event in an individual do not follow the same time course. Unconscious to the individual, the components may follow their own dynamic process: Processes of the central nervous system may only take fractions of seconds and lead to a first, quick appraisal, whereas more complex, cognitive appraisals and subsequent behavior like an approach or facial expression may last several minutes. The stimulation of any given component and its progression are different and depend on each other. The reciprocal dependence of the emotional components and their different dynamics create recursive effects between the consecutive early and late emotional appraisals, the emotional reaction, and the action tendencies. The consecutive elements influence, in the form of feedback, the emotional evolution with regard to the chronological progression and the emotional quality (Colombetti, 2009).

4.2 Cognitions: Sequential check theory

The currently influential component process models (CPMs) describe the five emotional components listed in Section 4.1 and their high interdependency (Scherer, 2001). These models are based on the assumption that perceived emotional stimuli of the social and physical environment require cognitive appraisal, and try to further specify the connection between a specific stimulus and the pattern of the resulting physiological, expressive and motivational change in order to map an integration of the entire emotional process. The cognitive component in which the appraisal or assessment of inner and/or outer stimuli takes place plays an important role.

According to Leventhal and Scherer (1987), cognitions check sequentially every inner and outer emotional stimulus. This cognitive process, described as Sequential Check Theory (SCT), should be understood as a part of the dynamic CPM, where the cognitive appraisal takes place. The sequential check theory tries to explain the different emotional states of an individual as the result of a specific stimulus evaluation check and makes predictions about subsequent response patterns in the individual organic sub-systems. The stimulus evaluation is divided into four main steps that are necessary for an adaptive response to an emotional stimulus: Relevance, Implication, Coping Potential, and Normative Significance. The assessment is subjective and does not have to match the objective characteristics of a given situation. Furthermore, it should be noted that the assessments take place both subconsciously and consciously (see Figure 2).

The relevance assessment in the *first step* determines how important the event is for the individual. Both the external and the internal environments are monitored with regard to the occurrence of potentially emotional stimuli that require allocation of attention, further information processing, and possibly adaptive reactions. The

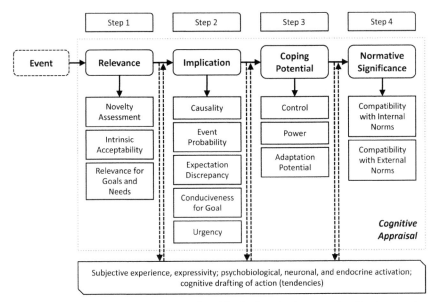

Figure 2. Stepwise Cognitive Appraisals with checks of relevance, implication, coping potential, and normative significance as a result of emotional events according to in the Sequential Check Theory.

relevance detection is divided into an appraisal of novelty (within the meaning of sudden onset and familiarity of the events), intrinsic acceptability, and relevance for goals and needs.

The *second step* comprises the assessment of the implication, and deals with an estimation of consequences and future developments, whereby the causes and the consequences of the event are considered. This process is divided into the following five assessment dimensions: Causal attribution, meaning who did what and why?; the probability with which certain consequences are expected; the discrepancy or consistency with expectations that the individual has regarding the situation (created by the event) at this point in time; the positive or negative effect it has on the individual's goals and needs; and the urgency that depends on the importance of the events for the individual as well as chronological reaction contingencies.

Step three is responsible for evaluating the coping potential, which is assessed in order to change an event and its consequences or to successfully adapt to unchangeable events. The control over the event or its consequences, the power (of the individual to be able to align the results of the event with his/her own interests, when he/she has some measure of control) and the adjustment potential to the event play an important role. The *fourth step* checks the normative significance, in which the individual assesses how the majority of other group members will assess his/her actions. A distinction is made here between internal and external standards. Internal standard checks determine the extent to which an action corresponds, for example, to one's personal self-image. In the external standard check, the compatibility of an action with perceived standards is reviewed.

Leventhal and Scherer (1987) postulated that the evaluation steps follow a fixed sequential order. This sequence supposition is explained from system economics and logical dependencies. However, the behavioral and subjective results of every step influence the next cognitive evaluations which, in turn, influence the evaluation process in the individual steps (shown by dashed arrows in Figure 2).

Since it can be shown that moods, emotions, motives, individual personality differences and even cultural values and group pressure greatly influence the result of cognitive evaluations, these determining factors must also be taken into account by the modeling of user emotions in companion sytems.

Table 3.   Emotion processing system on the sensory-motor, schematic, and the conceptual level, according to Leventhal and Scherer (1987, p. 17).

	Novelty	Pleasantness	Goal/Need Conductiveness	Coping Potential	Norm/ Self- Compatibility
Sensory- motor Level	Sudden, intensive stimulation	Intrinsic preferences/ aversions	Basic needs	Available strength	Emphatic adaptation
Schematic Level	Familiarity: Pattern comparison	Learned preferences/ aversions	Acquired needs and motives	Body schemata	Self/Social schemata
Conceptual Level	Expectations: Cause/ Effect, Probability	Remembered, anticipated or deducted positive- negative assessments	Conscious goals and plans	Problem- solving ability	Self-ideal, moral evaluation

Leventhal and Scherer (1987) present an emotion-processing system, in which the evaluation process takes place on three different levels: The sensory-motor, the schematic, and the conceptual level (see Table 3). These three levels could also be aligned with the evaluation processes of the sequential check of emotional stimuli. On the sensory-motor level, the assessment of events takes place mainly on a subconscious level based on intrinsic functions and reflexes. On the schematic level, social, individually learned patterns are used to evaluate the event, which are mostly automatic and therefore rather subconscious. A situational similarity recognized by the companion could help here, in case of a conflict between the situation and the user (motives), to adequately solve it. The sensory-motor and schematic level of cognitive processing are assessable by extended measurement of psychobiological sensors of the motor, autonomic and central nervous system and by automated classification procedures. The last conceptual level now allows information stored in the memory to be intentionally used for the evaluation in a reflexive, conceptual-symbolic process in humans. On this level, it is appropriate to have the companion system run semantic analyses to anticipate the user's emotional and dispositional state.

The cognition-theoretical formulations (Leventhal and Scherrer, 1987) differentiate appraisals by complexity, cognitive content and the level of control between automated and intellectually derived. The automated appraisals are only cognitive in as much as all higher functions of the brain can be referred to as cognitive. These appraisals are not necessarily conscious; however, its result could be an emotional, subjective experience. Even evaluations of goal achievement or impairment can run in an automated and unconscious manner, but will create a subjective feeling. It should be mentioned that appraisal theories, which consider cognitive stimuli evaluation as the main causal factor for emotions, are not generally accepted. According to Frijda (Frijda and Zeelenberg; 2001; Frijda, 2007), evaluations are the result of a monitoring process of actions and intentions. He believes that such evaluations do not cause any emotions, but rather, are cognitive accessory phenomena of the situation or the action, but not causal for emotions. Whether it is accessory phenomena or causal triggers, it can only be answered for human-companion interactions if researching the subjectively felt emotions (feelings) and the other components of the emotional behavior with regard to their chronological development.

4.3 Dispositions: Motives, action tendencies, and various personality traits

4.3.1 Motives

A user's needs, which are relevant for the interaction between humans and companions, can be categorized by deficits and the needs for growth (deficit needs and growth needs within the meaning of Maslow (1943)). The basic needs on the highest level characterize the need for respect by others, social acceptance, and self-realization (also refer to personal growth as per Bandura (2001) and Aubrey (2010)). These include[5]:

- Improving self-awareness and self-knowledge
- Building or renewing identity
- Developing strengths or talents/potential
- Enhancing lifestyle or the quality of life
- Improving health
- Enhancing personal autonomy
- Improving social abilities

Such needs can govern the intensive use of complex and socially competent companions. With regard to deficit needs, companion technologies can compensate for a lack of safety and order. Maslow (1943) also lists belonging and **attention as important deficit needs.** Schuler and Prochaska (2000) distinguish between differentiated primary motives, social motives (dominance, competition and status orientation) and performance motives (e.g. commitment, the willingness to put forth effort, and persistence). There is a respective measuring instrument for this empirical motivation model (*Leistungsmotivationsinventar [performance motivation inventory], LMI by* Schuler and Prochaska, 2000). In the available test version, the scales included in the LMI refer to self-assessments and therefore allow documentation of the cognitive context that a user brings with him/her when engaging in a dialog with a companion. At the same time, the scales can also be used to describe behavior characteristics in the HCI because they describe action tendencies. Two scales from the total of 17 personality variables of the performance motivation are particularly relevant for tutorial applications. Persistence is defined as perseverance and the use of energy with which tasks are handled. Confidence in success describes the optimistic attitude toward difficult tasks that the abilities, skills, and knowledge will successfully lead to the desired goal.

4.3.2 Action Tendencies

Action tendencies are the result of imbalances of emotions and the behavioral consequences of motives. For a companion technology, it is therefore important to recognize changes of emotional components and the subjective experience of the balance. Vigilance, selective attention, approach/avoidance, interest, frustration and conflict/ambivalence are companion-relevant action tendencies.

---

[5] http://en.wikipedia.org/wiki/Personal_development

Vigilance refers to the sustained attention (German: Wachheit) during a certain period of time. Vigilance is the requirement for conscious information processing. It correlates in a causal-functional fashion with the stimulation of the central nervous system. The two poles on the vigilance continuum are high activation, e.g. extreme stress or startle and slow wave sleep. *Vigilance* refers also to the ability to respond to accidental, low-threshold, and seldom events in a meaningful manner. The vigilance stages can be measured continuously with the electroencephalogram.

Selective attention is the limited ability to simultaneously pay attention to multiple stimuli or sensory modalities. The reason for this limited ability is the assumption of limited information-processing capacity of an individual. The selective attention (also concentration) describes the focused attention on certain stimuli, mostly provided within the context of a task, while other stimuli can be ignored. Eye tracking would be an appropriate measure.

Avoidance is an action tendency to withdraw from a situation or action. It is triggered by a (conscious or unconscious) assessment of the situation as unpleasant, dangerous, or threatening. Also, a threat to one's self-worth or the anticipation of effort can trigger avoidance. Avoidance of a behavior triggered by (anticipated or imaginary) ideas can protect from unpleasant states, but also prevents new and positive experiences. Avoidance behavior is behavior that is learned through a combination of traditional and operant conditioning or by learning from role models. The self-reinforcement of avoidance behavior by negative reinforcement turns avoidance behavior into a stable behavior pattern.

Interest is a form of selective attention, which is referred to as the cognitive participation and attention to certain topics, tasks (for example, the reading of information) and content areas. It classifies the interests of a person for certain things (e.g. professional interests, hobbies, or political interests). Modern interest theories and research approaches (Krapp, 2002) describe a person-object concept, in which the degree of interest is defined by the subjective appreciation of an object area. This term is particularly relevant for tutorial systems because interest is defined there as the emotional, motivational, and cognitive interaction between a person and his/her object areas. Lack of interest can therefore be described as distraction, lack of selective attention, etc.

Frustration is created when a person is prevented from reaching a goal because of real (external) or imagined (internal) reasons. The intensity of the frustration depends on the attractiveness of the goal and the motivation to reach the goal. An emotional response to frustration may be anger and regression (helplessness). A subsequent action tendency may be either approach or avoidance.

Conflict and ambivalence refer, in the man-companion interaction context, to competing motives or action tendencies that can be triggered by ambiguous or several stimuli. The desired objectives or goals (real or imaginary) have either an appetence (the individual pays attention to it) or an aversion (the individual does not pay attention to it). Since an objective may trigger several appetences or several aversions or both, the result for two objectives/goals, according to Miller (1959), is as follows:

1. Appetence-appetence conflict: Both objectives/goals are considered positive,

2. Aversion-aversion conflict: Both objectives/goals are considered negative or

3. Appetence-aversion conflict (ambivalence conflict): Both objectives/goals are considered both, positive and negative.

Flow: If a dialogue behavior is mainly controlled by external stimuli (references, reward), inner involvement decreases. To maintain a difficult, task-related interaction with a technical system, the joy the activity brings should be the motivator. Such states are often referred to as flow (Keller et al., 2011).

4.3.3 Stable personality traits: Optimism, hardiness and a sense of coherence, NEO-FFI, emotion regulation, and attribution style

Personality traits describe individual differences that affect emotions and dispositions because they affect the perception of internal and external events and their cognitive and emotional processing. Processing introversion correlates with the intensity of the psychological stimulation during negative emotions, the tendency not to show emotions (suppression in ERQ) influences the reduction, the facial expression, as well as increased psychophysiological reactivity (Traue and Deighton, 2007). All cognition-related personality characteristics such as need for cognition, attribution style, etc. will impact stimuli and coping appraisals. The sociability scale of the NEO-FFI has a strong impact on the social action tendencies, and scales such as optimism, hardiness, coherence, etc. are important in coping with stressful interactions (Traue et al., 2005).

Optimism is referred to as the positive general belief that one has enough resources to cope with stress. It is not important that such "subjective optimism" is justified, but the mobilization of behaviors and cognitive patterns enables the respective individuals to cope with difficulties.

These assumptions are based on research regarding self-perception and personality traits that show that mild and permanently positive illusions about one's own person and overestimation of one's own control of situations has a positive impact on self-confidence and the manner in which challenges are handled (Maruta et al., 2002).

In life event research, these abilities to meet stressful situations with *resistance* are defined with the key word **hardiness**. Hardiness is understood as the cooperation of three attitude and behavior patterns: *Control, Challenge* and *Commitment.* This kind of ideal-typical person always trusts in his/her own abilities, even under difficult life circumstances. Such a person considers life a challenge, in which every change can also be considered an opportunity and is mostly without any ambivalence both in his/her private and professional life, has few doubts, and is usually very dedicated and motivated.

The term **coherence feeling** is used for a global orientation that expresses to what extent an individual has a generalized, lasting and dynamic feeling of trust that his/her own inner and outer environment is predictable and that things will, in all likelihood, will develop in the manner that can be reasonably expected. Antonovsky (1987) proposes three components that relate to each other: *Comprehensibility* refers to the extent to which stimuli, events or developments can be perceived as structured, orderly and predicable. *Manageability* refers to the extent to which an individual perceives appropriate personal and social resources that can help cope with internal and external requirements. *Meaningfulness* finally refers to the extent to which an individual perceives his/her life as meaningful. In particular, the meaningfulness component puts the coherence feeling in a closer relation with emotional behavior because the assignment of situational meaning is a central emotional process.

The term **personality** refers to all mental characteristics of an individual that it shares with others or in which it differs from others. Widely accepted are five-factor models of personality: Extroversion, neuroticism, openness, conscientiousness and agreeability. These are described as stable, independent and fairly culture-stable factors. Extroversion is characterized by an outward-looking attitude. Individuals scoring high on the extroversion scale can be described as active, social, cheerful and/or talkative. Neuroticism, which is also referred to as emotional instability, describes the experience of and coping with negative emotions. Individuals scoring high on the neuroticism scale often experience fear, nervousness, stress, sadness, insecurity and embarrassment. The *openness* factor describes the degree to which an individual shows interest in and seeks new experiences. Individuals scoring high on this factor are often characterized as artistically inclined, imaginative, inquisitive and intellectual. *Conscientiousness* describes the degree of reliability, organization, deliberateness and efficiency an individual displays. The *agreeability* factor mainly describes to what extent an individual is altruistic. The more agreeable an individual is, the more empathetic, understanding and cooperative that individual can be described. The five factors can be measured with a standardized questionnaire, the NEO-FFI (Costa and McCrae, 2002).

The term **emotion regulation** refers to the ability to influence one's emotions in an active and targeted manner and not to interpret them as the consequence of another person's actions or the environment, which one cannot control. Emotion regulation consists of the following steps: the experience feeling must be detected, followed by a reflection about which response would be appropriate in order to avoid any reflexive or impulsive actions. Individuals with good emotion regulation show indications for mental diseases less often.

The **attribution style** defines which type of cause attribution an individual performs in order to explain his or her own behavior or the behavior of others. Different researchers have proposed different dimensions describing the attribution. The most frequently used dimensions are the distinction between internal/external and stable/variable.

The **need for cognition** is a personality attribute that describes how often and how much an individual likes to think about a topic. Individuals scoring high for this attribute enjoy thinking intensively about various situations and topics. Opinions are formed by way of an intensive review of the arguments. An exchange of opinions may therefore lead to a stable change in opinion. Individuals scoring low for this attribute generally use peripheral attributes such as attractiveness, credibility, etc., but the quality of the arguments seems rather unimportant. When such individuals change their opinion, the status is unstable, which is why it seems to be much more difficult to predict such an individual's behavior than for an individual with a high need for cognition. This construct can be measured with a standardized questionnaire, the need for cognition scale (Bless et al., 1994), and will be addressed in further detail in the operationalization section.

4.5 How to embed the process model in a companion system

At the beginning of an emotional event, there is an exogenic or endogenic stimulus. The individual confronted with the event is in a predefined state. This state is a result of the social context, prior cognitive activities (priming), the motivational situation and personality (Garcia-Pieto and Scherer, 2006). It switches like a filter between exogenic and endogenic stimulation and that furthermore modulates the emotional response. Figure 3 shows the progression and the structural connections for a single emotional behavior sequence consisting of the stimulus and the response. Since the emotional response itself acts within seconds as an endogenic trigger stimulus, it is reflected as a response.

The entire emotional process including its detailed recursion is also shown in Figure 3. In this chart, the initially simple (primary appraisal) and later more complex (secondary appraisal) cognitive responses are shown. These cognitive appraisals relate to the emotional stimuli, the coping competency and personality characteristics that tend not to change (for example, expressive suppression). Significance or meaning (stimulus appraisal) is attributed in several steps: First, the newness factor is assessed by comparing the event with memories in the working memory without any further cognitive involvement. If an orientation reaction takes place, it is a stimulus that is perceived as new. The stimulus is then assessed as positive or negative, depending on its relevance for the individual (preferences need no interferences, Zajonc, 1980). This primary appraisal or relevance detection process (Scherer, 2001) triggers the actual emotional response with behavioral, cognitive and psychobiological components. These may be primary emotions or, if less discrete, a shifting of the core affect in the three-dimensional space spanned by the dimensions of valence, arousal and dominance.

The emotional response is able to interrupt the current process. A strong fear response, for example, leads to a freezing of all movement. This emotional response is experienced as subjectively gestalt-like, whereby individual components most certainly can be perceived in a differentiated manner. Once formulated, the emotional response and the triggering event are subjected to an iterative, cognitive process for implications, coping potential, and ultimately, normative standards. The secondary appraisal is complex because it evaluates the individual emotional response by interoception (awareness of bodily responses) of the physiological activation, uses experience-based memories for coping strategies and because it must evaluate the necessity of adherence to norms. During this appraisal step, the individual checks whether the emotional stimulus is conducive or an obstacle to achieving a goal. The result may be fear and anger as a response to the interruption of a planned chain of actions. If the stimulus is conducive to reaching the goal, the individual might experience satisfaction or joy. In a last step, the individual's coping capabilities regarding situations are reviewed with regard to his or her goals and plans. The basis for this assessment is a causal attribution, i.e. the determination of what caused a certain stimulus. Without this causal attribution, it is often not possible to assess coping capabilities. If the individual cannot cope with the respective stimulus constellation without putting his or her important goals at risk, the result is anger or, in the event of habitual insufficiency, helplessness or depression. Finally, the relevance for the individual's self-image is processed. In an unfavorable case, it coincides with feelings of embarrassment, shame or guilt. Also included in this complex stimulus processing is information about the external stimulus, aspects of the self-image and especially social norms. In total (possibly after some back and forth iteration), this complex cognitive assessment leads to a determination of action tendencies and ultimately actions. In this process, the importance of the triggering stimulus may have changed.

The perceptions of the emotional response managed are not always identical and sometimes are even conflicting action tendencies. An emotional anger response may, for example, trigger action tendencies to show the anger in one's face and body language. The individual perceives this response and action tendency simultaneously, which can block his or her action tendency, depending on the social norm an individual has internalized (also refer to Traue and Deighton (2007)). These considerations lead to the recursion (Figure 3) of the various components (Scherer, 2003b).

The linear process character of an isolated emotional event (feed forward) starts from the stimulus and then proceeds via the primary appraisal and differentiation as well as the emotional response and the control of the action tendency and evaluation of behavior options

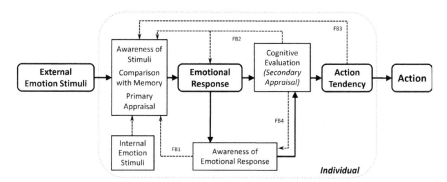

Figure 3. Recursive effects between emotion stimulus, emotion response, cognitive appraisal and action tendencies (adapted from Traue, 1998; Scherer, 2003b).

for the behavior. The first feedback (FB1) influences the primary stimulus appraisal by perceiving the response itself (e.g. the greater the anger, the more upsetting the situation). The second feedback (FB2) can influence the emotional response itself (e.g. keep cool, a little nervousness is okay), and the third feedback (FB3) affects the primary stimulus appraisal by assessing coping abilities or desirability (e.g. to make an omelet, you have to break a few eggs). Finally, the resulting action tendency may also alter the appraisal of the individual emotional response by feedback loop (FB4) (e.g. it annoys me, but I am not willing to do anything about it).

> *A man should take as companion one older than himself.   African.*

5. Modeling of Emotions and Disposition with COMPLEX

Based on a model used to simulate the interaction of artificial agents (SIMPLEX, Simulation of Personal Emotion Experience, Kessler et al., 2008), an expanded model was developed, which can map and formalize the interaction of a human user with a companion system (COMPLEX, Companions Personal Emotion Experience). Aside from the dynamic mapping of emotional and dispositional states within a technical-cognitive system, the emotional responses of users can be simulated and predicted in consideration of internal and external events, so that the functionality of the companion system can be enhanced in a meaningful manner.

External events (for example from the environment) are subjected to an individual appraisal process, based on the respective response to the event, in consideration of the available knowledge base (user or domain knowledge) and then individual goals can be determined. In addition to the appraisal of external events, internal events (such as psychophysiological parameters) can serve as input signals for the appraisal process. To customize the model, the values determined with such assessment processes are specifically modified on the basis of variable and stable personality traits (e.g. NEO-FFI, emotion regulation, etc.) as well as the current mood. Personality (long-term), action tendency (more medium-term), mood (medium-term) and emotions (short-term) consequently represent different semantic and temporal levels in the emotion model that interact with each other in a realistic fashion.

The special modeling challenge lies in the mapping of non-linear intensity curves (which can differ depending on the emotion or disposition). Aside from the general progression of short-term (emotional) states, the temporal characteristics of changed mid-term to long-term parameters (e.g. mood, action tendency) and/or their reciprocal effect is relevant for predicting emotional behavior. The technical implementation of COMPLEX is shown in Figure 4.

The theoretical basis for the appraisal process currently used in COMPLEX is the OCC model (Ortony et al., 1988). It is based on the assumption that (discrete) emotions are the direct outcome of an individual appraisal process, which appraises an event and/or an action based on three aspects: (1) consequences of the event (for one's own goals), (2) appraisal of the action on the basis of individual standards and (3) certain aspects of objects. These three aspects are further differentiated by the idea of several agents involved in the interaction because the relationship between agents must also be taken into account in the appraisal process. For example, if Person A, who is friends with Person B, fails an exam, which is important to the goal "graduation", the model would generate the emotion *'pity'* for Person A (based on the positive relationship and because an important goal was negatively impacted). For Person C, who has a negative relationship with Person A, the OCC model would, however, predict the emotion *'gloating'*.

To implement these appraisal processes in COMPLEX, it is first of all important to map the individual variables (events/actions, goals, etc.) and their interplay:

Relationships between agents are described by a value ranging from −1 to +1 ($rel_{Ag1, Ag2} = [-1, 1]$). The "1 reflects a maximum negative and +1 reflects a maximum positive relationship. The relationship is

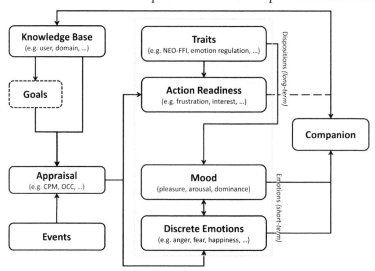

Figure 4. Overview of the COMPLEX model.

directional and therefore does not necessarily apply in the opposite direction. For initialization purposes, if two agents do not know each other, a neutral initial value of 0 is assumed.

The importance of the goal is especially relevant for the individual **goals**. It is represented for each defined goal with a value between 0 (no relevance or non-existent) and 1 (maximum relevance) (rel_{Goal} = [0, 1]).

Currently, COMPLEX only defines the actions of individual agents as **events**. These events are categorized depending on the individual values of each agent with regard to their praiseworthiness (praiseworthiness$_{Action}$ = [−1, 1]). Consequently, this variable defines whether an agent feels "ashamed" for performing an action, or feels "pride" in the other case. The *'praiseworthiness'* variable may range between "1 (low value) and +1 (high value). Internal events (e.g. psychophysiological parameters of the user) are (not yet) taken into account in the current implementation of COMPLEX.

How an event is assessed within COMPLEX also depends on its **consequences** (consequence$_{Goal, Group, Prospect}$ = [−1, 1]). Consequence is understood as the influence on an individual goal. The *consequence* variable may range between "1 (hindering the goal) and +1 (facilitating the goal). In addition, the consequences of an action are appraised separately for different groups (Group = [Self, Other, Concerned]) and may also occur in the future and not at the time an action is carried out (Prospect = [true, false]). Handing in a paper, for example, may not lead to a feeling of deep satisfaction, but initially only the hope that it is accepted. Satisfaction comes later, once the paper has been accepted.

Consequently, an event or an action may have numerous consequences. The definition of these individual consequences is very difficult to implement in COMPLEX because the variables are domain-specific and must be defined for every application domain. Especially problematic are the variables *'praiseworthiness'* and *'consequence'* because they cannot be defined globally for a domain, but are based on individual values (such as, for example, stable or variable personality traits) and are therefore agent-specific.

The direct output of the appraisal process, which uses the OCC model, are discrete emotions. In terms of time, however, these last milliseconds to seconds are therefore not long enough to model the emotional experience over longer periods of time. COMPLEX tries to solve this problem with the help of additional variables that last longer, for example, moods (medium-term) or personality (long-term). Both the emotional state and the mood are internally mapped as coordinates in the three-dimensional core affect space (valence, aroausal, dominance, VAD). Stable personality traits are taken into account in the current implementation of COMPLEX in a rather rudimentary fashion (variable personality traits are currently not taken into account at all). Depending on the respective personality structure (NEO-FFI), individual starting points are determined in the VAD space for the basic mood (Mehrabian, 1996). These anchor points are firmly located in the VAD space and serve to slowly attract the mood, if no other emotional response is "active" or influences the mood. The speed with which the mood is drifted back to this starting value in the VAD space is described with the help of the formula $I * \cos\left(\sqrt{\dfrac{d}{m}} * t\right)$

(spring model: I = Starting Intensity, d = Spring Constant, m = Mass, Becker-Asano, 2007). The mood is deflected by the output of the OCC model (emotions). To do so, the discrete emotions of the OCC appraisal are first mapped in the VAP space and averaged. Consequently, the COMPLEX moves both in the discrete and the dimensional space. The result of the averaging is one single point in the VAD space, which serves to manipulate the agent's current mood with regard to valence, arousal and dominance. Taking valence as an example, this is done as follows: $V_{mood\_new} = V_{mood\_old}$ + neuroticism * $V_{emotion}$. Arousal and dominance are deflected accordingly. A high score on the neuroticism scale, for example, would then lead to a "faster" change of the values within the VAD space, which means that the individual is emotionally less stable.

The modular structure of COMPLEX makes it possible to exchange individual components at any time in new implementations. The OCC model that is currently used for the appraisal process can be replaced by implementation of other theories, such as the component process model (see Figure 2).

It is better to travel alone than with a bad companion.   Senegalese

6. Measurements of Emotions, Dispositions and Various Personality Traits

The component process model makes it possible to describe emotions and dispositions in the human-companion interaction in a structural and dynamic manner. This description is initially phenomenological because neither model contains any formal or statistic descriptions of the dynamic relationships between the various components in the process. It is necessary for the operationalization of the individual components to measure and model concrete man-companion interactions (see Table 3). There are, of course, definitions and measuring techniques for all emotional components that lead to scale values and variables. Usually, these operationalizations are not, however, designed for real-time collection. Most suitable are behavior data (video), speech data (audio) and psychobiology data, which can be gathered on a continuous basis during the man-companion interaction. These measurements are not very reactive, i.e. the measurement itself only has little influence on the communicative process. While the measurements themselves are unproblematic, the analysis of the data in real time is a major challenge.

Table 4.   Measurements of the emotions and dispositions in the component process model of the emotional behavior based on emotion components.

Measurements Emotions and Dispositions	Subjective Experience	Facial Expressiveness	Psychobiology	Psychomotor Behavior: Gestures, body movements, attention, focus
Moods, core affect	SAM, affect grid, interview	FACS, Ratings, RTAutomatic recognition of facial expressions	Partially through ANS pattern, voice parameters	
Novelty and Valence (N&V)	SAM, affect grid, interview	FACS, EM-FACS, RTAutomatic recognition of facial expressions	P300 (EEG), EDA, EMG, voice parameters	Head movement (OR), defensive reaction, response time, eye-tracking
Discrete Emotions (VAD, DES)	SAM, affect grid, Differential Emotion Scale (DES), interview, semantic	FACS, EM-FACS, ratings, response time	Partially through ANS pattern, voice parameters	Ratings, automatic gesture and body movement detection, localization
Action tendencies	Interview, semantic		Partially through EMG, hemispheric shifts in the EEG power spectrum, voice parameters	Eye-tracking, automatic gesture and body movement detection
Motives	LMI, interview	Rare expressions (contempt)		Eye-tracking
Personality Traits	Personality scales, Interview		Partially through ANS and CNS patterns	

6.1 Measures of subjective experience

The **Self-Affective Manikin** (Bradley and Lang, 1994) can determine the variables valence, arousal and dominance and assess them on a scale from 1 to 9. The rating scale for valence means: "1" is absolutely negative, "5" is neutral and "9" is absolutely positive; for arousal, it means: "1" is absolutely relaxed, "5" is average arousal and "9" is high arousal; and for dominance, "1" means absolute control. The **Differential Emotion Scale** (DES; Izard et al., 1974) consists of 10 emotion categories (interest, joy, sadness, anger, fear, guilt, disgust, surprise, shame, shyness) with three emotion adjectives each.

The **Performance Motivation Inventory** (Schuler and Prochaska, 2000) integrates several dimensions of the performance-oriented personality: Perseverance, dominance, commitment, confidence in success, flexibility, flow, fearlessness, internality, compensatory effort, pride in performance, willingness to learn, preference for difficulties, independence, self-control, status orientation, competition orientation and level of ambition. The analysis may be dimension-specific or as a general value. The results are presented in the form of a profile.

Need for Cognition: The Need for Cognition concept can be individually determined with the German version "Skala zur Erfassung von Engagement und Freude bei Denkaufgaben". The scale is subdivided into three factors: 1. Pleasure from engaging in thinking and solving brain teasers, 2. Positive self-assessment of one's own cognitive abilities and 3. Brooding and conscientiousness (Bless et al., 1994).

The **NEO-FFI** is a proven method to measure five different personality traits: Agreeability, openness, extroversion, neuroticism and conscientiousness. The NEO-FFI consists of 60 questions, 12 for each trait. This model is a data-based, cross-sectional and empirically proven model. The following internal consistencies are provided for the NEO-FFI: Neuroticism = .79, extroversion = .79, openness = .80, agreeability = .75, conscientiousness = .73. The output format is a five-point Likert scale.

The **ERQ** (Abler and Kessler, 2009; based on Gross and John, 2003) makes it possible to scientifically research emotion regulation processes. Preferences for two frequently used strategies can be identified: suppression and reappraisal. To determine these two parameters, 10 items each are provided, which can be rated from 1 ("do not agree at all") to 7 ("agree completely"). The German version reaches an internal consistency of $r = .74$ for "suppression" and $r = .76$ for "reappraisal".

6.2 Measures of facial expressiveness

Human codings: The most commonly used system is the Facial Action Coding System, an anatomy-based method to describe visually distinguishable facial movements. The FACS does not interpret the facial expression, but detects inseparable facial movements, the so-called 44 facial action units (AUs). Individual muscles create the basis for AUs, but some AUs are produced by groups of muscles. Some muscles can even produce several different AUs. With the help of the 44 AUs, it is possible to describe all emotional (EmFACS) and non-emotional movements (such as communication signals) of the face. The analysis is based on a video recording. Furthermore, the intensity and temporal resolution can be captured (onset, apex, offset).

The Active Appearance Model (Cohn, 2010) and CERT (Bartlett et al., 2008) are able to automatically recognize the AUs. Such automated video processing systems for the recognition of facial movements are based on the gathering and classification of features (Wimmer and Radig, 2007). The efficiency of these methods is significantly increased by taking into account the modeling of the facial geography and thus the separation of dynamic (contours, coloring and contrasts) and static features (Skelley et al., 2006).

6.3 Measures of psychobiological activity

The psychobiological emotion recognition can be subdivided into traditional psychological research (Stemmler and Wacker, 2010; Kreibig et al., 2007) and the area of affective computing (Picard et al., 2001). It has been noticed, however, that the two areas are converging (Kolodyazhniy et al., 2011; Walter et al., 2010; Hrabal et al., 2012). In the basic research area and the affective computing area, discrete (fear, anger, joy, etc.) and dimensional (valence, arousal, dominance) models are used. In both research areas, however, the dimensional approach has become more prevalent. Both areas use similar parameters: blood volume pulse (BVP), skin conductance level (SCL), respiration (RSP) and electromyography (EMG). Von Kolodyazhniy and coworkers (2011) added additional parameters. Especially reliable are the correlations between the dimensions valence and corrugator and/or Zygomaticus EMG (Tan et al., 2012), as well as between arousal and SCL or BVP, respectively. In this process, predominantly action potential and frequency analyses are determined. Frequency analyses are generally suitable for real-time applications.

The main difference between basic research and affective computing is the following: In the basic research domain, averages and standard deviations are formed for signals, and signal increases or decreases calculated with these. To do so, statistical methods such as *t*-tests and variance analysis are used and effect strengths calculated (Kreibig et al., 2007, 2010; Stemmler and Wacker, 2010; Schupp et al., 2004; Bradley, 2009). In the affective computing area, generally spoken, the raw signals are initially subjected to a (1) **pre-processing** and then a (2) **feature extraction** ($f_\#$). The number of extracted features ranges from 13 (Haag et al., 2004) to 110 (Kim and André, 2008).

There is no consensus yet about which feature extraction is better. As an example, the extraction of Gu et al. (2008) will be mapped. Gu et al. used the following formulas for corrugator and Zygomaticus EMG, BVP, SCL, temperature (TMP) and electrocardiogram parameters (ECG) for each of the six parameters and extracted 36 features:

- The mean of $x(n)$
- The standard derivation of $x(n)$
- The mean of the absolute values of the first differences of $x(n)$
- The mean of the absolute values of the first differences of normalized $x(n)$
- The mean of the absolute values of the second differences of $x(n)$
- The mean of the absolute values of the second differences of the normalized $x(n)$

What follows is an automatic (3) **feature selection**. Gu et al. (under review) were able to show with the Sequential Floating Forward Search (SFFS) algorithm that, with regard to valence, the accuracy and robustness reaches their highest levels at 10 features and starts to decrease at 20 features, but arousal already reaches its highest level at 5 features and starts to decrease at 22 features. The problem with the feature selection is that these features are individual-specific and trans-situational dependent. Nevertheless, Kolodyazhniy et al. (2011) selected features that are inter-individually and trans-situationally robust: SCL, Corrugator-EMG, Zygomathicus-EMG, pCO_2 (end-tidal carbon dioxide partial pressure) and PEP (pre-ejection period).

The last step is the (4) **classification** (LDP, SVM, MLP, etc.) or hybrid classification, respectively. Overall, however, based on current findings, it can be said that psychobiological signals have the advantage that they can be obtainable in a permanent fashion and regardless of the location. It is absolutely necessary, however, that sensors are "comfortable". The psychobiological gathering of data tends to require individual-specific processing (Walter et al., 2013a, Böck et al., 2012). A trans-situational, robust feature selection still currently presents a problem (Walter et al., 2013a).

6.4 Measurements of prosody and paralinguistic parameters

Verbal communication can be analyzed in a hierarchically ordered manner. After processing the acoustic information with a signal-theoretical approach, prosodic and linguistic language information is processed separately. The prosodic information contains references about the emotional and motivational content of the act of communication. These correspond with or differ from the linguistic emotion- and cognition-related information from the semantics. On the highest level, the prosody and semantics of the spoken information can be interpreted as possible intentions. Within the context of this framework, reference is made only to emotion-related information.

The analysis of emotional language produces a multitude of prosodic characteristics, which differ significantly from modal and/or unemotional language. Among others, features such as fundamental frequency progressions, sound pitch progressions, energy progressions, speech pause frequency and pattern, the stretching of words and syllables, frequency changes and voice qualities were identified (Scherer, 2003a; Yanushevskaya et al., 2007). Researchers working in the area of automatic emotion recognition in speech use these features (Borst et al., 2004). In addition, there are many para-linguistic events that transmit emotionally colored information, such as laughing (Scherer et al., 2011). As already mentioned above, semantic and speech-content phenomena are also researched in addition to the prosodic and para-linguistic information hidden in speech (Schuller et al., 2003; Scherer et al., 2012b).

6.5 Semantics

Content-analytical processes, for some time now, have been based on key words or phrases. More complex processes carry out syntactic and semantic analyses. In certain areas of application with a limited speech scope, the results are robust and can be applied in practice. The text-based detection of emotional information is a proven method: Harvard Dictionaries, LIWC (Pennebaker and Francis, 1996) and LEAS (Kessler et al., 2010) and is easy to use, once words have been identified. Since natural speech rarely follows grammar rules, computer-linguistic methods often face significant problems in the identification of meanings. The automated content analysis can be rendered less ambiguous with decision-theoretical methods.

So far, however, there is an insufficient number of language databases with natural emotionality in the language. First analyses, however, show promising emotion classification methods (Gnjatović and Rösner, 2010). Emotion-related annotated data is classified through a dynamic recognition process based on Hiden Markov models, neuronal networks, and kernel-based processes as well as support vector machines (SVM) and kernel logistic regression (KLR). The language-independent classification of basic emotions lies at approximately 75%, and the language-dependent classification at 93%, which is very precise, with regard to artificially created language material with emotional prosody (Wagner, 2005), if the language signals come from a defined interaction and are recorded without any acoustic interference.

6.6 Psychomotor functions: Gestures, body movements and attention focus

The automatic recognition of gestures generally takes place in three steps: (1) the object detection, (2) the chronologically recursive filtering (tracking) and (3) the classification and/or verification (Bar-Shalom and Li, 1995). (1) In the detection step, the measured data is individualized, which means that object hypotheses with individually measured data are extracted from the measured data, which then must be allocated to the known objects. (2) The chronological filtering then, in the second step, uses a multi-instance filter approach, in which every object is individually assigned a dynamic filter. For chronological filtering processes, approximations of the generally recursive Bayes estimation are used (Bar-Shalom and Li, 1995). (3) In the area of gesture recognition, numerous classification methods have been developed ((Morguest, 2000), (Corradini and Gross, 2000) and (Barth and Herpers, 2005)). Once the static and dynamic gestures have been recognized, they can be classified. For static gestures, model-based processes (Stenger et al., 2001), Active Shape Models (Wimmer and Radig, 2007), or feature-based processes in connection with a classification algorithm (e.g. Bayesian Networks (Ong and Ranganath, 2003) or artificial neuronal network, 2001) can be used. In the dynamic methods, the focus is on the analysis of a chronological sequence of individual images to determine the movement trajectory. To do so, image and pattern recognition methods for time-dependent features must be used in order to analyze the information from the video sequence. The typical sequential process flow for gesture classification includes segmentation, feature recognition and feature extraction.

Good company makes short miles.   Dutch.

7. Multi-modal Assessment and Multi-modal Fusion for Emotion and Disposition Recognition

Recognizing the users' emotional and dispositional states can be achieved by analyzing different modalities, e.g. analyzing facial expressions, body postures, and gestures, or detecting and interpreting paralinguistic information hidden in speech (see section on measurements above). In addition to these types of external signals, psychobiological channels can provide information about the user's current emotional state (honest signals sensu Pentland and Pentland, 2008). Although emotion recognition is often performed on single modalities, particularly in benchmark studies on acted emotional data, for the recognition of more naturalistic emotions, multi-modal events or states need to be considered, and principles of multi-modal pattern recognition become increasingly popular (Caridakis et al., 2007; Walter et al., 2011).

Basically, any multi-modal classification problem can be treated as a uni-modal one, just by extracting relevant data or feature vectors from each modality and concatenating them in a single vector that is then applied as input vector to a single monolithic classifier. This fusion scheme is called *data fusion, early fusion* or *low-level fusion.* The opposite of data fusion is *decision fusion, late fusion* or *high-level fusion.* This means that information of different modalities is processed separately until the classifier decisions were computed. After that an aggregation rule is applied combining the entire bunch of decisions into a final overall decision. All these different notions are reflecting the processing level (data/decision, early/late, or low/high) where information fusion takes place. In addition to these two principles, *feature (level) fusion* or *intermediate (level) fusion* or *mid (level) fusion* is a common fusion scheme. This notion is used to express the fact that information sources are fused after computing some type of higher-level discriminative features, e.g. action-unit intensities, statistics of spoken words, speech content (Schwenker et al., 2006).

Besides the spatial fusion types of different modalities, in multi-modal data streams the integration of temporal information is required (Dietrich et al., 2003). In human-computer interaction scenarios of typical events in the environment, the user's states or actions cannot be detected or classified on the basis of single video frames or short-time speech analysis windows (Glodek et al., 2011b). Usually such events or states are represented through multi-variate time series and thus fusion in these applications almost always means both spatial and temporal information fusion. The simplest temporal fusion scheme is chunking by filtering, averaging, or static decision fusion such as (weighted) voting. Chunking assumes that entries of the series are independent and can be computed separately; this assumption might be true for the classification of a global emotional state, but in case of data, such as actions or user dispositions, the sequential nature has to be explicitly modeled through recurrent neural networks or hidden Markov models (see also Chapter 4 in this book; Bishop, 2006; Glodek et al., 2011a).

REFERENCES

Abler, B. and H. Kessler. 2009. Emotion Regulation Questionnaire—Eine Deutsche Version des ERQ von Gross and John. *Diagnostica*, **55(3):** 144–152.

Antonovsky, A. 1987. Unraveling the Mystery of Health—How People Manage Stress and Stay Well. Jossey-Bass Publishers, San Francisco.

Asendorpf, D. 2011. Netz auf Rädern. DIE ZEIT, March 10th, 40 p.

Aubrey, B. 2010. Managing Your Aspirations: Developing Personal Enterprise in the Global Workplace. McGraw-Hill Education, Singapore.

Bandura, A. 2001. Social cognitive theory: An agentic perspective. *Ann. Rev. Psychol.*, **52:**1–26.

Bar-Shalom, Y. and X.R. Li. 1995. Multitarget-Multisensor Tracking—Principles and Techniques. YBS, Urbana, IL.

Barth, A. and R. Herpers. 2005. Robust head detection and tracking in cluttered workshop environment using GMM, pp. 442–450. In DAGM symposium, LNCS 3663. Springer, New York, Berlin.

Bartlett, M. G. Littlewort, E. Vural, K. Lee, M. Cetin, A. Ercil and M. Movellan. 2008. Data mining spontaneous facial behavior with automatic expression coding. Lecture Notes in Computer Science 5042, Springer, Berlin, pp. 1–21.

Baum, S.R. and M.D. Pell. 1999. The neural bases of prosody: Insights from lesion studies and neuroimaging. *Aphasiology*, **13:**581–608.

Becker-Asano, C. 2008. WASABI: Affect Simulation for Agents with Believable Interactivity. Dissertation der Universität Bielefeld.

Bierhoff, H.-W. 2011. Attribution: In Lexikon der Psychologie. http://www.wissenschaft-online.de/abo/lexikon/psycho/1584.

Bishop, C. 2006. Pattern Recognition and Machine Learning. Springer, New York, Berlin.

Biundo, S. and A. Wendemuth. 2010. Von kognitiven technischen Systemen zu Companion-Systemen. *Künstliche Intelligenz*, **24(4):**335–339.

Bless, H., M. Wänke, G. Bohner and R.F. Fellhauer. 1994. Need for Cognition: Eine Skala zur Erfassung von Engagement und Freude bei Denkaufgaben. *Zeitschrift für Sozialpsychologie*, **25:**147–154.

Böck, R., K. Limbrecht, S. Walter, S. Glüge, D. Hrabal, A. Wendemuth and H.C. Traue. 2012. Intraindividual and interindividual multimodal emotion analyses in human-machine-interaction. Proceedings of 2012 IEEE Conference on Cognitive Methods in Situation Awareness and Decision Support, pp. 59–64.

Borst, M., G. Langner and G. Palm. 2004. A biologically motivated neural network for phase extraction from complex sounds. *Biological Cybernetics*, **90**:98–104.

Bradley, M.M. 2009. Natural selective attention: Orienting and emotion. *Psychophysiology*. **46(1)**:1–11.

Bradley, M.M. and P.J. Lang. 1994. Measuring emotion: The self-assessment manikin and the semantic differential. *J. Behav. Ther. Exp. Psychiatry*, **25**:49–59.

Bradley, M.M. and P.J. Lang. 2008. The International Affective Picture System (IAPS) in the study of emotion and attention, pp. 29–46. In: J.A. Coan and J.J. Allen (eds.), Handbook of Emotion Elicitation and Assessment. Oxford University Press, Oxford.

Bryson, J.J. 2010. Robots should be slaves, pp. 63–74. In: Y. Wilks (ed.), Close Engagements with Artificial Companions: Key Social, Psychological, Ethical and Design Issues. John Benjamins, Amsterdam.

Caridakis, G., G. Castellano, L. Kessous, A. Raouzaiou, L. Malatesta and S. Asteriadis. 2007. Multimodal emotion recognition from expressive faces, body gestures and speech, pp. 375–388. In: C. Boukis, L. Pnevmatikakis and L. Polymenakos (eds.), Artificial Intelligence and Innovations 2007: From Theory to Applications. Springer, Boston, Berlin.

Cohn, J.F. 2010. Advances in behavioral science using automated facial image analysis and synthesis. IEEE Signal Processing Magazine, **128(6)**:128–133.

Colombetti, G. 2009. From affect programs to dynamical discrete emotions. *Philosophical Psychology*, **22(4)**:407–445.

Corradini, A. 2001. Real-time gesture recognition by means of hybrid recognizers, pp. 34–46. Proceedings of the International Gesture Workshop on Gesture and Sign Languages in Human-Computer Interaction, USA.

Corradini, A. and H. Gross. 2000. Camera-based gesture recognition for robot control, pp. 133–138. In Proceedings of IJCNN'2000.

Costa, P.T. and R.R. McCrae. 2002. Revised NEO Personality Inventory (NEO PI-R) and NEO Five-Factor Inventory (NEO-FFI) 2002. Psychological Assessment Resources, Odissa.

Cowie, R. and R.R. Cornelius. 2003. Describing the emotional states that are expressed in speech. *Speech Communication*, **40**:5–32.

Deighton, R.M. and H.C. Traue. 2006. Emotionale Ambivalenz, Körperbeschwerden, Depressivität und soziale Interaktion: Untersuchungen zur deutschen Version des Ambivalence over Emotional Expressiveness Questionnaire (AEQ-G18. *Zeitschrift für Gesundheitspsychologie*, **14**:158–170.

Dietrich, C., G. Palm and F. Schwenker. 2003. Decision templates for the classification of bioacoustic time series. *Information Fusion*, **4**:101–109.

Ekman, P. and W. Friesen. 1978. *Facial Action Coding System: Investigator's Guide*. Palo Alto: Consulting Psychologists Press.

Frick, R.W. 1985. Communicating emotion: The role of prosodic features. *Psychol. Bull.*, **97**:412–429.

Friederici, A.D. and K. Alter. 2004. Lateralization of auditory language functions: A dynamic dual pathway model. *Brain Lang*, **89**:267–276.

Frijda, N.H. 1988. The laws of emotion. *Am. Psychol.*, **43**:349–358.

Frijda, N.H. 2007. The Laws of Emotion. Erlbaum, Mahwah New Jersey and London.

Frijda, N.H. and M. Zeelenberg. 2001. Appraisal: What is the dependent? pp. 141–155. In: K.R. Scherer, A. Schorr and T. Johnstone (eds.), Appraisal Processes in Emotion: Theory, Methods, Research. Oxford University Press, New York.

Garcia-Prieto, P. and K.R. Scherer. 2006. Connecting social identity theory and cognitive appraisal theory of emotions, pp. 189–208. In: R. Brown and D. Capozza (eds.), Social Identities: Motivational, Emotional, Cultural Influences. Psychology Press, New York.

Glodek, M., L. Bigalke, M. Schels and F. Schwenker. 2011a. Incorporating uncertainty in a layered HMM architecture for human activity recognition, pp. 33–34. Proceedings of the 2011 Joint ACM Workshop on Human Gesture and Behavior Understanding. 2011. ACM.

Glodek, M., S. Tschechne, G. Layher, M. Schels, T. Brosch, S. Scherer, M. Kächele, M. Schmidt, H. Neumann, G. Palm and F. Schwenker. 2011b. Multiple classifier systems for the classification of audio-visual emotional states, pp. 359–368. In: S.D'Mello et al. (eds.), ACII 2011, Part II, LNCS 6975, Springer, Heidelberg.

Gnjatović, M. and D. Rösner. 2010. Inducing genuine emotions in simulated speech-based human-machine interaction: The NIMITEK Corpus. *IEEE Transactions on Affective Computing,* **1:**132–144.

Gross, J.J. and O.P. John. 2003. Individual differences in two emotion regulation processes: Implications for affect, relationships, and well-being. *J. Pers. Soc. Psychol.,* **85:**348–362.

Gu, Y., S.L. Tan, K.J. Wong, M.H.R. Ho and L. Qu. 2008. Emotion-aware technologies for consumer electronics. *IEEE International Symposium on Consumer Electronics*, Vilamoura, pp. 1–4.

Guralnik, D.B. 1970. *Webster's New World Dictionary*. Foster, Toronto/Canada.

Haag, A., S. Goronzy, P. Schaich and J. Williams. 2004. Emotion recognition using biosensors: First step towards an automatic system. Affective Dialogue Systems, Tutorial and Research Workshop, Kloster Irsee, Germany, pp. 36–48.

Hrabal, D., S. Rukavina, K. Limbrecht, S. Walter, V. Hrabal, S. Gruss and H.C. Traue. 2012. Emotion Identification and Modelling on the Basis of Paired Physiological Data Features for Companion Systems. Proceedings of the MATHMOD 2012—7th Vienna International Conference on Mathematical Modelling. Vienna, Austria.

Izard, C.E., F.E. Dougherty, B.M. Bloxom and N.E. Kotsch. 1974. The Differential Emotions Scale: A Method of Measuring the Meaning of Subjective Experience of Discrete Emotions. Nashville: Vanderbilt University, Department of Psychology.

James, T.W., G.K. Humphrey, G.S. Gati, S.J.S. Ravi, R.S. Menon and M.A. Goodale. 2000. The effects of visual object priming on brain activation before and after recognition. *Current Biol.,* **10:**1017–1024.

Keller, J., S. Ringelhan and F. Blomann, 2011. Does skills-demands compatibility result in intrinsic motivation? Experimental test of a basic notion proposed in the theory of flow-experiences. *Journal of Positive Psychology*, **6(5):**408–417.

Kernberg, O.F. 1992. Objektbeziehungen und Praxis der Psychoanalyse. Klett-Cotta Stuttgart.

Kessler, H., A. Festini, H.C. Traue, S. Filipic, M. Weber and H. Hoffmann. 2008. SIMPLEX—Simulation of Personal Emotion Experience, Affective Computing. pp. 255–271. In: Jimmy Or (ed.), Affective Computing: Emotion Modelling, Synthesis and Recognition, I-Tech Education and Publishing.

Kessler, H., H.C. Traue, M. Hopfensitz, C. Subic-Wrana and H. Hoffmann. 2010. Level of Emotional Awareness Scale-Computer (LEAS-C): Deutschsprachige digitale Version. *Psychotherapeut, Bnd 55, Heft,* **4**:329–331.

Kim, J. and E. André. 2008. Emotion recognition based on physiological changes in music listening., *IEEE Transactions on Pattern Analysis and Machine Intelligence*, **30**:2067–2083.

Kohut, H. 1987. The Kohut Seminars on Self Psychology and Psychotherapy with Adolescents and Young Adults. W.W. Norton & Company, New York.

Kolodyazhniy, V., S. Kreibig, J.J. Gross, W. Roth and F.H. Wilhem. 2011. An affective computing approach to physiological emotion specificity: Toward subject-independent and stimulus-independent classification of film-induced emotions. *Psychophysiology*, **48**:908–922.

Krapp, A. 2002. Structural and dynamic aspects of interest development: theoretical considerations from an ontogenetic perspective. Learning and Instruction. **12**:383–409.

Kreibig, S.D., F.H. Wilhelm, W.T. Roth and J.J. Gross. 2007. Cardiovascular, electrodermal, and respiratory response patterns to fear and sadness-inducing films. *Psychophysiology*, **44**:787–806.

Kreibig, S.D., T. Brosch and G. Schaefer, 2010. Psychophysiological response patterning in emotion: Implications for affective computing, pp. 105–130. In: K.R. Scherer, T. Baenziger and E. Roesch (eds.), A Blueprint for an Affectively Competent Agent: Cross-Fertilization Between Emotion Psychology, Affective Neuroscience, and Affective Computing. Oxford University Press, Oxford.

Kuhl, J. 2001. Motivation und Persönlichkeit. Interaktionen psychischer Systeme. Hogrefe, Göttingen.

Leventhal, H. and K. Scherer. 1987. The relationship of emotion to cognition: a functional approach to semantic controversy. *Cognition and Emotion*, **(1)**:3–28.

Loewenstein, G. and J.S. Lerner, 2003. The role of affect in decision making, pp. 619–642. In: R.J. Davidson, K.R. Scherer and H.H. Goldsmith (eds.), Handbook of Affective Sciences. Oxford University Press, Oxford.

Maruta, T., R.C. Colligan, M. Malinchoc and K.P. Offord. 2002. Optimism-pessimism assessed in the 1960s and self-reported health status 30 years later. *Mayo. Clin. Proc.*, **77**:748–753.

Maslow, A.H. 1943. A theory of human motivation. *Psychol. Rev.*, **50**:370–396.

Mehrabian, A. 1996. Analysis of the big-five personality factors in terms of the PAD temperament model. *Austral. J. Psychol.*, **48**:86–92.

Miller, N.E. 1959. Liberalization of basic S-R concepts: Extension to conflict behavior, motivation, and social learning. In: S. Koch (ed.), Psychology: A Study of Science (Vol. 2. McGraw-Hill, New York).

Morguet, P. 2000. Stochastische Modellierung von Bildsequenzen zur Segmentierung und Erkennung dynamischer Gesten. Ph.D. thesis, TU München, München, Germany.

Nass, C. and S. Brave. 2005. Wired for Speech: How Voice Activates and Advances the Human-Computer Relationship. MIT Press, Cambridge, MA.

Ong, S. and S. Ranganath, 2003. Classification of gesture with layered Meanings. Proceedings of International Gesture Workshop Singapore, pp. 239–246.

Ortony, A., G.L. Clore and A. Collins. 1988. The Cognitive Structure of Emotions. Cambrige: Cambrige University Press.

Pennebaker, J.W. and M.E. Francis. 1996. Cognitive, emotional, and language processes in disclosure. *Cogn. Emot.*, **10**:601–626.

Pentland, A. and S. Pentland. 2008. Honest Signals: How They Shape Our World. Bradford Books. Cambrige: MIT Press.

Picard, R. 2003. Affective computing: Challenges. *International Journal of Human-Computer Studies*, **59**:55–64.

Picard, R.W., E. Vyzas and J. Healey. 2001. Toward machine emotional intelligence: Analysis of affective physiological state. *IEEE Transactions on Pattern Analysis and Machine Intelligence*, **23**:1175–1191.

Puchalska-Wasyl, M. 2007. Types and functions of inner dialogues. *Psychology of Language and Communication*, **11**:43–62.

Scherer, K.R. 2001. Appraisal considered as a process of multilevel sequential checking, pp. 92–120. In: K.R. Scherer, A. Schorr and T. Johnstone (eds.), Appraisal Processes in Emotion: Theory, Methods, Research. Oxford University Press, New York.

Scherer, K.R. 2003a. Vocal communication of emotion: A review of research paradigms. *Speech Communication*, **40**:227–256.

Scherer, K.R. 2003b. Introduction: Cognitive components of emotion, pp. 563–571. In: R.J. Davidson, K.R. Scherer and H.H. Goldsmith (eds.), Handbook of Affective Sciences. Oxford University Press, Oxford.

Scherer, S., E. Trentin, F. Schwenker and G. Palm. 2009. Approaching emotion in human computer interaction. In: International Workshop on Spoken Dialogue Systems. (IWSDS'09), pp. 156–168.

Scherer, S., M. Glodek, M. Schels, M. Schmidt, G. Layher, F. Schwenker, et al. 2012a. A generic framework for the inference of user states in human computer interaction: How patterns of low level communicational cues support complex affective states. *Journal on Multimodal User Interfaces*, **6**:117–141.

Scherer, S., M. Glodek, F. Schwenker, N. Campbell and G. Palm. 2012b. Spotting laughter in naturalistic multiparty conversations: A comparison of automatic online and offline approaches using audiovisual data. ACM Transactions on Interactive Intelligent Systems: Special Issue on Affective Interaction in Natural Environments, **2**:111–144.

Scherer, S., M. Schels and G. Palm. 2011. How low level observations can help to reveal the user's state in hci. In S. D'Mello, A. Graesser, B. Schuller, and J.-C.M. (eds.), Proceedings of the 4th International conference on Affective Computing and Intelligent Interaction (ACII'11), volume 2, pp. 81–90. Berlin: Springer.

Schuler, H. and M. Prochaska. 2000. Entwicklung und Konstruktvalidierung eines berufsbezogenen Leistungsmotivationstests. *Diagnostica*, **46**:61–72.

Schuller, B., G. Rigoll and M. Lang, 2003. Hidden Markov model-based speech emotion recognition. *Proceedings IEEE International Conference on Multimedia and Expo*, pp. 401–404.

Schupp, H.T., B.N. Cuthbert, M.M. Bradley, C.H. Hillman, A.O. Hamm and P.J. Lang, 2004. Brain processes in emotional perception: Motivated attention. *Cogn. Emot.*, **18**:593–611.

Schwenker, F., C. Dietrich, C. Thiel and G. Palm. 2006. Learning of decision fusion mappings for pattern recognition. *International Journal on Artificial Intelligence and Machine Learning (AIML)*, **6**:17–21.

Skelley, J., R. Fischer, A. Sarma and B. Heisele. 2006. Recognizing expressions in a new database containing played and natural expressions. *Proceedings of 18th International Conference on Pattern Recognition (ICPR)*, **1**:1220–1225.

Sloman, A. 2010. Requirements for artificial companions: It's harder than you think, pp. 179–200. In: Y. Wilks (ed.), Close Engagements with Artificial Companions. J. Benjamins Publishing Company, Amsterdam.

Stemmler, G. and J. Wacker. 2010. Personality, emotion, and individual differences in physiological responses. *Biol. Psychol.*, **84**:541–551.

Stenger, B., P. Mendonca and R. Cipolla. 2001. Model-based 3D tracking of an articulated hand. *IEEE Conference on Computer Vision and Pattern Recognition*, **2**:310–315.

Tan, J.-W., S. Walter, A. Scheck, D. Hrabal, H. Hoffmann, H. Kessler and H.C. Traue. 2012. Repeatability of facial electromyography (EMG) activity over corrugator supercilii and zygomaticus major on differentiating various emotions. *Journal of Ambient Intelligence and Humanized Computing*, **3**:3–10.

Traue, H.C. and R.M. Deighton. 2007. Emotional inhibition, pp. 908–913. In: George Fink (Editor-in-Chief), Encyclopedia of Stress, Second Edition, Volume 1. Academic Press, Oxford.

Traue, H.C. and H. Kessler. 2003. Psychologische Emotionskonzepte, pp. 20–33. In: A. Stephan and H. Walter (eds.), Natur und Theorie der Emotion. mentis Verlag, Paderborn.

Traue, H.C., H. Kessler and A.B. Horn. 2005. Emotion, Emotionsregulation und Gesundheit, pp. 149–171. In: R. Schwarzer (ed.), Gesundheitspsychologie. Enzyklopädie der Psychologie. Hogrefe, Göttingen.

Turtle, S. 2010. In good company, pp. 3–10. In: Y. Wilks (ed.), *Close Engagements with Artificial Companions*. J. Benjamins Publishing Company, Amsterdam.

Wagner, J. 2005. Vom physiologischen Signal zur Emotion: Implementierung und Vergleich ausgewählter Methoden zur Merkmalsextraktion und Klassifikation. M.S. thesis, Universität Augsburg, Augsburg, Germany.

Wagner, J., J. Kim and E. André. 2008. From physiological signals to emotions: implementing and comparing selected methods for feature extraction and classification. *Proceedings of IEEE International Conference on Multimedia and Expo ICME*, pp. 940–943.

Walter, S., C. Wendt, J. Böhnke, S. Crawcour, J.W. Tan, A. Chan, K. Limbrecht, S. Gruss and H.C. Traue. 2013b. Similarities and differences of emotions in human–machine and human–human interactions: What kind of emotions are relevant for future companion systems? Ergonomics, Taylor & Francis, pp. 1–13.

Walter, S., D. Hrabal, A. Scheck, H. Kessler, G. Bertrain, F. Nodtdurft and H.C. Traue. 2010. Individual emotional profiles in Wizard-of-Oz-Experiments. In: F. Makedon, I. Maglogiannis and S. Kapidakis (eds.), *Proceedings of the 3rd International Conference on Pervasive Technologies Related to Assistive Environments.* **S:**58–63.

Walter, S., K. Jonghwa, D. Hrabal, S.C. Crawcour, H. Kessler, H.C. Traue. 2013a. Transsituational Individual-Specific Biopsychological Classification of Emotions. *IEEE Transactions on Systems, Man and Cybernetics: Systems,* **43(4):**988–995.

Walter, S., S. Scherer, M. Schels, M. Glodek, D. Hrabal, M. Schmidt, R. Böck, K. Limbrecht, H.C. Traue and F. Schwenker. 2011. Multimodal Emotion Classification in Naturalistic User Behavior. In J.A. Jacko (ed.). Human Computer Interaction, Part III, HCI 2011, LNCS 6763, Springer-Verlag Berlin, pp. 603–611.

Weiner, B. 1986. An Attributional Theory of Motivation and Emotion. Springer, New York, Berlin.

Wendemuth, A. and S. Biundo. 2012. A Companion Technology for Cognitive Technical Systems. In: Cognitive Behavioural Systems. A. Esposito, A.M. Esposito, A. Vinciarelli, R. Hoffmann, V.C. Müller (eds.), LNCS 7403, Springer Berlin Heidelberg, pp. 89–103.

Wilks, Y. 2010. Introducing artificial companions, pp. 11–22. In: Y. Wilks (ed.), Close Engagements with Artificial Companions. J. Benjamins Publishing Company, Amsterdam.

Wimmer, M. and B. Radig. 2007. Automatically Learning the Objective Function for Model Fitting. Proceedings of the Meeting in Image Recognition and Understanding (MIRU) in Hiroshima, Japan.

Witherington, D.C. and J.A. Crichton. 2007. Framework for understanding emotions and their development: Functionalist and dynamic systems approaches. *Emotion,* **7:**628–637.

Wolff, S., C. Kohrs, H. Scheich, A. Brechmann. 2011. Temporal contingency and prosodic modulation of feedback in human-computer interaction: Effects on brain activation and performance in cognitive tasks. Informatik, Lecture Notes in Informatics, Berlin, **192:**238.

Woolf, B.P. 2010. Social and caring tutors: ITS 2010 Keynote Address, pp. 5–14. In: V. Aleven, J. Kay and J. Mostow (eds.), *Intelligent Tutoring Systems, Part I, LNCS 6094.* Springer, Berlin, Heidelberg.

Yanushevskaya, I., M. Tooher, C. Gobl and A. Ní Chasaide. 2007. Time- and amplitude-based voice source correlates of emotional portrayals, pp. 159–170. In: A. Paiva, R. Prada and R.W. Picard (eds.), Affective Computing and Intelligent Interaction: Proceedings of the ACII 2007 (Vol. 4738). Springer-Verlag, Lisbon, Portugal.

Zajonc, R.B. 1980. Feeling and thinking: Preferences need no inferences. *American Psychologist,* **35(2):**151–175.

CHAPTER 6

French Face-to-Face Interaction: Repetition as a Multimodal Resource

Roxane Bertrand, Gaëlle Ferré and Mathilde Guardiola

1. Multimodal Analysis of Human Communication and Interaction

Human-human interaction implies numerous studies to significantly improve the efficiency, naturalness and persuasiveness in human-computer interaction (HCI) systems. But there is still inadequate knowledge on what and how cues interact in face-to-face interaction. The complexity of human-human interaction involving the description of verbal and non-verbal modalities still needs theoretical and empirical foundations. To achieve this goal, researchers need to develop resources and tools that enable them to take into account the different modalities. Verbal, vocal and gestural cues have been studied separately for a long time. This favored the precise description of the mechanisms and rules governing each domain. But today the question of how these various cues in the different modalities are connected, has become important for linguists.

In this study, we present the perspective adopted in the national OTIM project (Blache et al., 2009) which aimed to answer some of the issues raised in multimodality in French face-to-face interactions. To achieve the global aim of the project, i.e. to better understand how the different linguistic levels interact, several steps were necessary, among which the specification of a standardized way of representing multimodal information, the development of generic and reusable annotated resources based on the elaboration of a multimodal annotation schema, the development and/or the adaptation of different annotation tools (see http://www.lpl-aix.fr/~otim/ for details of conventions, tools and annotations).

Drawing on the Corpus of Interactional Data (CID, Bertrand et al., 2008), the project involved different steps, from the development of the various coding schemes in the different modalities to the annotation and analysis. The corpus itself is an audiovisual recording of 8 hours of French conversational dialogs. The recording of the corpus was born out of an interest in human interaction based on a very fine-grained analysis of each linguistic domain and their relationships. Such an analysis in the phonetic domain requires a semi-experimental setting with a high quality of recordings enabling the acoustic analysis of speech. In the same way, the gestural level requires a particular setting, both in terms of lighting, framing and placement of the speakers in respect to each other and to the camera. The various recordings should be comparable and the frame chosen for the recording should allow good visibility of fine movements as well as larger ones made by the speakers. At the same time, conversational analysis requires yet to consider other criteria such as the level of (in)formality, the symmetric or complementary status between participants, the absence of pre-determined discursive role of participants, the presence/absence of a third party to regulate turn-taking, etc. This corpus affords a good balance between the elicited and very controlled corpora usually used by phoneticians or prosodists until recently and 'natural' conversational data analyzed in the field of Conversational Analysis (Couper-Kuhlen and Selting, 1996) on which the present study on repetition is drawing.

In this latter framework, the authors claim that every aspect of talk-in-interaction is collaboratively accomplished through participants' ongoing negotiations in situ (Szczepek Reed, 2006:8). In the same way, the collaborative model of Clark (1996) defines conversation as a *joint-action* implying a coordination of actions by participants at the level of content and at the level of process. Joint-action is achieved through different phenomena in interaction, among which backchannel signals, but also collaborative or competitive turn completion. Repetition naturally contributes to the co-construction of interaction as it supposes that one of the participants is taking into account what has been produced by the other at a certain time. Repetition then supposes some kind of adaptation in between participants to an interaction.

The use of such terms as *adaptation* as well as *alignment* (Garrod and Pickering, 2004; Pickering and Garrod, 2006), *accomodation* (Giles et al., 1987) or *mimicry* (Kimbara, 2006 among others) to quote but a few studies, refers to convergence phenomena. In the Interactive-alignment Model of Dialogue (Pickering and Garrod, 2006), the alignment observed at one level is automatically extended to other levels, resulting in a similarity at the level of discourse or gesture and at the level of representations. For the authors, this alignment is the basis of successful communication in dialog. Adaptation refers to the fact that participants adapt their responses to the other interactant(s)' productions. In the Communication Accomodation Theory (CAT, Giles et al., 1987), adaptation and accomodation can be used indifferently. Speakers are tailored to their partners (*adaptive behavior*) to affiliate with their social status for example. Mimicry is a direct imitation of what the other participant produces (exact match at prosodic level, Couper-Kuhlen and Selting, 1996; exact or very close match at gesture level, Jones, 2006). A discussion about the relevance of one term or another is out of the scope here (Guardiola, in progress).[1] It can nevertheless be noted that the choice of a term varies according to the linguistic field (psycholinguistics, sociolinguistics, phonetics, …) but also the modality considered in the type of study (see Section 3).

In this chapter, after presenting the corpus as well as some of the annotations developed in the OTIM project, we then focus on the specific phenomenon of repetition. After briefly discussing this notion, we show that different degrees of convergence can be achieved by speakers depending on the multimodal complexity of the repetition and on the timing in between the repeated element and the model. Although we focus more specifically on the gestural level, we present a multimodal analysis of gestural repetitions in which we met several issues linked to multimodal annotations of any type. This gives an overview of crucial issues in cross-level linguistic annotation, such as the definition of a phenomenon including formal and/or functional categorization.

2. Corpus and Annotations

A multimodal analysis of interaction requires the encoding of many different pieces of information, from different domains, with different levels of granularity. All the information has to be connected and synchronized (with the signal for example). Different steps in the annotation were adopted in the OTIM project to achieve this goal. Before presenting the annotation process and some of the annotations

---

[1] For further details, see SPIM ANR-08-BLAN-0276-01: http://spim.risc.cnrs.fr/

used in this study on repetition, it is important to consider that the project aimed not only to provide and develop conventions and tools for multimodal annotation but also to define the organization of annotations in an abstract description from which a formal XML scheme could be generated (Blache and Prévot, 2010).

2.1 Corpus

For a few years, numerous programs have been conducted in different countries to provide large-scale spontaneous speech interactions involving the creation and development of resources (in terms of both corpora and annotations). Among others, one can mention the Map-Task corpus (Anderson et al., 1991) that is one of the first semi-elicited corpus and which has been reduplicated in many languages, the Columbia Game Corpus (http://www.cs.columbia.edu/speech/games-corpus/), the Buckeye Corpus (Pitt et al., 2005), the Corpus of Spontaneous Japanese (Furui et al., 2005), DanPASS (the Danish phonetically annotated spontaneous speech corpus, Grønnum, 2006), as well as corpora annotated at the gestural level such as the Göteborg Spoken Language Corpus (Allwood et al., 2000), the MIBL Corpus (Multimodal Instruction Based Learning, Wolf and Bugmann, 2006) or the D64 Corpus (Campbell, 2009), among others (for a more exhaustive list, Knight, 2011). The *Corpus of Interactional Data* (CID) (Bertrand et al., 2008) described here is an audiovideo recording of conversational French (eight dialogs of 1 hour each, 110.000 words). Participants were filmed by a single camera and recorded with a head-set microphone (one track per speaker, in order to enable the acoustic analysis of speech and segments produced in overlap by both speakers). Participants were asked to tell about conflicts or unusual events in their personal lives.

2.2 Corpus transcription

The first and most important step in the annotation process is transcription because most of the annotations in the different domains are based on this particular level.

In a preliminary stage, the speech signal was automatically segmented in inter-pausal units (IPUs), speech blocks surrounded by 200-ms silent pauses. The transcription process takes this series of IPUs as input. The transcription conventions adopted in the project derive from the ones defined by the GARS (Blanche-Benveniste and Jeanjean, 1987). They take into account some remarkable and frequent phonetic phenomena: non-standard elisions, phoneme substitution or additions, assimilation phenomena, word truncation, silent pauses, filled pauses as well as some specific phenomena such as the pronunciation of schwas in Southern French and laughters. From this initial transcription, two versions were generated: (i) a *standard orthographic transcription* from which the orthographic tokens are extracted to be used for semantics, syntax and discourse analysis and their related tools (POS tagger, parser, etc.) and (ii) a *phonetic transcription* from which the phonetic tokens are used in the next steps of *grapheme-phoneme conversion* and alignment presented below.

The enriched orthographic transcription is time-consuming: three passes have been made for each speech file. In the first one, the entire transcription was made by one transcriber. The second and third passes involved a correction of this first transcription. However, it guarantees a faithful transcription and improves the phoneme/signal alignment.

The grapheme-phoneme converter is a dictionary and rule-based system (Di Cristo and Di Cristo, 2001); it takes a phonetic token sequence extracted from the transcription as input and provides a sequence of phonemes as output. From this, the aligner assigns each phoneme its time localization. This aligner is HMM-based (Brun et al., 2004), and relies on acoustic models based on standard French. The alignment is done for each IPU separately. In a first pass, labeling was automated. A second pass involving hand-correction of vowel boundaries was conducted on two speakers. From the time-aligned phoneme sequence and the enriched orthographic transcription, the orthographic tokens are also time-aligned.

From this tokenization and its alignment on the signal, a wide range of annotations have been conducted in each of the different domains: prosody (phrasing, pitch contours), morphosyntax and syntax, discourse and interaction (discursive units, reported speech, disfluencies, backchannel signals, etc.[2]). Not all the annotations will be fully described here, since there have been many in several linguistic fields and not all of them are relevant to the present study.

2.3 Morphosyntactic annotations

Morphosyntactic annotations were done in two steps. In a first stage, the enriched orthographic transcription was filtered of information to which no morphosyntactic category could be assigned, such as laughter or disfluencies, in order to form the input for a modified version of the syntactic parser for written French text (Blache and Rauzy, 2008). This was then modified in order to account for the characteristics

---

[2] More details are provided on http://aune.lpl.univ-aix.fr/~otim/

of spoken French. In a second stage, the output of the parser was manually corrected for the totality of the CID. The annotation process is time-consuming whether it is manual or automatic. The manual annotation requires several annotators (either expert or not, sometimes both) and tests of labeling consistency to measure inter-annotator agreement. The automatic annotation is less time-consuming but also requires evaluation between the different tools or involves manual corrections, which enable to evaluate the performance of the parser (only 5% of error rates).

2.4 Prosodic annotations

The prosodic level can be annotated in a manual or an automatic way depending on whether we observe rather phonological (more abstract) phenomena or phonetic parameters. In OTIM, we focused on the prosodic phrasing which corresponds to the structuring of speech material in terms of boundaries and groupings. The manual annotation was very time-consuming but was necessary to improve the knowledge of prosodic domains in French. In a first stage, such a manual annotation made by experts enabled us to test the robustness of annotation criteria. A previous study involved two experts; results have shown a very good inter-coder agreement and kappa score for the higher level of constituency (IP) (Nesterenko et al., 2010). In a second stage, the elaboration of a guideline for transcribing prosodic units in French by naïve annotators enabled us to test the reduplicability of these annotation criteria. Naïve transcribers have to annotate four levels of prosodic break defined in terms of a ToBI-style annotation[3] (0 = no break; 1 = AP break; 2 = ip break; 3 = IP break) in Praat (Boersma and Weeninck, 2009). The global aim is to develop a phonologically based transcription system for French that would be consistent enough to be amenable to automatic labeling. One of the steps is to compare the manual annotations. Another step consists in improving existing automatic tools (such as *Intsint* for example, Hirst et al., 2000) by comparing the output of different annotation tools and manual expert/naïve annotation (Peskhov et al., 2012).

At last, another aspect of prosodic annotation concerns the intonation contours associated to intonational or intermediate phrases (levels 3 and 2 above). Pitch contours are formally and functionally defined (Portes et al. (2007) for details). Intonation contours were coded for six speakers.

---

[3] http://www.ling.ohio-state.edu/~tobi/ame_tobi/annotation_conventions.html

2.5 Gesture annotations

Ninety minutes of the CID involving six speakers were coded for gestures. We manually annotated hand gestures, head and eyebrow movements as well as gaze direction with Anvil (Kipp, 2001).

Different typologies have been adopted for the classification of hand gestures, based on the work by Kendon (1980) and McNeill (1992, 2005). The formal model we use for the annotation of hand gestures is adapted from the specification files created by Kipp (2004) and from the MUMIN coding scheme (Allwood et al., 2005). Both models consider McNeill's research on gestures (1992, 2005).

The changes made to existing specification files only concerned the organization of the different information types and the addition of a few values for a description adapted to the CID. For instance, we added a separate track 'Symmetry'. In case of a single-handed gesture, we coded it in its 'Hand_Type': left or right hand. In case of a two-handed gesture, we coded it in the left Hand_Type if both hands moved in a symmetric way or in both Hand_Types if the two hands moved in an asymmetric way. For each hand, the scheme and annotation file in Anvil has 10 tracks.

2.5.1 Functional categories

The gesture types we annotated are mostly taken from McNeill's work. Iconics present "images of concrete entities and/or actions", whereas *Metaphorics* present "images of the abstract", they "involve a metaphoric use of form" and/or "of space" (McNeill, 2005:39). *Deictics* are pointing gestures and *Beats* bear no "discernible meaning" and are rather connected with speech rhythm (McNeill, 1992:80). *Emblems* are conventionalized signs and *Butterworths* are gestures made in lexical retrieval. *Adaptors* are non-verbal gestures that do not participate directly in the meaning of speech since they are used for comfort. Although they are not linked to speech content, we decided to annotate these auto-contact gestures since they give relevant information on the organization of speech turns. For gesture phrases, we allowed the possibility of a gesture pertaining to several semiotic types using a boolean notation.

2.5.2 Descriptive annotations

A gesture phrase (i.e. the whole gesture) can be decomposed into several gesture phases, i.e. the different parts of a gesture such as preparation, stroke (the climax of the gesture), hold and retraction (when both hands return to rest) (McNeill, 1992). The scheme presented

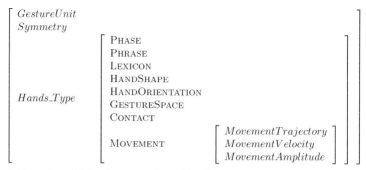

Figure 1. Formal model for the annotation of hand gesture.

in Figure 1 also enables us to annotate gesture lemmas (Kipp, 2004:237), the shape and orientation of the hand during the stroke, suppress gesture space (where the gesture is produced in space in front of the speaker's body, McNeill, 1992:89), and contact (hand in contact with the body of the speaker, of the addressee, or with an object). We added three tracks to code the hand trajectory (adding the possibility of a left-right trajectory to encode two-handed gestures in a single Hand_Type, and thus save time in the annotation process), gesture velocity (fast, normal or slow) and gesture amplitude (small, medium and large). A gesture may be produced away from the speaker in the extreme periphery, while having a very small amplitude if the hand was already in this part of the gesture space.

Head and eyebrow movements, as well as gaze direction and global facial expressions (laughters and smiles) were annotated as well, although not all the items projected in the coding scheme provided in Figure 2 were noted.

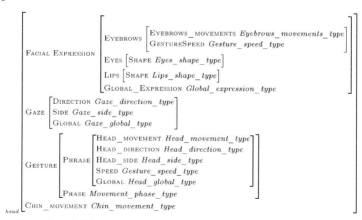

Figure 2. Formal model for the annotation of head and eyebrow movements, gaze and facial expressions.

3. Repetition

3.1 Theoretical background: A general definition

To take a rather objective term, "repetition" in interaction has been observed by many researchers working in different fields. Chartrand and Bargh (1999), but also Garrod and Pickering (2004) argue that repetition is needed to make conversations easier or more fluent and that speakers align "their representations at different linguistic levels at the same time" (2004:9), thus reducing the processing load for each participant in a conversation. Several terms have been coined to refer to repetition that do not, however, necessarily refer to the same process. Working on gesture and sound repetition of adults by infants, Jones (2006:3) distinguishes between emulation, a repetition of an "outcome produced by a model, with no requirement that the actual motor behavior should match that of the model", and mimicry (the most widely used term in multimodal studies), "a behavior that matches or closely approximates the movements of another". She also goes against the general view that mimicry is innate and contends instead that it is a learned behavior that arises in infants around 18 months and that is the result of infants being mimicked by their caregivers. An infant's response to an adult protruding their tongue with the same movement would not be imitation according to her but rather a general response to any interesting stimulus. The debate is out of the scope of this chapter, but Garrod and Pickering (2004) adopt a similar view when they describe conversations amongst adults, as they establish a sequential link between primed representations—what is called by Chartrand and Bargh (1999) the chameleon effect, i.e. a perception-behavior link—which leads to imitation, which in turn leads to alignment of representations (op. cit., 9), or what is termed elsewhere *convergence*. They illustrate this sequential process with 'yawning': if one sees someone yawn, one yawns in return (primed representation), but one also tends to feel more tired or bored (imitation and alignment of representations with the initial yawner). They also mention that the whole process is unconscious and largely automatic but "is also conditional to the extent that it can be inhibited when it conflicts with current goals and purposes, or promoted when it supports those goals" (op. cit., 10). At last, both Garrod and Pickering (op. cit.) and Shockley et al. (2009) mention that alignment does not mean that speakers have to be in agreement in a conversation and that it is rather a process speakers use to simply understand each other. Tabensky (2001:217), however, states that what she calls echoing "can be merely a sign of co-presence, and not necessarily an indication of understanding or alignment with the speaker's proposition".

There is more consensus on the fact that a conversation can be considered as a joint action (Garrod and Pickering, 2004; Holler and Wilkin, 2011; Kimbara, 2006; Shockley et al., 2009; Tabensky, 2001, to cite but a few studies) which entails the co-construction of meaning by all the participants to the interaction as suggested in the introduction.

3.2 *Behavior and gesture repetition*

There has already been quite a large body of work on the role of behavior and gesture-pattern repetition and on the conditions for their emergence. Lakin et al. (2003) observe that some situations activate a desire to affiliate in the participants to an interaction and thus encourage mimicry. This work is derived from Chartrand and Bargh (1999) who noted the social role of mimicry. On experimental data, they observed that postures and adaptors (in their study, the shaking of one's foot) were regularly mimicked by the participants, and that when the confederate mimicked the participants, the latter felt greater empathy with the model. Working on posture and gaze, Shockley et al. (2009) find that similar gaze patterns emerge in participants together with the increase of joint understanding. Also working on experimental data, they observe that participants adopt more postural coordination when they see the same words on a screen than when the words are different. Mol et al. (2009) go further on experimental data as well. They find that reproduced iconic gestures are not just imitation: only gestures that are consistent with verbal content are copied. Much in the same vein, Holler and Wilkin (2011) find that mimicked hand gestures in experimental conditions play an active role in the grounding process and help create mutually shared understanding. Their classification of mimicked gestures is both semantic and formal. To count as repeated, a gesture has to represent the same meaning or have the same referent, use the same mode of representation and have the same overall form. Drawing on data from a joint narration task, Kimbara (2006:45) adds that temporal proximity together with co-referentiality between a gesture and its repetition show "realizations of a shared image construal". Besides, she observes that not all the features of gestures are repeated, but a subset has to be present in the repetition for the gesture to be considered as mimicked. From her study, she concludes that hand-gesture mimicry creates *gesture catchments* (McNeill, 2001) across speakers.

In a later study, Kimbara (2008) notes that gesture repetition is not a chance phenomenon. In experimental conditions, she notices that participants produce gestures which are more similar in terms of handshape when they can see each other than when they cannot. Parrill and Kimbara (2006), also working on experimental data, note that observing mimicked gestures induces more mimicry in the participants. They consider a gesture is repeated when two of the following features are reproduced: motion, handshape or location. Hand-gesture features are also central in Mol et al. (2012) who state that imitators in laboratory speech are influenced in their mimicry by features of the original gesture, for instance handedness. They show as well that participants are influenced in their repetition of hand gestures by the cognitive perspective adopted by the confederate (like the description of items on a map from a vertical or a horizontal viewpoint).

Instead of focusing on exact matches between the verbal and gestural productions of participants to conversations in three languages, Tabensky (2001) describes what she calls *rephrasings*, namely how speakers mimic some semantic features while adding new features at the same time. She is also concerned with temporal alignment of the productions and what forms a language unit. From her corpus, she observes that in some instances, the semantic features which formed a package in speech and/or gesture in the original production are separated into different units in the rephrasing (a process she terms *separation*), whereas in other instances, semantic features expressed in several units in the original production are merged into a single unit in the rephrasing (a process she terms *fusion*).

At last, von Raffler-Engel (1986) observes full or partial gesture imitation in *transfers*, namely the gestures made by an interpreter into another language in consecutive translation. She describes gesture repetition in terms of "repetition of parts to the whole" and determines a series of *components* which must be proportional to the model for a gesture to be considered as repeated but need not be identical: muscular tension, gesture duration and movement extension, a series of components that we also consider in the present study. Other gesture characteristics have to be identical to the model to give an impression of sameness. With this in mind, she notices that in many instances, interpreters retake the gestures they observe in the speaker they translate, instead of changing the original gestures in the re-packaging of information involved in a translation. Yet, she mentions that the gestures produced in the translation were judged natural enough by native speakers of this language, so that no culturally inappropriate gesture was copied into the translation.

3.3 Verbal other-repetition

Verbal other-repetitions (henceforth OR) consist in repeating a word or a sequence of words that have been previously uttered by another interactant. This process leads to a lexical similarity of the participants' speech that can be analyzed as a means to align with the interlocutor. OR have been identified as forming an important mechanism in face-to-face conversation through their discursive or communicative functions (Norrick, 1987; Tannen, 1989, 2007; Perrin et al., 2003). According to Tannen (2007), participants notably use lexical repetition to show their involvement in the interaction. She argues that repetition is useful at several levels of verbal communication: production (easier encoding), understanding (easier decoding), connection (better cohesion in discourse), and interaction (repetition maintains the link between participants). Repetition can also be considered as a *specific* form of feedback (in the sense of Bavelas et al., 2000). Perrin (2003) proposes a four-function typology for other-repetitions, nearly corresponding to backchannel functions: taking into account, confirmation request, positive reply and negative reply. More largely, repetition functions as a device for getting or keeping the floor (Norrick, 1987).

3.4 Prosodic repetition

In a similar way to the verbal or gestural level, the main issue raised by prosodic repetition is to know when one speaker's repetition of a prosodic pattern can be considered as mimicry (Couper-Kuhlen and Selting, 1996:366). More recently, this question is addressed by Szczepek Reed (2006) through the notion of prosodic orientation that refers to the "interactional orientation whereby (...) speakers display in their sequentially "next" turns an understanding of what the "prior" turn was about" (Hutchby and Wooffitt, 1998:15). Szczepek Reed defines several types of prosodic orientation such as the prosodic matching (copy) of the previous speaker's prosodic design, the complementation of a prior turn with a second structurally related prosodic design or a continuation of the previously unfinished prosodic pattern. Gorisch et al. (2012) report some works on pitch matching and interactional purpose. To the authors' credit, they provide precise definitions of different terms often used in the same way. They propose to consider that prosodic matching is used for continuing the project in hand, aligning or affiliating with the previous speaker. In line with Stivers (2008) and Barth-Weingarten (2011), they distinguish between the terms *alignment* and *affiliation* that have been used until recently in an indefinite way. Alignment refers then to the endorsement of the sequence/activity in progress and contrasts with the notion of affiliation which refers to the endorsement of the previous speaker's evaluative positioning, or stance (cited by Gorish et al., 2012:7). More specifically, Stivers (2008) uses the term of alignement to describe actions by a second speaker which support the activity being undertaken by the first speaker. In this way, the production of backchannel signals in conversation can be considered as adapted and expected responses from the listener during conversation (Bertrand et al., 2007; Heldner et al., 2010 among others) and more particularly in a storytelling activity in which the main speaker (narrator) is indeed ratified as main speaker by the listener (Stivers, 2008). At last, Gorish et al.'s study also constitutes a first attempt to develop a method that enables the measurement of the acoustic similarity (in terms of f0 and intensity) of pitch contours in naturally occurring data. Similar parameters were considered by De Looze et al. (2011) in a study on prosodic convergence in spontaneous conversations. In Gorish et al.'s study, the prosodic matching observed is then considered as a resource used to demonstrate alignment with the prior action. In a similar way, Bertrand and Priego-Valverde (2011) have shown that a copy of some prosodic cues by both participants could be a resource to demonstrate orientation to a humorous utterance expressed by the speaker. A series of prosodic matching repetitions by both participants is leading to the creation of a short sequence called *joint fantasy* (Kotthoff, 2006).

4. Identification of Multimodal Repetition

4.1 Gesture repetition

The criteria we adopted for the repetition of hand gestures were very much inspired from von Raffler-Engel (1986). For a hand gesture to be considered as repeated, gesture phrase, lexicon and movement trajectory have to be identical, that is the functional category of the gesture, whereas other descriptive features like gesture space, tension, amplitude and velocity do not have to be strictly identical to the model for the gesture to be considered as a repetition of the model. The criteria for the repetition of head movements and gaze orientation are stricter than for hand gestures since we considered a gesture was repeated if the movement in the repetition was strictly identical or mirrored between the repetition and the model.

4.2 *Lexical repetition*

In the same way as for gesture repetition, for a lexical repetition to be considered as repeated, different formal and functional criteria are involved.

First of all, we proposed to formally define verbal repetition as the production of a word or phrase that has already been uttered by another speaker. We specify that too frequent words cannot be considered as repeated, in order to avoid 'accidental' similarities in discourse.

Annotations concerning lexical other-repetition were made in two steps: a first automatic output, followed by a manual correction by two experts. The first automatic stage, based on the transcription of tokens, allows the detection of potential other-repetitions. It is based on a set of rules and on relevance criteria themselves based on word frequency (for each speaker). The rules were elaborated during a previous study (Bigi et al., 2010).

A preliminary processing transforms words into lemmas, containing no morphological mark of conjugation or plural. A set of two rules is then applied on the data: Rule 1—An occurrence is accepted if it contains one or more 'rare' words (the rarity is measured on the vocabulary of the speaker who makes the repetition). Rule 2—An occurrence which contains at least five words is accepted if the order of words is strictly identical in both speakers' discourse.

The tool locates co-occurrences of relevant lemmas: the words which were uttered by a speaker in an IPU, and by the other speaker in a simultaneous or in a previous IPU.

Obviously, these formal criteria are not sufficient to only select occurrences of other-repetition. Following Perrin et al. (2003), a repetition has to have an *ostensive* character to be considered as a real repetition (intention of quotation). Then only an expert analysis enables to eliminate co-occurrences that are not other-repetitions. The tool, however, greatly reduces the amount of time necessary for the detection of repetitions. In a last step, two experts checked the speech segments identified by the tool as possible repetitions on the basis of formal criteria. Three hundred and fifty consensual cases were then retained in the CID.

4.3 *Prosodic repetition*

In the same way as for gesture or lexical repetition, for a prosodic pattern to be considered as repeated, the repetition and the model have to be identical, either at a phonetic level (duration, fundamental frequency) or at a phonological level (phrasing, pitch contours, or rhythm pattern). Following Szczepek Reed (2006), we use the term *prosodic matching* to refer to the repetition of a prosodic pattern.

5. Analysis

First, it must be noted that *other-repetition* is not quite frequent in our corpus when considering gesture. Although we have seen in the introduction and Section 3 that conversations constitute joint activities in which meaning is co-constructed by participants, this co-construction does not necessarily entail gesture repetition. Gesture repetition is therefore probably dependent on many other factors like topics, collaborative tasks, but also the conversational history of the participants. Indeed, Tabensky (2001) mentions that gesture repetition occurs pretty much at the beginning of the conversations she works on, in which participants are not acquainted with each other. In our corpus, participants know each other quite well and therefore have a long conversational history which could explain why less gestural adjustment is needed between them. Tabensky (op. cit.: 232–233) also states that "textual repetition of words and retakes with small adjustments are generally not accompanied by gesture", which is also what we find since there is a much vaster quantity of verbal repetition and hardly any at all involve gesture repetition. However, we will see in this section that when gesture repetition is present, it can be quite complex especially when related to verbal and prosodic repetition in a multimodal perspective.

5.1 Repetition as a confirmation request

In example 1 below, speaker A describes the hard work he had to put in to redecorate a room in his house since the walls were coated with several layers of wallpaper and paint. An approximate translation of the example is given in italics and the transcription conventions are provided at the end of this chapter.

Example 1

　1 A: c'était tapissé peint [alors c'était l'enfer quoi] *the wall was papered, painted, so it was a nightmare*

　2 B: [ah ouais avant que t'enlèves les couches] *oh yes, before you can take off all the layers*

　3 A: oh putain j'ai mis j'ai mis un mois quoi /mh/ *oh fuck it took it took me a month /mh/*

4 A: enfin bon c'est un mois en faisant que le week-end si tu veux /ah ouais/ mais *well a month but I was only working on week-ends you know /oh yeah/*

5 A: tu vois là tout à la ra[clette] *you see, all of it with a scraper*
       {Gesture 1 ——————————————————————————}

6 B: [c'était peint sur la ta]pisserie [truc comme ça] *it was painted on the paper and stuff*
       {Gesture 2 ——————————————————————————}

7 A: [et ouais] *yes it was*

Figure 3a illustrates the gesture made by speaker A while saying *all of it with a scraper*. He produces an iconic gesture as if he was holding a scraper, a gesture in which his right hand goes up and down twice. Speaker B repeats the iconic gesture adopting a mirror perspective in a double up and down movement of his right hand with the same hand shape, although the gesture is not as high in the gesture space as Gesture 1 and a bit more towards his side instead of being in front of his chest. Whereas Gesture 1 lasts 1.60 s, Gesture 2 in Figure 3b is slightly shorter with a duration of 1.52 s, yet the difference in length between the two gestures is not dramatic. What is interesting though is their temporal alignment. Speaker B prepares for the gesture just

Figure 3a. Iconic gesture produced by speaker A (Gesture 1) in example 1.

Figure 3b. Iconic gesture reproduced by speaker B (Gesture 2) in example 1.

as speaker A is preparing for his second "scraping" gesture and the stroke of speaker B's gesture is beginning three frames before the end of speaker A's stroke. This means that at the same time as he is producing his gesture, speaker B is attending to speaker A's own gesture. He cannot know at the beginning of his gesture that the model will stop after the second "scraping movement".

The example illustrates a certain multimodal parallelism between gesture and speech: the case does not correspond to the fusion nor to the separation described by Tabensky (2001) since the repetition contains two pieces of information pertaining to two different modalities. The iconic gesture repeats the information of the scraper, whereas the utterance produced by B is not an exact repetition of what was said by speaker A, although some information is similar. At the prosodic level, however, there is a similarity. The two utterances considered here form one intonational phrase (IP) each (the IP of speaker B starting before the end of speaker A's turn, with an overlap of 0.655 s). Both IPs present similar configurations. We can note three things that are particularly striking though: first of all, speaker A produces a slightly emphatic accent on "tout" (*all*) which is realized as a slight reinforcement of the initial plosive /t/. The same reinforcement is met in the initial plosive of the emphatic word "peint" (*painted*) for speaker B. Towards the end of the IP we can see a similar list pitch contour even if the second utterance (B) could be considered as a confirmation request. Speaker B seems indeed to ask confirmation that he understood well when saying "it was painted on the wallpaper" as the first verbal mention of the utterance by speaker A (line 1) did not make it explicit that the coat of paint had been applied onto the wallpaper (the utterance "the wall was papered, painted" could be understood as a chronological description of two actions with no link between them, not necessarily as the wallpaper being painted).

Nevertheless, the prosodic matching is also expressed by the strong lengthening associated with the last syllables of words "raclette" (*scraper*) and "tapisserie" (*wallpaper*), which was described by Portes et al. (2007) as the main cue of the list contour. And at last, speaker A adopts a flat trailing contour around 135 Hz on the whole phrase which is also copied by speaker B with the same f0 height, although speaker B generally has a much lower voice than speaker A.

The match which occurs both at the gestural and at the prosodic level is interesting in two respects: first, considering the fact that the two utterances do not constitute the same kinds of speech acts —speaker A's utterance is a statement, whereas speaker B's could be a confirmation request—the two utterances would probably have had completely different prosodic contours in another context. Then, because of the overlapping speech, it means that speaker B is copying prosodic information while speaker A is still speaking and this exactly matches the pattern we have for gesture since there was also a gesture overlap in between the model and the copy.

5.2 *Repetition as a hedge*

Just before the extract below, two male participants were discussing a school experience one of them had. His teacher was very strict and forbade the children to leave class. Once, he needed to go to the bathroom, didn't dare to ask the teacher and messed himself. As his mother actually worked in the school as a teacher, he went to see her. In the example, speaker A, after acknowledging the narrative with a backchannel, asks if the mess showed in a verbally elliptical utterance ("parce que t'étais tout", *because you were all*). The question is, however, not exactly elliptical as it is completed by a gesture, which is repeated by speaker B in his answer.

Example 2

1 A: parce ce que t'étais tout (0.075) *because you were all*
{gesture 1 ————————}

2 B: non ça se voyait peut être /non/ je me rappelle plus trop / ouais/ mais je crois pas que ça se voyait mais bon euh @ ça ça devait sentir tu vois @ *and then, no perhaps it didn't show / no/, I don't quite remember /yeah/, but I think it didn't show, but uh it must have smelt you see*

{gesture 2 —}{gesture 3 ————————————————}{gesture 4 ——————————————————————-}

Figure 4a illustrates the metaphoric gesture produced by speaker A who starts with both hands slightly rising from his lap and places them in the lower periphery, palms oriented towards his body. He then extends his hands away from his body thus representing the extent of the mess. The whole gesture from the beginning of the preparation phase to the end of the retraction lasts 0.96 s. One frame before the end of the retraction of Gesture 1, speaker B initiates a repetition of the gesture (Figure 4b), yet the two gesture strokes are not in overlap. The difference between the two gestures as illustrated in the figures is that Gesture 2 is much shorter (0.60 s) than Gesture 1, the movement is not as ample and the fingers are much more relaxed than those of Gesture 1.

What is interesting, however, is that immediately after Gesture 2, B produces an emblem (Gesture 3)—which is of no particular interest

Figure 4a. Metaphoric gesture produced by speaker A (Gesture 1) in example 2.

Figure 4b. Metaphoric gesture reproduced by speaker B (Gesture 2) in example 2.

Figure 5. Second reproduction of metaphoric gesture by speaker B (Gesture 4) in example 2.

here—without any retraction of Gesture 2, and then produces Gesture 4 without retracting his hands. Gesture 4 happens to be a second repetition of Gesture 2. This time, it is slightly longer than the first repeat (0.92 s) and the gesture features are more similar to the original production of the gesture as illustrated in Figure 5. In this second repetition, the amplitude of the movement is slightly larger than in the first repetition and finger tension is also greater. What is different from the original gesture is the position of the hands: the palm of the left hand is facing up instead of the hand being on its side.

What can be added in this example is that both Gesture 1 and Gesture 2 are not redundant with the message content. When speaker A produces Gesture 1, he is anticipating some assumption on the part of speaker B who was narrating what happened at school. The gesture in this context completes what is left unsaid in the elliptical utterance, probably out of decency. Although speaker B repeats speaker A's gesture (with Gesture 2), he contradicts the verbal assumption, so that the gesture which was consistent with the initial verbal message is repeated (as such gestures tend to be repeated as pointed out by Mol et al., 2009), but then becomes quite inconsistent with the answer. A gesture linked to the syntactic negation would rather have been expected here. In this example, the prosodic level is in accordance with the content of the utterances. In the first turn, speaker A formulates the elliptical question with a low and trailing pitch and a very strong lengthening on the last word that is typically used in unfinished turns. By contrast, the next turn produced by B, starting with the answer "no", exhibits a rising-falling contour while at the same time speaker B is repeating the gesture previously produced by A in the first turn. Gesture 4 is also a repetition that comes together with the repetition of the contradiction and this looks as what has sometimes been called a *hedge*, i.e. a way of softening a contradiction, contradictions being generally not preferred by interactants. It is interesting to note that the same rising-falling contour is again produced by speaker B on "voyait" (*showed*) (which is also the second repetition of this word) as speaker B is once again repeating speaker A's gesture. Therefore, we can say that there is a complete dissociation between the double repetition of the other participant's gesture by speaker B, and the contrast expressed both in verbal content with two negations and prosody with the self-repetition of the contrastive prosodic contour. Speaker B then develops with "mais ça devait sentir" (*but it must have smelt*), an utterance which is later repeated as well as a self-confirmation.

5.3 Cross-repetition

In example 3 below, the two speakers are discussing the arrangements speaker A will make to look after the baby his wife is expecting.

Example 3

    1 A: bien par exemple t'façon Laure elle est prof *well, for example, anyway, Laure is a teacher*

    2 A: donc elle travaille pas tu vois /ouais/ tout le tem[ps] *so she doesn't work, you see /yeah/ all the time*

    3 B: [to]ut le temps ouais *all the time yeah*

4 A: puis à ce moment là les matinées où où elle est au au co- si elle doit aller au collège (0.62) *so then on the mornings when when she is at at scho- if she must go to school*

5 A: [bien moi moi je reste ici je prends euh enfin je m'en occupe] *so I I stay here, I take uh, well I look after it* {Gesture 1}{Gesture 2}{Gesture3}{Gesture 4 ——————}

6 B: [ouais toi tu restes ouais ouais vous euh] *yeah yeah you stay yeah yeah you uh* {Gesture 5}

Example 3 is slightly different from what we have seen above. As he utters *so I*, speaker A produces a metaphoric gesture (Gesture 1) which is an asymmetrical double-handed gesture (the left hand moves a bit more than the right one), and which consists of his hands being oriented palm up at the beginning of the stroke. Then he rotates his wrists, so his hands are on the side and opens them a little again. Speaker B has a similar movement of his right hand only although the configuration of his fingers is different (Gesture 5). The twisting movement is actually what makes the gesture look as a repetition of A's metaphoric. This is shown in Figures 6a and 6b.

Figure 6a. Two-hand metaphoric gesture produced by speaker A (Gesture 1) in example 3.

Figure 6b. Single-hand metaphoric gesture produced by speaker B (Gesture 5) in example 3.

In terms of temporal alignment, Gesture 5 starts 0.88 s after Gesture 1, yet, contrary to what we saw in example 1, the repeat is much longer than the model (0.88 s vs. 0.36 s) and is also more complex as well. Whereas Gesture 1 is only composed of a stroke because it is part of a series of gestures which we will not describe here since they are not relevant to the present study, Gesture 5 contains a preparation and a retraction. If we consider the stroke only, then the repeat is shorter than the model as it lasts 0.32 s.

One may consider that there is a redundancy between gesture and verbal repetition in this example with no new information added. However, the repetition plays a role in the message structure as it is the global pattern which is repeated including words and gesture and which has a function of backchannel. The whole extract is very collaborative: speaker A produces some argumentation as to who will take care of the baby and speaker B collaborates to the argumentation first producing the backchannel signal "ouais" (*yeah*), then repeating "tout le temps" (*all the time*) and repeating his own *yeah* again. His whole utterance forms a complex backchannel with a function of acknowledgement. At the prosodic level, the model and the repeat are clearly distinct at least because of the location in the IP and the discursive function of each one. A produces "tout le temps" in the end of the IP with a major terminal rising contour (about 100 Hz) followed by a high plateau while B produces "ouais tout le temps ouais" as a single IP with a minor rise on "temps" (around 30 Hz). At this point, the two speakers seem to be reaching the end of a conversational sequence which could be the reason why their overall pitch is so low. The configuration of this repetition exhibits a compressed span as it is often the case in backchannels. Concerning the gesture repetition (Gesture 5), although it is quite clear in the video that speaker B is repeating speaker A's gesture, the timing in the verbal modality is rather a repetition of B's speech by A. In fact, B's "restes" (*[you] stay*) begins 0.197 s before A's "reste" (*[I] stay*), so that when speaker A is beginning to utter "reste", he has enough acoustic material to know what is being said by B. B's "restes" is much longer so that speech rate is not similar for the two speakers. The lengthening on "restes" by speaker B directly corresponds to the lengthening of his copied gesture. However, according to the location of the word "reste" in the two repetitions, their contour is not quite the same (see Figure 7). Whereas speaker B's contour is a low plateau at the end of an IP, followed by another IP ("ouais") A's "reste" is in the middle of the IP that ends on "ici" expressed with a rise. The configuration of the repetition that also functions as an acknowledgement is in accordance with the previous verbal repetitions of this sequence.

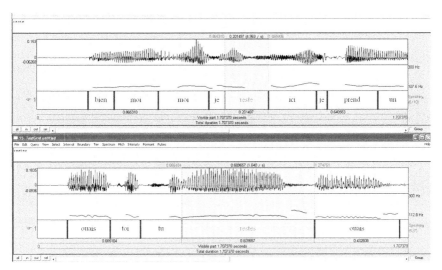

Figure 7. Pitch curves (Hz) of the utterances produced by speaker A (top) and speaker B (bottom) in example 3.

This analysis provides a good example of cross-repetition: whereas one of the participants is repeating the other's gesture, the other participant repeats the first one's speech in terms of verbal content. Prosody is matching between the two speakers only in the fact that they both use compressed span as projecting the end of a conversational sequence.

5.4 Posture-match: A case of extreme convergence?

The example below illustrates the social role of posture coordination between participants in a way quite similar to the observations made by Chartrand and Bargh (1999) and Shockley et al. (2009). It occurs just after the beginning of an interaction between two female participants. They have been urged to speak about unusual things that might have happened to them and at the very beginning of the recording they were thinking about what to say and each of them was turned away from the other, looking up while discussing the meaning of the word "unusual" as illustrated in Figure 8a. At the beginning of example 4, they both turn their heads towards each other with their chin slightly raised without changing the orientation of their body (Figure 8b) and both encourage the other to come up with a narrative.

Example 4

    1 A: insolite (0.674) euh si le p- *unusual um yes the p-*

2 B: [bon je vois que tu es tellement à court d'idées allez vas-y tu démarres vas-y vas-y @] *so I see you're really lacking an idea, here you go, you begin, here you go, here you go*
{Both A and B turn their head towards each other in exact synchrony ————————————————————————

3 A: [si si vas-y vas-y j'ai j'ai un truc qu'est qui était] extrêmement marrant *yes yes, here you go, here you go, there is something that is that was extremely funny*

————————————————————————————————}

When considering the head orientation of the two participants in example 4, one is compelled by the exact match both before and after they turned their heads towards each other. In determining the presence of repetition, simultaneity plays as important a role in gesture as in the other modalities. One cannot talk of gesture repetition in this example because the change in head direction starts at exactly the same time for each participant. However, there is a strong convergence in both the visual and verbal modality as not only do both participants turn their heads towards each other at the same time but they also speak in overlap repeating the phrase "vas-y" (*here you go*) several times both in self- and other-repetitions. Prosodically, this overlapping sequence presents a high pitch and intensity for both participants. This characteristic is known to indicate a competitive sequence to gain the floor (French and Local, 1986). To describe what exactly happens in this sequence, we can say that there is a real adjustment between speech turns. For each speaker, we observe a similar phrasing in three units. For A: "allez vas y vas y // tu démarres// vas y vas y" and for B: "si si// vas y vas y// j'ai un truc //", the second IP for B being a repetition of A's first IP and A's last IP being the repetition of B's second one. This precise timing inside the overlapping sequence provides evidence that both speakers are in a legitimous position to take the floor according to the rules governing the organization of turn-taking (Sacks et al., 1974). After a certain time lag, both participants to the interaction are entitled to take the turn at speech and are therefore potential next speaker. They then start speaking at the same time because the time lag is shared by both participants, a process which is described as case of *blind-spot overlap* (Jefferson, 1987). The effect is to achieve some sort of 'social convergence' insofar as conversation can be seen as a social activity governed by a certain number of rules of politeness. Politeness does not only involve what is said and in what manner but also involves behavior patterns like gaze alternation in between speakers and listeners as well as body orientation towards the co-participant.

Figure 8a. Posture of the two participants in line 1 of example 4.

Figure 8b. Posture of the two participants in lines 2 and 3 of example 4.

Example 4 can be contrasted to later moments in the same interaction where the two speakers are not involved in the interaction to the same degree as illustrated in Figure 9. Their body is not oriented towards the co-participant and they do not gaze at each other. At these moments, the previous topic was finished and they had not started a new topic yet. They nevertheless repeat phrases such as "à part ça" (*apart from this*) and "et sinon" (*and otherwise*) which carry little semantic content. The repeated phrases are similar from a lexical and prosodic viewpoint (echo utterances) and they seem to be the only link left between the two speakers, playing a role in the regulation (in terms of cohesion, Tannen, 2007) of the interaction. The repetitions show an interactional alignment at the level of forms, but also at a meta-interactional level (both speakers express convergence in their search for a new topic).

Figure 9. Postural misalignment.

6. Conclusion

One of the issues raised in the field of multimodality is precisely how the verbal, the vocal and the visual modalities articulate with one another in the construction of interaction. We know that information is conveyed not only through words and sentence types at the semantic and syntactic levels, but also through prosodic phrasing and contours used by the speaker. It has been shown more recently that co-speech gestures also participate in the conveying of semantic information, and that they play a role in the organization of discourse by speakers. At last, much like what happens in the verbal and vocal modalities, they reveal something of the interpersonal relationship between participants to an interaction. It would, however, be simplistic to suggest that in any utterance, exactly the same information is conveyed in the three modalities at the same time. It cannot be expected therefore that when information is repeated by a participant to an interaction, all of the information will be copied. Rather, the participant is more liable to copy different pieces of the message: part of what was said (semantic information) and/or part of its format (prosodic and gestural information). And since the main role of repetition, as seen in previous studies described in Section 3 of this chapter, is to help participants to an interaction achieve some sort of convergence, it is to be expected that depending on the amount of information repeated by a participant, the degree of convergence will be lower or higher.

In order to test this, we analyzed some examples with a focus on gesture repetition. The examples were drawn from the Corpus of Interactional Data (CID) recorded at Aix en Provence. It comprises a series of video recordings of unprepared dialogs in French which were transcribed and annotated in several linguistic domains, including gesture for part of the corpus.

The examples confirmed results from previous studies showing that gesture repetition does not have to be strictly identical to be considered as repetition and that it is rather what makes the semantics of the gesture (namely the type and direction of movement, general hand shape) which has to be copied, whereas other features are not strictly necessary in the repeat (gesture speed or gesture space for instance). These may be considered as variable features of the gesture. It became apparent as well that although the copy goes towards a reduction of the model in most cases, it sometimes happens that the copy is an improved version of the model, both in terms of length and structure. Gesture repetition may be used to accompany a confirmation request on the part of one of the participants, and therefore as a means to achieving a convergence which is not yet there. In some cases, when the speaker repeats a gesture whereas the prosodic pattern and the verbal message are in contradiction with what was said by the other participant, the gesture repetition may be seen as a means to fake convergence. This reveals how important convergence is to participants in an interaction. It also reveals that, although co-speech gesture is sometimes considered as forming a single idea unit with speech (McNeill, 1992), there must be some kind in independence between the different modalities, for them to be repeated or not independently from each other.

We saw as well that timing between model and repetition is of extreme relevance in terms of convergence. When two gestures are produced in complete overlap (and therefore cannot be termed model and repetition), convergence between interactants is at its highest. These particular occurrences of gesture match between participants are also generally accompanied by verbal and prosodic matches.

Beyond the study of repetition, we presented here the more global perspective of the OTIM project which aims to create resources in terms of multimodal corpus and annotations. Thanks to the annotations now available we can investigate numerous phenomena in conversation, that we can compare and that we hope to be able to analyze shortly in a more systematic way, thanks to the adaptation of tools, and automatization in gesture annotation.

Acknowledgements

This research is supported by the French National Research Agency (Project number: ANR BLAN0239). The OTIM project is referenced on the following webpage: http://aune.lpl.univ-aix.fr/~otim/.

REFERENCES

Allwood, J. , L. Cerrato, L. Dybkjaer, K. Jokinen, C. Navarretta and P. Paggio. 2005. The MUMIN Multimodal Coding Scheme, NorFA yearbook 2005. HYPERLINK "http://www.ling.gu.se/~jens/publications/B%20files/B70.pdf"

Allwood, J., M. Bjornberg, L. Gronqvist,. E. Ahlsén and C. Ottesjö. 2000. The Spoken Language Corpus at the Department of Linguistics, Göteborg University. *Forum: Qualitative Social Research*, **1(3)**:1–20.

Anderson, A.H., M. Bader, E. Gurman Bard, E. Boyle, G. Doherty, S. Garrod, S. Isard, J. Kowtko, J. McAllister, J. Miller, C. Sotillo, H.S. Thompson and R. Weinert. 1991. The HCRC map task corpus. *Language and Speech*, **34**:351–366.

Barth-Weingarten, D. 2011. Double sayings of German JA—More observations on their phonetic form and alignment function. *Research on Language & Social Interaction*, **44(2)**:157–185.

Bavelas, J.B., L. Coates and T. Johnson. 2000. Listeners as co-narrators. *Journal of Personality and Social Psychology*, **79**:941–952.

Bertrand, R., G. Ferré, R. Espesser, S. Rauzy and P. Blache. 2007. Backchannels revisited from a multimodal perspective. In *Proceedings of AVSP (Auditory Visual Speech Processing)*, Hilvenbareek, Pays-Bas, [On CD-rom].

Bertrand, R., P. Blache, R. Espesser, G. Ferré, C. Meunier, B. Priego-Valverde and S. Rauzy. 2008. Le CID—Corpus of Interactional Data—Annotation et Exploitation Multimodale de Parole Conversationnelle. *Traitement Automatique des Langues*, **49**:105–133.

Bertrand, R. and B. Priego-Valverde. 2011. Does prosody play a specific role in conversational humor? *Pragmatics and Cognition*, **19(2)**:333–356.

Bigi, B., R. Bertrand and M. Guardiola. 2010. Recherche automatique d'hétéro-répétitions dans un dialogue oral spontané. In: *Proceedings of XVIIIèmes Journées d'Étude sur la Parole*, Mons (Belgium), Cederom.

Blache, P. and S. Rauzy. 2008. Influence de la qualité de l'étiquetage sur le chunking: une corrélation dépendant de la taille des chunks, Proceedings of the TALN conference, Avignon, France, pp. 290–299.

Blache, P., R. Bertrand and G. Ferré. 2009. Creating and exploiting multimodal annotated corpora: The ToMA project. In: M. Kipp et al. (eds.), Multimodal Corpora. From Models of Natural Interaction to Systems and Applications. Springer-Verlag, Berlin, Heidelberg, pp. 38–53.

Blache, P. and L. Prévot. 2010. A Formal Scheme for Multimodal Grammars. Proceedings of COLING-2010, Beijing, China, pp. 63–71.

Blanche-Benveniste, C. and C. Jeanjean. 1987. *Le français parlé : transcription et édition*, Publication du *Trésor de la langue française*. INALF, Didier Érudition.

Boersma, P. and D. Weenink. 2009. Praat: doing phonetics by computer (Version 5.1.05) [Computer program]. Available: Retrieved May 1, 2009, from HYPERLINK "http://www.praat.org/" http://www.praat.org/

Brun, A., C. Cerisara, D. Fohr, I. Illina, D. Langlois, O. Mella and K. Smaili. 2004. Ants: le système de transcription automatique du Loria. *Actes des XXVᵉ Journées d'Etudes sur la Parole*, Fès, Morocco, pp. 101–104.

Campbell, N. 2009. Tools and resources for visualising conversational-speech interaction. In: Kipp, M.; Martin, J.-C.; Paggio, P.; Heylen, D. (eds.), Multimodal Corpora: From Models of Natural Interaction to Systems and Applications. Heidelberg: Springer, pp. 176–188.

Chartrand, T.L. and J.A. Bargh. 1999. The chameleon effect: the perception-behavior link and social interaction. *Journal of Personality and Social Psychology*, **76(6)**:893–910.

Clark, H.H. 1996. Using Language. Cambridge, UK: CUP.

Couper-Kuhlen, E. and M. Selting. 1996. Towards an interactional perspective on prosody and a prosodic perspective on interaction. In: E. Couper-Kuhlen and M. Selting (eds.), Prosody in Conversation. Cambridge: Cambridge University Press, pp. 11–56.

De Looze, C., C. Oertel, S. Rauzy and N. Campbell. 2011. Measuring dynamics of mimicry by means of prosodic cues in conversational speech. In: Proceedings of ICPhS XVII, Hong Kong, pp. 1294–1297.

Di Cristo, A. and P. Di Cristo. 2001. Syntaix, une approche métrique-autosegmentale de la prosodie. *Traitement Automatique des Langues*, **42(1)**:69–111.

French, P. and J. Local. 1986. Prosodic features and the management of interruptions. In: C. Johns-Lewis. (ed.), Intonation in Discourse. San Diego: College-Hill Press, pp. 157–180.

Furui, S., M. Nakamura, T. Ichiba and K. Iwano. 2005. Analysis and recognition of spontaneous speech using corpus of spontaneous Japanese. *Speech Communication*, **47**:208–219.

Garrod, S. and M.J. Pickering. 2004. Why is conversation so easy? *TRENDS in Cognitive Sciences*, **8(1)**:8–11.

Giles, H., A. Mulac, J. Bradac and P. Johnson. 1987. Speech accomodation theory: The first decade and beyond. In: M.L. McLaughlin (ed.). *Communication Yearbook*, 10, London, UK: Sage Publications, pp. 13–48.

Gorisch, J., B. Wells and G.J. Brown. 2012. Pitch contour matching and interactional alignment across turns: An acoustic investigation. *Language and Speech*, **55**:57–76.

Grønnum, N. 2006. DanPASS—a Danish phonetically annotated spontaneous speech corpus. In: *Proceedings of LREC 2006*. Genoa, Italy: 5th LREC conference.

Guardiola, M. in progress. Contribution multimodale à l'étude de phénomènes de convergence en interaction, Ph.D. Thesis, Aix-Marseille Université.

Heldner, M., J. Edlund and J. Hirschberg. 2010. Pitch similarity in the vicinity of backchannels. In Proceedings Interspeech 2010, Makuhari, Japan. pp. 3054–3057.

Hirst, D., A. Di Cristo and R. Espesser. 2000. Levels of description and levels of representation in the analysis of intonation. In: M. Horne (ed.), Prosody: Theory and Experiment. Dordrecht, The Netherlands: Kluwer, pp. 51–87.

Holler, J. and K. Wilkin. 2011. Co-speech gesture mimicry in the process of collaborative referring during face-to-face dialogue. *Journal of Nonverbal Behavior*, **35**:133–153.

Hutchby, I. and R. Wooffitt. 1998. Conversation Analysis. Cambridge, UK: Polity Press.

Jefferson, G. 1987. Notes on "latency" in overlap onset. In: G. Button, P. Drew and J. Heritage (eds.), Interaction and Language Use. Special issue of Human Studies, pp. 153–183.

Jones, S.S. 2006. Infants learn to imitate by being imitated. In: Proceedings of International Conference on Development and Learning (ICDL), Bloomington, IN: Indiana University, pp. 1–6.

Kendon, A. 1980. Gesticulation and speech: Two aspects of the porcess of utterance. In: M.R. Key (ed.), The Relationship of Verbal and Nonverbal Communication, Mouton: The Hague, pp. 207–227.

Kimbara, I. 2006. On gestural mimicry. *Gesture*, **6(1)**:39–61.

Kimbara, I. 2008. Gesture form convergence in joint description. *Journal of Nonverbal Behavior*, **32**:123–131.

Kipp, M. 2001. Anvil—A generic annotation tool for multimodal dialogue. In: Proceedings of 7th European Conference on Speech Communication and Technology (Eurospeech), Aalborg, Denmark, pp. 1367–1370.

Kipp, M. 2004. Gesture Generation by Imitation—From Human Behavior to Computer Character Animation. Boca Raton, Florida, Dissertation.com.

Knight, D. 2011. The future of multimodal corpora. *RBLA, Belo Horizonte* **11(2)**:391–415.

Kotthoff, H. 2006. Oral genres of humor: On the dialectic of genre knowledge and creative authoring. Interaction and Linguistic Structures, No. 44. http://www.uni-potsdam.de/u/inlist/issues/44/index.htm

Laforest, M. 1992. Le back-channel en situation d'entrevue, in *Recherches Sociolinguistiques* 2, Québec : Université Laval.

Lakin, J.L., V.E. Jefferis, C.M. Cheng and T.L. Chartrand. 2003. The chameleon effect as social glue: Evidence for the evolutionary significance of nonconscious mimicry. *Journal of Nonverbal Behavior*, **27(3)**:145–162.

McNeill, D. 1992. Hand and Mind. What Gestures Reveal about Thought. Chicago: The University of Chicago Press.

McNeill, D.. 2001. Growth points and catchments. In: C. Cavé et al. (eds.), *Oralité et Gestualité (ORAGE): Interactions et comportements multimodaux dans la communication*. L'Harmattan, Aix-en-Provence, pp. 25–33.

McNeill, D. 2005. Gesture and Thought. Chicago, London: The University of Chicago Press.

Mol, L., E. Krahmer, A. Maes and M. Swerts. 2012. Adaptation in gesture: Converging hands or converging minds? *Journal of Memory and Language,* **66(1)**:249–264.

Mol, L., E. Krahmer and M. Swerts. 2009. Alignment in Iconic Gestures: Does it make sense? In: Proceedings of AVSP 2009—International Conference on Audio-Visual Speech Processing, University of East Anglia, Norwich, UK, pp. 1–8.

Nesterenko, I., S. Rauzy and R. Bertrand. 2010. Prosody in a corpus of French spontaneous speech: perception, annotation and prosody ~ syntax interaction. Proceedings of Speech Prosody 2010, May 11–14: Chicago, United States of America.

Norrick, N. 1987. Functions of repetition in conversation. *Text, Interdisciplinary Journal for the Study of Discourse*, **7(3)**:245–264.

Parrill, F. and I. Kimbara. 2006. Seeing and hearing double: The influence of mimicry in speech and gesture on observers. *Journal of Nonverbal Behavior,* **30(4)**:157–166.

Perrin, L., D. Deshaies and C. Paradis. 2003. Pragmatic functions of local diaphonic repetitions in conversation. *Journal of Pragmatics*, **35**:1843–1860.

Peshkov, K. et al. 2012. Quantitative experiments on prosodic and discourse units in the corpus of Interactional Data, Seinedial, 16th Workshop on the Semantics and Pragmatics of Dialogue, Paris, pp. 19–21.

Pickering, J. and S. Garrod. 2006. Alignment as the basis for successful communication. *Research on Language and Computation*, **4**:203–228.

Pitt, M.A., K. Johnson, E. Hume, S. Kiesling and W. Raymond. 2005. The Buckeye corpus of conversational speech: Labeling conventions and a test of transcriber reliability. *Speech Communication*, **45**:89–95.

Portes, C., R. Bertrand and R. Espesser. 2007. Contribution to a grammar of intonation in French. Form and function of three rising patterns. *Cahiers de linguistique française*, **28**:155–162.

von Raffler-Engel, W. 1986. The transfer of gestures. *Semiotica*, **62(1–2)**:129–145.

Sacks, H., et al. 1974. A simplest systematics for the organization of turn-taking for conversation. *Language*, **50(4, part 1)**:696–735.

Szczepek Reed, B. 2006. Prosodic orientation in English Conversation. Basingstoke, UK: Palgrave MacMillan.

Shockley, K., D.C. Richardson and R. Dale. 2009. Conversation and coordinative structures. *Topics in Cognitive Science*, **1**:305–319.

Stivers, T. 2008. Stance, alignment, and affiliation during storytelling: When nodding is a token of affiliation. *Research on Language and Social Interaction*, **41(1)**:31–57.

Tabensky, A. 2001. Gesture and speech rephrasings in conversation. *Gesture*, **1(2)**:213–235.

Tannen, D. 1989, 2007. Talking Voices: Repetition, Dialogue, And Imagery In Conversational Discourse. Cambridge: CUP.

Wolf, J.C. and G. Bugmann. 2006. Linking Speech and Gesture in Multimodal Instruction Systems. In: Proceedings of IEEE RO-MAN 2006. Plymouth, UK, pp. 141–144.

CHAPTER 7

The Situated Multimodal Facets of Human Communication

Anna Esposito

1. Introduction

Humans interact with each other through a gestalt of emotionally cognitive actions which involve much more than the speech production system. In particular, in human interaction, the verbal and nonverbal communication modes seem to cooperate jointly in assigning semantic and pragmatic contents to the conveyed message by unraveling the participants' cognitive and emotional states and allowing the exploitation of this information to tailor the interactional process. These multimodal signals consist of visual and audio information that singularly or combined may characterize relevant actions for collaborative learning, shared understanding, decision making and problem solving. This work will focus on the visual and audio information including contextual instances, hand gestures, body movements, facial expressions, and paralinguistic information such as speech pauses, all grouped under the name of nonverbal data, and on the role they are supposed to play, assisting humans in building meanings from them.

Just to give an example, I am reporting a dialogue I had with a friend of mine a few days ago. She asked me about a common friend and at a certain point she said: "*I can never forget when his wife gave birth to his last child. I was in the hospital because of my daughter's car accident, and then I heard this tic, tic, tic ... I turned round and saw him, his wife and his little daughter all together ...".* For someone that was not seeing her while speaking, figuring out what she meant with this *"tic, tic, tic ...",* can be a bit difficult, even though it would be possible to appropriately guess what it was with a little imaginative effort and by exploiting the context she indirectly provided in her sentence. For me, speaking face to face (I must say body-to-body to be appropriate) there was no need to guess or ask for clarifications, since through her gestures, body movements, and facial expressions, I immediately understood that she was referring to the noise of the wheelchair while running through the corridor of the hospital.

There were several communicative signals that should be accounted for while the abovementioned information exchange took place, all contributing to my comprehension:

a) The context she was referring to;
b) Her vocal intonation and pauses in mimicking the wheelchair noise;
c) Her hand gestures while mimicking it;
d) Her body movements;
e) Her facial expressions;
f) Her speech;
g) More.

I refer to these as the situated multimodal facets of human communication and I want, in the discussion below, to consider their contribution to human information exchanges. Therefore, in the following, I will dedicate the first section of this paper to the importance of context, the second one to pausing strategies, the third one to hand gestures and their role in communication. The fourth section is dedicated to the perception of facial expressions, in particular "facial emotional expressions". I will neglect the speech produced since it is language dependent and can be formulated through different speech motor programs, and the "more" item, since it includes my personal background and knowledge, as well as, my cultural specificity and the social specific communication rules adopted in my Mediterranean community. This is not to say that these aspects are less important, and surely they deserve to be accounted for, nevertheless they are not the focus of this paper.

Several themes have emerged during my own research on human multimodal communication, and I highlight these here. These are: the importance of context; pausing strategies when speaking; gestures; and facial expressions. I will review each of these, with the overall aim of summarizing and highlighting our work in humans for the broader multimodal communication community.

2. The Importance of Context

The interaction between our sensory-motor systems and the environment (which includes people and things) dynamically affects/enhances our social perception and actions/reactions. In the terminology of cognitive psychology, this is known as embodiment. Cognitive processes, such as inference, categorization and memory, are embodied, with individual choices, perception and actions emerging dynamically from this interaction. As such, also the processing of human communication involves embodiment (Esposito et al., 2009). Traditional models of cognition did not account for such concepts, instead asserting that cognitive processes are independent of their physical instantiations. As a consequence, cognition was supposed to be based on amodal representations and mental operations are performed by a central processing unit that exploits the sensory (input) and motor (output) subsystems for collecting and sending representations of the external world, and executing commands (Block, 1995; Dennett, 1969; Fodor, 1983; Newell and Simon, 1972; Pylyshyn, 1984). Recently, however, new cognitive models have been proposed, which account for embodied knowledge acquisition and embodied knowledge use (Barsalou et al., 2003; Smith and Semin, 2004; Barsalou, 2008), and experimental data have been provided that support this idea. For example, Schubert (2004) showed that the visual act of making a fist influenced men's and women's automatic processing of words related to the concept of strength. Montgomery et al. (2007) and Kröger et al. (2010) showed that recognizing and understanding a gesture is wedged in the perceiver's own motor action inventory.

My own experiments on context aimed to investigate visual context effects on the perception of musical emotional expressions. The results of these were partially published in Esposito et al. (2009). A set of such experiments (unpublished) involved four groups of subjects, each composed of 38 participants, equally balanced between males and females and aged from 18 to 30 years. The material consisted of musical and visual stimuli. There were eight 20-second-long musical pieces representative of happiness, sadness, fear and anger (two for each emotion) already assessed as able to arouse the abovementioned emotional feelings (Esposito and Serio, 2007; Nawrot, 2003). The visual stimuli consisted of 15 color images, 5 judged to arouse a positive, 5 a negative, and 5 a neutral feeling according to the emotional dimension of valence (Russell, 1980). The negative, positive and neutral images were previously assessed by 16 independent subjects and received an average recognition score of 84%, 95% and 74% for arousing a positive, negative and neutral feeling, respectively. Participants were asked to label the eight musical pieces played according to valence value (*positive, negative, I don't know*), under the following experimental conditions:

- Group 1) Only Audio
- Group 2) Positive context—each of the eight musical pieces played against a positive visual stimuli background
- Group 3) Negative context—each of the eight musical pieces played against a negative visual stimuli background
- Group 4) In a neutral context—each of the eight musical pieces played against a neutral visual stimuli background.

The results show that music alone, or music combined with congruent visual stimuli, is effective in raising emotional feeling (Figure 1). In addition, a χ^2 analysis with Yeats correction[1] shows that judgments on valence values do not significantly enhance the perception of fear and sad musical pieces when played against a negative visual stimuli background. Instead happy musical pieces are significantly affected by a negative visual stimuli background (χ^2 Yeats = 24.106, Yeats-p-value = 0.000005 for piece 1; χ^2 Yeats = 15.256, Yeats-p-value = 0.0004, for piece 2). These effects are weak for angry musical pieces since significant differences for the valence values, when the audio alone and the audio played with negative visual stimuli were compared, were found only for piece 2 (χ^2 Yeats = 12.026, Yeats-p-value = 0.002 for angry musical piece 2).

These results seem to fit into the scheme proposed by Partan and Marler (1999, 2005) to categorize redundant or non-redundant multimodal signal composites taking into account the behavioral response elicited from the receiver (either human or non-human). This categorization proposes that the simultaneously occurring of two redundant signal components may elicit from the recipient either an identical or enhanced response with respect to the one elicited from the single component. In contrast, the simultaneously occurring of two non-redundant signal components << *"can each continue to have an effect (independence), or one can overshadow or change the effect of the other (dominance or modulation, respectively); ... or their combination may elicit*

---

[1] The χ^2 analysis was separately made for each musical piece in order to be able to assess also their different valence strength.

an entirely new result (emergence)" >> (see Partan and Marler, 2005, pp. 234-235). According to this scheme, for the data reported in the present paper, the sad, fear and angry musical pieces played against the negative visual stimuli background are redundant multimodal signal composites (from the valence point of view, i.e. they have a negative valence) and the expected receiver's response, as predicted by the Partan and Merler model, should be identical or enhanced. This is true in our data, except for the angry musical piece 2. Instead, the happy musical pieces played against the negative visual stimuli background are ambiguous and contrastive (from the valence point of view) non-redundant signal composites and they produced a modulated (lowered in intensity from a valence point of view) recipient's response. It is worth to note that both the recipient's response and the redundancy/non-redundancy of the signal composites depend on the assigned task (i.e. attribute a valence value to the combined stimulus). It can be hypothesized that changing the task will change the information value (redundant/non-redundant) of the signal components and therefore the recipient's response.

It is important to note that the happy musical pieces, when played with the positive context, do not show enhancement effects with respect to the songs played alone. This is also true when the angry, sad and fear musical pieces (that can be considered to arouse a negative valence (see Figure 1)) are played with a negative context.

In a subsequent experiment, the images were substituted by a written text consisting of three different sentences assessed by a previous selected group of 30 independent subjects as arousing happiness (sentence 1, 80.0% of subjects), anger (sentence 2, 76.6% of subjects) and sadness (sentence 3, 76.6% of subjects). The three sentences (Text, T) were congruently and incongruently paired (producing a total of nine paired stimuli) with the three musical pieces listed below (Melodies, M) and previously assessed in terms of emotional labels by a group of 20 adults (10 males and 10 females) with no musical education (Esposito et al., 2009).

- *Beethoven's Symphony No.9 in Re Minor Op. 125*, previously assessed by 72.2% of the subjects as an angry stimulus;
- *Adagio for Strings from Platoon by Barber*, previously assessed by 90% of the subjects as a sad stimulus;
- *Beethoven's Symphony No.6 Pastoral selection from Awakening of happy feelings on arriving in the country*, previously assessed by 90% of the subjects as a happy stimulus.

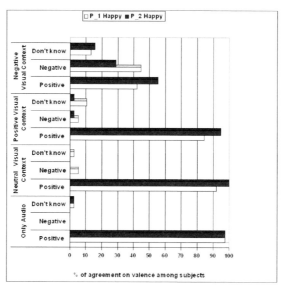

Figure 1. Percentage of agreement, among subjects, on valence values (*positive, negative, I don't know*), in the four experimental conditions (happy music only, happy music played with neutral, positive, and negative visual stimuli, respectively).

The resulting nine paired stimuli were then presented through a Power Point presentation to 42 subjects equally balanced for gender and aged between 19 and 31 years (mean = 23.98, SD = 3.49). The subjects were asked to judge, while reading the sentences, the emotional quality of the paired melody attributing to it one of the following labels: *happiness, sadness, anger, I don't know, another emotional label.*

Results show that subjects assign congruent labels to congruent pairs, whereas they are significantly less likely to do so when the Melody (M) does not match the Text (T) (Table 1).

As Table 1 illustrates, subjects' identification of the emotional feeling associated with the musical pieces is significantly higher for congruent paired stimuli with respect to the mismatched ones. The congruent data in Table 1 fit the "identical" or "enhanced" behavior suggested by the Partan and Merler model, since, when the paired stimuli are congruent (same emotional text, same emotional melody), the subject's response intensity is slightly enhanced (for happiness) or slightly weakened (for sadness and anger). The slightly different recipient's response from the expected "equivalent" or "enhanced" behavior (in particular for the sad melody that is better recognized alone (90% of correct identification) than when paired with the sad text (78.6% of correct identification)) suggests that other variables may

Table 1. Percentage of label attribution to each emotional Text (T), Melody, and Text (T)-Melody (M) stimulus pair (unpublished data).

Mode	Stimuli	Happy	Anger	Sad	Another emotion	I don't know
Single	Happy (T)	**80**	na	na	na	na
Single	Happy (M)	**90**	na	na	na	na
1	Happy (T) Happy (M)	**92.90**	0.00	0.00	2.38	4.76
2	Happy (T) Anger (M)	**38.10**	**31.00**	4.75	9.52	16.70
3	Happy (T) Sad (M)	30.95	2.38	**35.71**	26.19	4.76
Single	Anger (T)	na	**76.6**	na	na	na
Single	Anger (M)	na	**72.2**	na	na	na
4	Anger (T) Anger (M)	2.38	**69.00**	11.90	7.14	9.52
5	Anger (T) Sad (M)	0.00	26.20	**54.80**	9.52	9.52
6	Anger (T) Happy (M)	**31.00**	9.52	7.14	16.70	35.70
Single	Sad (T)	na	na	**76.6**	na	na
Single	Sad (M)	na	na	**90**	na	na
7	Sad (T) Sad (M)	2.38	0.00	**78.60**	16.70	2.38
8	Sad (T) Anger (M)	4.76	**47.62**	**23.81**	16.67	7.14
9	Sad (T) Happy (M)	**47.62**	2.38	**11.90**	9.52	28.57

play a role in such assessments, as for example individual differences among subjects in personality and experience.

To evaluate the effects of the text on the assessment of the melody, an ANOVA analysis was performed for each mismatched stimulus pair with gender as a between subject variable and emotional labels (congruent melody label, congruent text label, and a label named "other" that grouped all the remaining subject's response) as a within subject variable. No significant effects of the text and/or the melody were found for the paired stimuli **Happy(T)Angry(M)** ($F_{2,80}$ = .208, p = .8130), **Happy(T)Sad(M)** ($F_{2,80}$ = .068, p = .9339), and **Sad(T)Angry(M)** ($F_{2,80}$ = 2.059, p = .1343). Significant effects were found for the stimuli **Sad(T)Happy(M)** ($F_{2,80}$ = 5.294, p = .0069), **Angry(T)Happy(M)** ($F_{2,80}$ = 9.823, p = .0002) and **Angry(T)Sad(M)** ($F_{2,80}$ = 4.884, p = .01). Post-hoc tests revealed that the significant differences were due to a different emotional labelling of the melody for males and females. In particular for the **Angry(T)Happy(M)** stimulus, the label attribution was significantly different both for males ($F_{2,80}$ = 3.363, p = .04), and females ($F_{2,80}$ = 8.673, p = .001), with a male preference for the "HAPPY" label and a female preference for the "Other" label. Male subjects significantly differed from females in attributing the label "SAD" to the **Angry(T)Sad(M)** ($F_{2,80}$ = 4.186, p = .02) stimulus and the label "HAPPY" to the **Sad(T)Happy(M)** ($F_{2,80}$ = 7.227, p = .001) stimulus.

In contrast to the data illustrated in Figure 1, where statistically significant effects of context on the melody assessment are observed only for happiness, in this new context, neither the text, nor the melody predominates on the subject's assessment. For some pairs, the melody assessment was affected by the written text (see the paired stimuli 6 and 9), for others it was the opposite (see the paired stimuli 5 and 8).

Subjects are more oriented to the melody label (see the paired stimuli 4, 5, 7, 8 in Table 1) than to the text label (see the paired stimuli 6, 9 in Table 1) but the percentage of correct label attributions is always significantly lower than that attributed to the single text or to the single melody (see Table 1). Comparison between the two different experiments shows greater effects of written texts than visual images in affecting the emotional decoding process. However, the significance of these results for understanding multimodal integration in human communication remains unclear. It is possible to speculate that the results are affected by the subject's task that was changed from a dimensional (the valence) to a basic emotional label approach (or to a basic emotion assumption). In our view, the two approaches involve different evaluation processes. The dimensional is more immediate and instinctive than the basic emotion approach that requires a cognitive assessment of the linguistic labels. More experimental data are indeed needed to interpret these differences.

Similar effects in different contexts have been described by other authors (Bargh et al., 1996; Russell, 2003; Esposito, 2009), suggesting that contextual interactions at organizational, cultural and physical levels play a critical role in shaping individuals' social conduct, providing a means in the ability to render the world sensible and interpretable during everyday activities. New cognitive models must account for embodied knowledge acquisition and use. To this respect, the current literature has coined the term "embodied cognition" usually employed in a multiplicity of domains ranging from psychology, human-computer interaction, affective computing, sociology, neuroscience, up to cognitive anthropology. Embodied cognition theories explain cognition and consequently perception and action either from a radical position that has made them directly dependable from the body a creature possesses (Gallese, 2008) or as grounded in multiple ways, including context, culture consequence, and *"on occasion bodily states* (Barsalou, 2008)". On this ground, it can be suggested that the physical, organizational and social context functions as a signaling modality in human communication. Studies on human information processing must assess how humans integrate and synchronize visual, tactile and audio somatic sensory inputs in a unique percept (especially appreciated in noisy environments with corrupted and degraded signals) and combine them into a reasoned understanding.

It could be argued that the experiments reported in this section relate to how auditory information is interpreted alone, or in combination with visual congruent and incongruent stimuli and that a change in the context, and therefore a context effect on the decoding and interpretation of a given stimulus is better observed when the same signal is differently interpreted in different circumstances. In this line, the current literature refers to "cognitive bias" (Mathews et al., 1997 for human) and "judgment bias" (Mendl et al., 2009 for human and non-human animals). These experiments tested how ambiguous stimuli (with a valence value that may result ambiguous according to the past experience of the human or non-human animal involved in the judgment) are differently interpreted according to the experimented emotional state (Wheeler, 2009, 2010 for predator calls given by capuchins in feeding/non-feeding contexts). However, according to the task assigned, our subjects were asked to concentrate and assign a valence value or a label to the melody. No mention was made on the visual cues. In this sense, the visual cues have been considered the surrounding task environment and therefore the context, and it cannot be considered a multimodal signal integration. The reported data cannot be linked to cognitive and judgment bias experiments due to the lack of a pre-existing (i.e. an anxiety disorder) or an induced (induced depression) bias.

3. The Significance of Pausing Strategies

Speech is characterized by the presence of silent intervals (empty pauses) and vocalizations (filled pauses) that do not have a lexical meaning. Such pauses play a role in controlling the speech flow and are a multi-determined phenomenon attributable to physical, socio-psychological, communicative, linguistic and cognitive causes.

Physical pauses are normally attributed to breathing or the momentary stoppage of the breath stream caused by a complete closure of the glottis and/or the vocal tract in producing phonemes such as stop consonants. Communicative and socio-psychological pauses are meant to build up tension or generate the listener's expectations about the rest of the story, assist the listener in his/her understanding of the speaker, or to interrupt for questions and comments, as well as to signal anxiety, emphasis, syntactic complexity, degree of spontaneity, gender, and educational and socio-economical information (Bernestein, 1962; Goldman-Eisler, 1968; Abrams and Bever, 1969; Kowal et al., 1975; Green, 1977; O'Connel and Kowal, 1983). Linguistic pauses are more likely to coincide with discourse boundaries, realized as a silent interval of varying length, at clause and paragraph level (Brotherton, 1979; Gee and Grosjen, 1984; Rosenfied, 1987; Grosz and Hirshberg, 1992). This is particularly true for narrative structures where it has been shown that pausing marks the boundaries of narrative units (Chafe, 1980; Rosenfield, 1987; O'Shaughnessy, 1995; Oliveria, 2000, 2002). Lastly, several cognitive psychologists have suggested that pausing strategies will surface in the speech stream as the end product of a "planning" cognitive process that reflects the complexity of the neural information processing. The length of the pause is a sign of the cognitive effort related to lexical choices and semantic difficulties for generating new meanings (Goldman-Eisler, 1968; Butterworth, 1980; Chafe, 1987; Brennan, 2001).

Investigations into the role of speech pauses have been neglected by the recent literature leaving open several scientific questions. These include, how children exploit pausing to shape their discourse structure, and to signal the cognitive effort associated with the amount of *"given"* and *"added"* information they are conveying. Shedding light on this topic would help in the comprehension of the role that this paralinguistic information is supposed to play, assisting humans in building up meanings from them, suggesting appropriate mathematical models for implementing more friendly human-machine interfaces, and in particular, more efficient automatic speech recognition systems. Here, I report the results of a series of analyses on narrations made by 10 female and 4 male Italian children (mean age = 9 ± 3 months).

These children were asked to narrate episodes of a 7-minute animated color cartoon "Sylvester and Tweety" they had just seen (Esposito, 2006; Esposito and Esposito, 2011 for procedural details). The cartoon is of a familiar type to Italian children, involving a cat and a bird. The listener was familiar to the child (either the child's teacher or other children). The children were recorded while narrating the cartoon and the videos were analyzed frame by frame in slow motion to assess the association between utterance meaning and empty speech pauses.

Empty speech pauses, defined as a silent interval longer than 0.150 s, were divided into three categories according to their duration: a) short—from 0.150 up to 0.500 s long; b) medium—from 0.501 up to 0.900 s long; c) long—more than 0.900 s long. *Added* verbal material was defined as *any verbal material that produces a modification in the listener's conscious knowledge*, and *given* verbal material as intended not to produce such a modification (Chafe, 1974). The label *unclassified* was attributed to filled pauses (such as *"ehmm"*, *"uhh"*), and/or short interruptions (such as *"ap*"*) following empty pauses. Empty pauses identifying changes in scene, time and event structures were labeled *changes*, a *clause* defined as *"a sequence of words grouped together on a semantic or functional basis"*, and a *paragraph* as *"a sequence of clauses connected together by the same subject or scene"*.

The analyses of the children's narrations showed that according to their duration, each speech pause category plays a different role in the discourse organization. Pauses of short and medium length generally appear after utterances containing *added* information. Long pauses play the functional role of delimiting paragraphs and identifying changes in scene, time and event structures (Table 2).

Table 2.   Percentage and absolute occurrences of pauses of short, medium and long length associated with *given, added, unclassified* information and *changes of scene* in the discourse structure. The percentage is computed over the number of pauses in each duration range.

10 Female Children	Short EP		Medium EP		Long EP	
Given	9	4%	2	3%	0	
Added	225	88%	67	89%	36	77%
Unclassified	21	8%	6	8%	11	23%
Changes	51	20%	61	81%	45	96%
4 Male children						
Given	5	6%	0		0	
Added	65	86%	37	88%	13	81%
Unclassified	6	8%	5	12%	3	19%
Changes	6	8%	14	33%	15	94%

The data reported in Table 2 suggest that pauses belonging to different duration ranges may play a different role in structuring narrations, but more data and experiments are needed to confirm this hypothesis. Most empty pauses are associated with the production of *added* information, independently of the duration category to which they belong. The majority of pauses of long (96% for female and 94%

for male) and medium length (81% for female, 33% for male children) are followed by changes of scene that signal both the production of a greater amount of added information and boundaries in the discourse structure. Only a low percentage of short pauses (20% for females and 8% for males) serve this purpose, which probably reflects the role of pausing due in cognitive efforts to recall from memory and lexicalize concepts not yet known by the listener. It can be hypothesized that more cognitive effort is involved in providing more *added* information (a change in the scene requires a change in the discourse topic), resulting in longer pausing times. The data also suggest a predictive scheme for the alternating pattern of cognitive rhythm in the production of spontaneous narratives. In this alternating pattern, long pauses account for the highest percentage of paragraphs or changes followed by medium pauses. Even though more frequent than medium and long pauses, there is a low likelihood that short pauses may signal a change of scene (20% in female and 8% in male children) in the narration flow.

As in adults (Chafe, 1987), children use empty pauses of short, medium and long duration to cue both the *added* information they are providing to the listener and the cognitive effort they are making to recover and lexicalize it. Pausing strategies act as cues indicating to the listener the novelty of the received information and its cognitive cost, and allow him/her to carry out the appropriate inferences. Pausing strategies are also used as a linguistic tool for discourse segmentation, marking word, clause and paragraph boundaries. Previous works (Esposito et al., 2004; Esposito, 2005, 2006) showed that 56% of child pauses indicate a major transition in the speech flow helping to plan the message content for the continuation of the discourse, suggesting a common pausing strategy for structuring the discourse. In addition, the distribution of pauses of short, medium and long length is consistently adopted among the subjects suggesting a coarse pause-timing model that speakers exploit to regulate speech flow and discourse organization. It would be interesting to make sense of how this model works by collecting more data. Nevertheless, the results would be of great utility in the field of human-machine interaction, favoring the implementation of more natural speech synthesis and interactive dialog systems.

4. The Role of Gestures

We are not conscious of gesturing and I was extremely surprised when my colleagues in the United States, in mimicking my speech, were exuberantly moving their arms and hands. Then, I learned that, according to a stereotype point of view, the nature of the Italian gestural repertoire is reputedly considered to be richer than those of other speaker groups and my gesturing while speaking was an exemplification of such a view. The careful analysis of video-recordings of other speaker groups and personal discussions with experts in the field (among those Sue Duncan, Adam Kendon and David McNeill) proved that (to a large extent) gesturing is common to any worldwide speaker and comes out (while speaking) as a "cognitive need" (this is a personal opinion) of the communicative act (the multimodal facets of communication), even though the gesture types and their distribution along the speech may depend both on the speaker and language.

Gestures have been categorized using several classification systems (e.g. Efron, 1941; Ekman and Friesen, 1969; McNeill and Levy, 1982; McNeill, 1992) that mainly differ in the number of enlisted gesture categories and several theories were advanced to explain their role in communication. There is considerable body of evidence supporting the idea that gestures share with speech similar semantic and pragmatic functions, i.e. neither gestures nor speech alone have a primary role in the communicative act (Kendon, 1980, 2004; McNeil, 1992, 2005; De Ruiter, 2000; Goldin-Meadow, 2003; Esposito and Marinaro, 2007, Esposito and Esposito, 2011). There is also a large research work investigating gestures in non-human primates showing that their deployment of hands and body movements is both flexible[2] and intentional[3] and also for them it emerges as a need of the communicative act to clearly shape thoughts, intentions and needs.

Nevertheless, in relation to human primates, there are data suggesting that they are secondary to speech, being of support to the speaker's effort to encode his/her message (Freedman, 1972; Rimé and Schiaratura, 1992; Rausher et al., 1996; Krauss et al., 2000; Morsella and Krauss, 2005a, 2005b).

The idea that gestures support lexical retrievals and that in general their role in interactive communications is to assist the listener, or the speaker, or both, appears to be widespread. This is because during our daily interactions, awareness of the semantic content of our verbal messages is favored by the continuous auditory feedback. On the contrary, our gesticulation, posture and facial expressions are visible

---

[2] Non-human primates change their gestural actions in order to clarify the meaning of their messages when misunderstood, and flexibly create new gestural signals avoiding repetitions of behaviors that fail the communicative goal (Cartmill and Byrne, 2007; Cartmill et al., 2012).

[3] Non-human primates mainly exploit silent visual gestures when the interactant can easily see and is attentive to them, while tactile gestures are often used when this is not the case (Genty and Byrne, 2010; Genty et al., 2009).

only to our interlocutors. Most of our gesturing is made without awareness of attributing a semantic meaning to it, and additionally, successful human interactions are possible in situations where the interlocutors cannot see each other (on the telephone for example, Short et al., 1976; Williams, 1977). Moreover, the understanding of a sign language requires to be a speaker of such a language (as for any oral language) and it is difficult (even though possible) to infer the meaning of a message through pantomime and mute gestural interactions (Freedman, 1972; Rimé, 1982; Krauss et al., 1991, 2000).

In contrast, other studies have shown that gestures can resolve speech ambiguities and facilitate comprehension in noisy environment (Kendon, 2004; Rogers, 1978; Thompson and Massaro, 1986) and in some interactional contexts are more effective than speech in communicating ideas (see the mismatches theory in Goldin-Meadow, 2003). More remarkably, it has been shown that gestures are in semantic coherence with speech (Kendon, 2004; McNeill, 2005), coordinated with tone units and prosodic entities, such as pitch-accented syllables and boundary tones (Yasinnik et al., 2004; Shattuck-Hufnagel et al., 2007; Esposito et al., 2007) and are loosely synchronized with speech through pausing strategies (Butterworth and Beattie, 1978; Esposito et al., 2002, 2001; Esposito and Marinaro, 2007). The abovementioned results suggest that gestures and speech are partners in shaping and giving kinetic and temporal (visual and auditory) dimensions to communication, and must be regarded as an expressive system that, in partnership with speech, provides communicative means for giving form to our thoughts.

Some of the abovementioned results extend to non-human primates, which have been shown to be extremely efficient in detecting gestural and vocal correspondences or discrepancies (Hauser et al., 1993; Partan, 2002; Ghazanfar and Logothetis, 2003) as well as to benefit of multimodal signals for basic evolutionary-survival-related needs such as courtship, aggression, affiliation, detection and localization of preys and predators (Lewis et al., 2001; Rowe, 2002; Schneider and Lewis, 2004). In order to evaluate the relative merits of the above theories, more data and investigations are needed.

4.1 Data supporting the speech and gesture partnership

In assessing the role of gestures in interactive communications, our personal interest was concentrated on integrating it with our research on pausing strategies (summarized above). Our research has suggested that pausing strategies are administered by clear cognitive processes that aim at structuring the discourse and account both for the listener's current knowledge and novelty of the information the speaker is transmitting. Assuming that speech and gestures are employed together in the transmission of information, our research questions were:

a) Do gestures have any gestural entity equivalent to speech pauses?
b) To what degree do these entities synchronize with speech pauses?

The synchrony of gestures with pausing strategies engages them in the recovery and lexicalization process needed to formulate the utterance plan and in the cognitive process undertaken to convey the *added* information to the interactional instance. The meaning of this synchrony will attribute to the conveyed message both temporal and spatial dimensions, making the utterance a more complex entity than the simple verbal and phonological phrase. An utterance, as reported in Kendon (2004), is a *"dynamic visible action"* and gestures as speech are an expressive resource that can take on different functions depending on the communicative demand.

To answer the first question, a pilot study (Esposito et al., 2001) analyzing video-recordings of narrative discourse data identified a gestural entity named "hold". A gestural "hold" is detected when the arms and hands engaged in gesticulation remain still in the gesture space for at least three video frames (i.e. approximately 100–120 ms) in whatever position. As defined in Esposito et al. (2002), a hold is an active configured gestural state where no intended (hand) motion is perceived. Note that the absence of movement is judged perceptually by an expert human coder. Therefore, the concept of hold is ultimately based on the receiver's perception. A hold may be thought to be associated with a particular level of discourse abstraction. In producing a sentence, the speaker may employ a *metaphoric gesture* (McNeill, 1992) with a hold spanning the entire utterance. The speaker may also engage in a slight oscillatory motion centered around the original hold, without any change in hand shape, or add emphatic *beats* (McNeill, 1992) coinciding with peaks of prosodic emphasis in the utterance. While the oscillatory motions and emphatic beats may sit atop the original hold, observers will still perceive the underlying gesture hold.

A careful review of speech and gesture data showed that in fluent speech contexts, holds appear to be distributed similarly to speech pauses and overlap with them, suggesting a loose synchronization between the two entities.

A subsequent study (Esposito et al., 2002) investigated the robustness of such synchronization looking for it in discourses elicited by different tasks and spoken in different languages (Italian and American English in such case).

The audio/video analysis of the recorded data proved that synchrony between holds and speech pauses exists independently of the tasks and the language, with variations that can typically be attributed to the language and cultural specificity. American English speakers equally distribute the time for holds and speech pauses over their fluent speech duration, while holds in Italian speakers are in general shorter than speech pauses. Also, American English holds overlap in similar amounts both with empty and filled pauses, whereas the percentage of Italian holds overlapping with empty pauses is significantly lower than that overlapping with filled pauses and Italian speakers pause less than Americans (Esposito et al., 2002).

Further support to the above data was found in Italian narrative discourse data from children (9 years old ± 3 months females) and adults (two males and two females, average age 28 years old ± 3 years) who participated in a similar type of elicitation (Esposito and Marinaro, 2007). Both adults and children were native speakers of Italian. The two groups of speakers produced a similar distribution of hold and speech pause overlaps. An ANOVA performed on the data collected from each group using hold and speech pause rates as within subject variables showed that children's holds and speech pauses were equally distributed along their narrations ($F_{1,10} = 1.09$, $\rho = 0.32$) while adults used holds significantly more frequently ($F_{1,6} = 11.38$, $\rho = 0.01$) than speech pauses. Rates were computed as the ratios between the number of holds and/or speech pauses over the length of the subject's narrations measured in seconds.

The differences between hold and speech pause rates found in adults and not in children may be explained by the fact that sophisticated utterances are a prerogative of the adult communication behavior. Adults exploit holds also as a strategy for structuring the discourse in absence of speech pauses, whereas children, being less skilled in assembling bodily and verbal information, tend to attribute the same functions to both holds and speech pauses.

Does this synchrony develop during language acquisition? In order to answer this question, a further study was conducted with three new groups of children (Esposito and Esposito, 2011):

- 8 females, of 9 ± 3 months years old;
- 5 males, and 5 females of 5 ± 3 months years old;
- 3 males, and 3 females of 3 ± 3 months years old.

As in the abovementioned studies (Section 3), the children told the story of the "Sylvester and Tweety" cartoon and their complete narrations were analyzed. The lengths of the narrations were different among children and none of the participants were aware that holds and speech pauses were of interest.

The Pearson correlation coefficient computed to measure the amount of information about speech pause frequency from the known hold frequency resulted in $\tau = 0.72$, and $\tau^2 = 0.52$ for 3 year olds; $\tau = 0.18$, and $\tau^2 = 0.03$ for 5 year olds; $\tau = 0.83$, and $\tau^2 = 0.70$ for 9 year olds; and $\tau = 0.88$, and $\tau^2 = 0.78$ for adults.

A high correlation between holds and speech pauses was found for all the groups, except for the 5-year-old children. Analyzing the distribution of hold and speech pause rates (as displayed in Figure 2, for the 3-, 5- and 9-year-old groups) it is possible to notice that the number and length of speech pauses in the 5-year-old children was significantly higher than the number and length of holds. In addition, speech pause rates are differently distributed with respect to the remaining groups, independently of the child gender. Five-year-old children remain often silent, independently of the length of their narration and the language skills, even though they possess a more sophisticated language competence with respect to the 3-year-old children. A possible explanation for this result can be attributed to the emergent consciousness of the self with respect to the others that children acquire around 5 years (Leslie et al., 2004) assigning a social value to their performance. This may have forced them to put extra care in reporting their narrations increasing the number of pauses (socio-psychological pauses) in their speech.

Despite the abovementioned inconsistency, all the reported data seem to prove that, at least some aspects of speech and gestures reflect a unified planning process, similar in all humans and independent of the particular language of expression and age. Acting in synchrony,

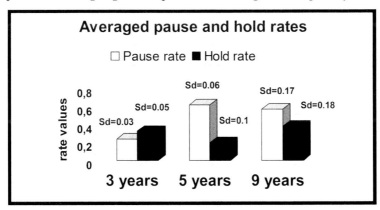

Figure 2. Averaged pause (white) and hold (black) rate values with the associated standard deviation (Sd) values along the speech for the three differently aged child groups.

speech and gestures are part of the same communicative plan, which exploits multiple communication devices to effectively combine the spatial and temporal features of the planned utterance (the internal representation of the communicative act), resolve verbal and nonverbal ambiguities efficiently, depict the social, physical and organizational context, and succeed in being appropriately decoded by the addressee. Without multiple communicative devices it would be hard to implement an effective interaction. It also seems evident that such synchrony disappears when there are no commonalities of goals in the interaction, reflecting both the cognitive and functional communication difficulties. It is not a case, for example, that when the speaker lies there is a loss of this synchrony, reflecting a disembodied communicative plan at the cognitive level.

5. Facial Expressions

Faces and consequently facial expressions play significant communicative functions in interaction. Changes in the facial muscles are exploited for changing the meaning of a sentence, controlling the conversational flow, expressing emotional states and guiding the speaker's intention. Things that can be signaled in facial expressions include disinterest, rising sympathy and empathy, distress, psychopathologies, cognitive activities, intentions, physical efforts and personality. According to Bavelas and Chovil (1997, p. 334), *"facial displays of conversants are actively symbolic components of integrated messages (including words, intonation and gestures)"*, i.e. they share with speech and gestures similar to semantic and pragmatic functions. In other words and in a way more close to the abovementioned discussion, facial expressions are co-involved in communication and participate in concert with speech and gestures to the communicative plan. This seems to be true also for non-human primate as the recent works of Slocombe et al. (2011) and Parr et al. (2010) aim to suggest. Bavelas and Chovil (1997, 2000) call them "facial displays" making it clear that they are used by the speaker to illustrate what she/he is saying and by the addressee as responses to the speaker in order to avoid, if not necessary, interruptions.

These facial displays *"do not express inner emotions; rather, they convey meaning to another person"* (Bavelas, 2012), "and to another non-human animal" (we would like to add) following some recent discussions on non-human primate multimodal communication (Partan, 2002; Parr, 2004; Slocombe et al., 2011; Parr et al., 2010).

The idea that interaction, and therefore face-to-face dialogue, is possible, with no exchange of emotional content, as well as that humans (and also non-human animals) can interact in a neutral interactional setting, is rather common and has been tacitly accepted to the point that facial emotional expressions have not been accounted into the set of "facial displays". According to Ekman (1989), *"affective facial displays"* are universal and must be considered independent of language (and then of communication?). Therefore, it seems that facial emotional expressions have nothing to do with "communication" and that in general emotions cannot be called in, within this context.

In the following, I will report data on the human ability to decode visual (generally facial and gestural) and vocal emotional information.

5.1 Weighing the verbal and nonverbal emotional information

Most of the current human psychology literature (with some very recent exceptions) on the recognition of emotional facial expressions refers to static photos of actors/actresses portraying the requested emotional face. As pointed out in Esposito (2007), these are static images, not embedded in a context, and not showing the dynamic features of facial muscle changes that characterize human-human interaction. To this aim, they can be considered qualitative targets that capture the apex of the expression, i.e. the instant at which the indicators of the selected emotion are most marked. Humans are good in attributing emotional labels to such facial expressions and in this task they perform much better than in the one where they are requested to assess vocal emotional expressions (Esposito, 2009). However, in interaction, facial emotional expressions are intrinsically dynamic and vary over time.

Is dynamic visual information emotionally richer than auditory information? To answer this question, it is worth reporting the results of a series of experiments based on emotional video-clips extracted from Italian and American English movies (Esposito, 2007, 2009). These experiments were aimed at evaluating the dynamic perception of six emotional states (happiness, sarcasm/irony, fear, anger, surprise and sadness) through the audio alone, mute video and audio/video combined. For each language (Italian and American English), 10 stimuli were collected for each emotional state (expressed by five different actors and five different actresses) for a total of 180 Italian stimuli (60 audio, 60 mute video, and 60 audio/video) and 180 American English stimuli. One hundred and eighty American and 180 Italian subjects, and each group was separated into six further groups of 30 subjects each tested on Italian audio alone, Italian video alone and Italian combined audio/video as well as on American English audio alone, American English video, and American English combined audio and video.

The percentage of correct label attribution is reported in Table 3 for American English and Italian subjects tested on American English emotional stimuli (audio, video, and combined audio/video) and in Table 4 for American English and Italian subjects tested on Italian emotional stimuli (audio, video, and combined audio/video).

The data were assessed statistically using an ANOVA with condition (audio alone, mute video, and combined audio/video) as a between-subject variable, and the emotion categories (the six under examination) as a within-subject variable.

Different ANOVAs were performed separately for each set of stimuli (American and Italian movies) and for each subject's country (America and Italy, Esposito, 2007, 2009; Esposito and Riviello, 2011).

Results showed that conditions (audio, mute video, and combined audio/video) always played a significant role for the subject's correct label attribution to the stimuli. However, significance was different

Table 3. Percentage of correct emotional label attributions for American and Italian subjects tested on American English stimuli played as audio alone, mute video, and combined audio/video.

Emotions	American Subjects on American English Stimuli			Italian Subjects on American English Stimuli		
	Audio	Video	Audio/Video	Audio	Video	Audio/Video
Happiness	33.3	57.3	60	45	61	60
Fear	76.7	65	81.7	66	68	79
Anger	86	89.3	93.3	79	81	92
Irony	44.3	56.3	62	33	62	57
Surprise	47.3	52.3	62	51	60	71
Sadness	52.3	62.3	68.3	56	55	69

Table 4. Percentage of correct emotional label attributions for American and Italian subjects tested on Italian stimuli played as audio alone, mute video, and combined audio/video (unpublished data).

Emotions	American Subjects on Italian Stimuli			Italian Subjects on Italian Stimuli		
	Audio	Video	Audio/Video	Audio	Video	Audio/Video
Happiness	45	68.7	72.33	48	61	50
Fear	45.33	58	64.67	63	59	48
Anger	76.67	76.67	91	77	68	60
Irony	49.67	55	64	75	49	64
Surprise	19.33	27	34.67	67	49	75
Sadness	56.67	70	79.67	37	37	59

depending on the native language (Italian and American English) of the subjects involved and the cultural context (American English and Italian movies) associated to the stimuli. For example, for the Italian subjects tested on the Italian stimuli, the combined audio/ video and the audio alone conditions conveyed the same amount of emotional information ($F_{1,8}$ = .004, p = .95), while such information was significantly different from the mute video and the combined audio/ video ($F_{1,8}$ = 10.414, p = .01) as well as from the mute video and the audio alone ($F_{1,8}$ = 13.858, p = .005) conditions. This was not the case for the American English subjects tested on the American English stimuli, where the combined audio/video and the mute video, as well as the mute video and the audio alone conditions conveyed the same amount of emotional information ($F_{1,8}$ = 2.323, p = .16; ($F_{1,8}$ = 1.696, p = .22) respectively) whereas such information was significantly different for the audio alone and the combined audio/video conditions ($F_{1,8}$ = 9.031, p = .01).

The data discussed above suggest that in decoding emotional states, humans do not sum up linearly the cues provided. The combined audio/video may convey the same amount of emotional information of the mute video or the audio alone, depending on the language, the expressed emotion and the cultural specific background. The visual channel may facilitate under certain conditions a cross-cultural decoding of emotional states but this is not always the case and cannot be applied to all emotional categories. For example, the decoding of irony is strictly language and cultural dependent. It has been suggested that the process to decode emotional information is driven by a "reducing cognitive load" strategy (Esposito, 2007, 2009) where in an attempt to reduce the amount of emotional cues to be processed for the task, humans select the cues that are more close to their personal/cultural competences, to their daily experience, and more congruent with the context of the contemporary interactional instance. Support to these speculations comes from literature on non-human animals where it has been shown that direct experience of male rhesus macaques with specific females allows them to interpret their facial signals more accurately (Higham et al., 2011). Unfortunately, to our knowledge, there are no data on human animals that report on similar experiments and this is surely another research aspect to be accounted for as a situated multimodal facet of communication.

6. Discussion

How can verbal and nonverbal emotional expressions be connected with the situated multimodal facets of human communication?

Communication is an emergent behavior and its success strongly depends on "readiness to act" and "prompting of plans" in order to appropriately handle the incoming communicative event, producing suitable mental states, feelings, and expressions, under conditions of limited time and resources.

In humans, emotions cannot be separated from communication, and communicative signals that bring meanings to the interactional instance also bring emotional contents. I have observed interactions in different languages, only among male or only female, or both, and I was always amazed by the amount of emotional content portrayed by the interactants through their body poses, facial, vocal, and gestural expressions, as well as head movements and gaze directions.

Communication is not neutral, it cannot be, since neutrality would not allow humans to rapidly and continuously adapt to the dynamical context of the communicative instance and the dynamic changes that it produces on our knowledge acquisition and use.

Communication is an embodied process that transforms the information and the meanings gathered in the course of everyday activities and builds up from them new knowledge and practical ability to render the world interpretable while interacting with others attuned to behavioral sequences that underpin collaboration.

Communication is a multimodal, multifunctional, multi-determined phenomenon and its components iterate its complexity. Therefore, in order to understand the multimodal facets of communication it is necessary to approach it holistically, including the emotional states that are an integrant part of such a process.

To do so, multiple theoretical investments are needed, ranging from mathematical models of interaction and dynamics of signal exchanges (in terms of shared meanings, emotional states, cultural differences, social, organizational and physical context effects), to social intelligence, behavioral analyses, action selection, and cognitive processes of cooperation, decision making and resource management.

The benefits of such an approach are threefold:

THEORETICAL: A holistic approach will allow the development of new mathematical models for representing data, reasoning, learning, planning, and decision making, as well as new individual/group behavior analysis models of social interaction in multilingual and cross-cultural contexts. This will produce new socio-psychological and computational approaches to model interaction, starting from existing cognitive frameworks such as the emotion model by Ortony et al. (1990) and from existing algorithmic solutions such as dynamic Bayesian networks (Dowe, 2010), long short-term memory networks (Hochreiter and Schmidhuber, 1997) and fuzzy models of computation (Tiwari, 2009).

TECHNOLOGICAL: The integration of cognitive theories, knowledge representation models and algorithms in a computational framework will enable rich models of human-machine interaction, and the development of novel machine learning techniques for the analysis of socially situated multimodal streams in human communication. The successful analysis, modeling, and understanding of context effects and social interactions will be of high relevance for Information Communication Technology (ICT) applications and crucial for implementing Behaving Human Computer Interactive systems that will be of public utility and profitable for a living technology that simplifies user access to future, remote and nearby social services encompassing language barriers and cultural specificity.

SOCIETAL: The implementation of Behaving Human Computer Interactive systems will produce applications such as: context-aware avatars replacing humans in high risk tasks, companion agents for elderly and impaired people, socially believable robots interacting with humans in extreme, stressful time-critical conditions like urban emergency, future smart environments, ambient assistive living technologies, computational intelligence in games/storytelling, embodied conversational avatars, and automatic healthcare and education services. Such applications will enhance quality of life in society and will change the individuals' social conduct, in typical as well as impaired circumstances.

To tackle this goal, research expertise in different yet complementary scientific fields and cultural-linguistic specificity must be accounted for, requiring a holistic scientific approach, cross-cultural collaborations and exchanges that should lead to the realization of the following objectives:

1. The implementation of neuropsychological, linguistic, and behavioral analyses of social signals exhibited in spontaneous dyadic/group interactions, taking into account different scenarios and the underlying perceptual, cognitive and emotional processes, to identify encoding features for annotation standards and common scientific frameworks;

2. The development of cognitive theories and mathematical models on how context influences interaction with the goal of better predicting and understanding the users' (individual and group) actions/reactions;

3. The delivery of algorithms and ICT applications for real-time cross-modal processing and fusion of multiple social signals in

order to supply encoded information at higher semantic and pragmatic levels;

4. The definition of models of interaction/communication and physical/cognitive constraints accounting for coupled mechanisms that rule Group Cognition and Collaborative Knowledge Building (Hargadon, 1999; Stahl, 2006);

5. The implementation of real-time, friendly, flexible, socially, and emotionally adaptable ICT interfaces coping with limited memory and bounded information.

I would call such systems BEINGS, i.e. BEhavINng coGnitive Systems.

7. Conclusions

The present paper addressed some aspects of the qualitative and quantitative features of communicative signals, in particular signals exchanged during human interactions. The general research idea portrayed is that communication is a *multimodal, multifunctional, multi-determined* phenomenon and its components iterate its complexity. Therefore, in order to understand the multimodal facets of human communication it has been proposed to approach its study holistically, accounting also for the cognitive emotional processes that are an integrant part of it. The benefits of this approach will be theoretical, technological, and societal, bringing to a new understanding of the user's needs that simplify the everyday-life man-machine interaction by offering natural ways of communication, motivating innovation and technological awareness.

Acknowledgement

I want to thank James HIGHAM from the University of Chicago. James has gone through this paper editing the English and providing really interesting and useful comments and suggestions. The paper would not have this form without his help. Thanks go to Maria Teresa RIVIELLO for sharing data to support the discussion. All the mistakes are of course mine.

REFERENCES

Abrams, K. and T.G. Bever. 1969. Syntactic structure modifies attention during speech perception and recognition. *Quarterly Journal of Experimental Psychology*, **21**:280–290.

Bargh, J.A., M. Chen and L. Burrows. 1996. Automaticity of social behaviour: direct effects of trait construct and stereotype activation on action. *Journal of Personality and Social Psychology*, **71**:230–244.

Barsalou, L.W., P.M. Niedenthal, A.K. Barbey and J.A. Ruppert. 2003. Social embodiment. In Ross B.H. (ed.). *The Psychology of Learning and Motivation*, San Diego, CA: Academic Press, pp. 43–92.

Barsalou, L.W. 2008. Grounded cognition. *Annual Review of Psychology*, **59**:617–645.

Bavelas, J.B. and N. Chovil. 1997. Faces in dialogue. In Russell J.A., Fernandez-Dols J.M. (eds.). The Psychology of Facial Expression. Cambridge University Press, Cambridge (UK), pp. 334–346.

Bavelas, J.B. and N. Chovil. 2000. Visible acts of meaning: An integrated message model of language in face-to-face dialogue. *Journal of Language and Social Psychology*, **19**:163–194.

Bavelas, J.B. 2012. See the website http://web.uvic.ca/psyc/bavelas/Facial_displays.html

Bernstein, A. 1962. Linguistic codes, hesitation phenomena, and intelligence. *Language and Speech*, **5**:31–46.

Block, N. 1995. The mind as the software of the brain. In Smith E.E., Osherson D.N. (eds.). Thinking, Cambridge, MA: MIT Press, pp. 377–425.

Brennan, S.E. 2001. How listeners compensate for disfluencies in spontaneous speech. *Journal of Memory and Language*, **44**:274–296.

Brotherton, P. 1979. Speaking and not speaking; process for translating ideas into speech. In Siegman, A., Feldestein, S. (eds.). Of Time and Speech, Hillsdale, NJ: Lawrence Erlbaum, pp. 179–209.

Butterworth, B.L. and G.W. Beattie. 1978. Gestures and silence as indicator of planning in speech. In Campbell, R.N., Smith, P.T. (eds.). Recent Advances in the Psychology of Language: Formal and Experimental Approaches, New York: Olenum Press, pp. 347–360.

Butterworth, B.L. 1980. Evidence for pauses in speech. In Butterworth, B.L. (ed.). Language Production, 1, Speech and Talk, London: Academic Press, pp. 155–176.

Cartmill, E.A. and R.W. Byrne. 2007. Orangutans modify their gestural signaling according to their audience's comprehension. *Current Biology*, **17**:1345–1348.

Cartmill, E.A., S. Beilock and S. Goldin-Meadow. 2012. A word in the hand: Action, gesture, and mental representation in human evolution. *Philosophical Transactions of the Royal Society, Series B.*, **367**:129–143.

Chafe, W. 1974. Language and consciousness. *Language*, **50**:111–133.

Chafe, W. 1980. The deployment of consciousness in the production of a narrative. In Chafe, W. (ed.). *The Pear Stories*, Norwood, NJ: Ablex, pp. 9–50.

Chafe, W. 1987. Cognitive constraint on information flow. In Tomlin R. (ed.). Coherence and Grounding in Discourse, John Benjamins, pp. 20–51.

Dennett, D.C. 1969. Content and Consciousness, London (UK): Routledge et Kegan Paul.

De Ruiter, J.P. 2000. The production of gesture and speech. In McNeill, D. (ed.). Language and Gesture, UK: Cambridge University Press, pp. 284–311.

Dowe, D.L. 2010. MML, hybrid Bayesian network graphical models, statistical consistency, invariance and uniqueness. In Bandyopadhyay, P.S. and Forster, M.L. (eds.). Handbook of the Philosophy of Science. Volume 7: Philosophy of Statistics, Elsevier BV (NL), **7**:901–982.

Efron, D. 1941. Gesture and Environment, New York, NY: King's Crown Press.

Ekman, P. and W. Friesen. 1969. The repertoire of non-verbal behaviour: Categories, origins, usage, and coding. *Semiotica*, **1**:49–98.

Ekman, P. 1989. The argument and evidence about universals in facial expressions of emotion. In Wagner, H., Manstead, A. (eds.). Handbook of Social Psychophysiology, Chichester: Wiley, pp. 143–164.

Erbaugh, M.S. 1987. A uniform pause and error strategy for native and non-native speakers. In Tomlin, R. (ed.). Coherence and Grounding in Discourse, John Benjamins (NL), pp. 109–130.

Esposito, A., K.E. McCullough and F. Quek. 2001. Disfluencies in gestures. In Proceedings of International Workshop on Cues in Communication, Hawai, December 9.

Esposito, A., S. Duncan and F. Quek. 2002. Holds as gestural correlates to empty and filled pauses. In Proceeding of the International Conference on Spoken Language Processing (ICSLP02), Denver, Colorado, **1**:541–544.

Esposito, A., M. Marinaro and G. Palombo. 2004. Children speech pauses as markers of different discourse structures and utterance information content. In Proceedings of the International Conference: From Sound to Sense: +50 Years of Discoveries in Speech Communication, MIT, Cambridge, USA, 10–13 June, H7–H12.

Esposito, A. 2005. Pausing strategies in children. In Proceedings of the International Conference on Nonlinear Speech Processing, Cargraphics, Barcelona, SPAIN, pp. 42–48.

Esposito, A. 2006. Children's organization of discourse structure through pausing means. In Faundez_Zanuy M. et al. (eds.). Nonlinear Analyses and Algorithms for Speech Processing, Springer Berlin Heidelberg, LNCS **3817**:108–115.

Esposito, A. 2007. The amount of information on emotional states conveyed by the verbal and nonverbal channels: Some perceptual data. In Stilianou, Y. et al. (eds.). Progress in Nonlinear Speech Processing, Springer Berlin Heidelberg, LNCS **4391**:245–268.

Esposito, A. and M. Serio. 2007. Children's perception of musical emotional expressions. In Esposito, A. et al. (eds.). Verbal and Nonverbal Communication Behaviours, Springer: Springer Berlin Heidelberg, LNCS, **4775**:51–65.

Esposito, A. and M. Marinaro. 2007. What pauses can tell us about speech and gesture partnership. In Esposito, A. et al. (eds.). Fundamentals of Verbal and Nonverbal Communication and the Biometric Issue. Amsterdam: IOS Press, NATO Publishing Series, pp. 45–57.

Esposito, A., D. Esposito, M. Refice, M. Savino and S. Shattuck-Hufnagel. 2007. A preliminary investigation of the relationships between gestures and prosody in Italian. In Esposito A. et al. (eds.). Fundamentals of Verbal and Nonverbal Communication and the Biometric Issue. Amsterdam: IOS Press, NATO Publishing Series, pp. 65–74.

Esposito, A. 2009. The perceptual and cognitive role of visual and auditory channels in conveying emotional information. *Cognitive Computation,* **1(2)**:268–278.

Esposito, A., D. Carbone and M.T. Riviello. 2009. Visual context effects on the perception of musical emotional expressions. In Fierrez J. et al. (eds.). Biometric ID Management and Multimodal Communication, Springer: Springer Berlin Heidelberg, LNCS, **5707:** 81–88.

Esposito, A. and A.M. Esposito. 2011. On speech and gesture synchrony. In Esposito A. et al. (eds.). Communication and Enactment, Springer: Springer Berlin Heidelberg, **6800:**252–272.

Esposito, A. and M.T. Riviello. 2011. The cross-modal and cross-cultural processing of affective information. In Apolloni, B. et al. (eds.). *Frontiers in Artificial Intelligence and Applications,* **226:**301–310.

Freedman, N. 1972. The analysis of movement behaviour during the clinical interview. In Siegmann A.W. and Pope B. (eds.). Studies in Dyadic Communication, New York: Pergamon Press, pp. 177–208.

Fodor, J.A. 1983. The Modularity of Mind. Cambridge, MA: MIT Press.

Gallese, V. 2008. Mirror neurons and the social nature of language: The neural exploitation hypothesis. *Social Neuroscience,* **3:**317–333.

Gee, J.P. and F. Grosjean. 1984. Empirical evidence for narrative structure. *Cognitive Science,* **8:**59–85.

Ghazanfar, A.A. and N.K. Logothetis. 2003. Facial expressions linked to monkey calls. *Nature,* **423:**937–938.

Genty, E. and R.W. Byrne. 2010. Why do gorillas make sequences of gestures? *Animal Cognition,* **13:**287–301.

Genty, E., T. Breuer, C. Hobaiter and R.W. Byrne. 2009. Gestural communication of the gorilla (*Gorilla gorilla*): Repertoire, intentionality and possible origins. *Animal Cognition,* **12:**527–546.

Green, D.W. 1977. The immediate processing of sentence. *Quarterly Journal of Experimental Psychology,* **29:**135–146.

Goldmar, Eisler F. 1968. *Psycholinguistic: Experiments in Spontaneous Speech.* London/New York: Academic Press.

Grosz, B. and J. Hirschberg. 1992. Some intentional characteristics of discourse structure. In Proceedings of International Conference on Spoken Language Processing, Banff, pp. 429–432.

Goldin-Meadow S. 2003. Gesture: How Our Hands Help Us Think. Cambridge, MA, USA.

Hargadon, A. 1999. Group Cognition and Creativity in Organizations. Stamford, Conn.: JAI Press.

Hauser, M.D., C.S. Evans and P. Marler. 1993. The role of articulation in the production of rhesus monkey, Macaca mulatta, vocalizations. *Animal Behaviour,* **45:**423–433.

Higham, J.P., K.D. Hughes, L.J.N. Brent, C. Dubuc, A. Engelhardt, M. Heistermann, D. Maestripieri, L.R. Santos and M. Stevens. 2011. Familiarity affects assessment of facial signals of female fertility by free-ranging male rhesus macaques. *Proceedings of the Royal Society* B, **278**:3452–3458.

Hochreiter, S. and J. Schmidhuber. 1997. Long short-term memory. *Neural Computation*, **9(8)**:1735–1780.

Kendon, A. 1980. Gesticulation and speech: Two aspects of the process of utterance. In Key, M.R. (ed.). The Relationship of Verbal and Nonverbal Communication, pp. 207–227. Mouton Publisher: The Hague (NL).

Kendon, A. 2004. Gesture: Visible Action as Utterance. Cambridge University Press, Cambridge (UK).

Kowal, S., D.C. O'Connell and E.J. Sabin. 1975. Development of temporal patterning and vocal hesitations in spontaneous narratives. *Journal of Psycholinguistic Research*, **4**:195–207.

Krauss, R.M., P. Morrel-Samuels and C. Colasante. 1991. Do conversational hand gestures communicate? *Journal of Personality and Social Psychology*, **61(5)**:743–754.

Krauss, R.M., Y. Chen and R.F. Gottesman. 2000. Lexical gestures and lexical access: A process model. In McNeill D. (ed.). Language and Gesture. Cambridge University Press, Cambridge (UK), pp. 261–283.

Kröger, B., S. Kopp and A. Lowit. 2010. A model for production, perception, and acquisition of actions in face-to-face communication. *Cogn. Process*, **11(3)**:187–205.

Leslie, A.M., O. Friedman and T.P. German. 2004. Core mechanisms in theory of mind. *Trends in Cognitive Science*, **8(12)**:528–533.

Lewis, E.R., P.M. Narins, K.A. Cortopassi, W.M. Yamada, E.H. Poinar, S.W. Moore, X.L. Yu. 2001. Do male white-lipped frogs use seismic signals for intraspecific communication? *American Zoologist*, **41**:1185–1199.

Mathews, A., B. Mackintosh and E.P. Fulcher. 1997. Cognitive biases in anxiety and attention to threat. *Trends in Cognitive Sciences*, **1(9)**:340–345.

McNeill, D. and E. Levy. 1982. Conceptual representations in language activity and gesture. In Jarvella R. and Klein S. (eds.). Speech, Place, and Action: Studies in Deixis and Related Topics, Chichester, England: John Wiley & Sons, pp. 271–295.

McNeill, D. 1992. Hand and Mind: What Gesture Reveal About Thought. Chicago: University of Chicago Press.

McNeill, D. 2005. Gesture and Thought. Chicago: University of Chicago Press.

Mendl, M., O.H.P. Burman, R.M.A. Parker and E.S. Paul. 2009. Cognitive bias as an indicator of animal emotion and welfare: Emerging evidence and underlying mechanisms. *Applied Animal Behaviour Science*, **118**:161–181.

Montgomery, K.J., N. Isenberg and J.V. Haxby. 2007. Communicative hand gestures and object-directed hand movements activated the mirror neuron system. *Soc. Cogn. Affect. Neurosci.*, **2(2)**:114–122.

Morsella, E. and R.M. Krauss. 2005a. Can motor states influence semantic Processing? Evidence from an interference paradigm. In Columbus A. (ed.). Advances in Psychology Research, New York: Nova Science. pp. 163–118.

Morsella, E. and R.M. Krauss. 2005b. Muscular activity in the arm during lexical retrieval: Implications for gesture-speech theories. *Journal of Psycholinguistic Research*, **34**:415–437.

Nawrot, E.S. 2003. The perception of emotional expression in music: Evidence from infants, children and adults. Psychology of Music, **31(I)**:75–92.

Newell, A. and H.A. Simon. 1972. Human Problem Solving. Oxford, England: Prentice Hall.

O'Connell, D.C. and S. Kowal. 1883. Pausology. *Computers in Language Research*, **2(19)**:221–301.

Oliveira, M. 2000. Prosodic features in spontaneous narratives. Ph.D. Thesis, Simon Fraser University.

Oliveira, M. 2002. Pausing strategies as means of information processing narratives. Proceeding of the International Conference on Speech Prosody, Aix-en-Provence, pp. 539–542.

Ortony, A., G.L. Clore and A. Collins. 1990. The Cognitive Structure of Emotions. Cambridge University Press, Cambridge (UK).

O'Shaughnessy, D. 1995. Timing patterns in fluent and disfluent spontaneous speech. Proceedings of ICASSP Conference, Detroit, pp. 600–603.

Parr, L.A. 2004. Perceptual biases for multimodal cues in chimpanzee (Pan trog-lodytes) affect recognition. *Animal Cognition*, **7**:171–178.

Parr L.A., B. Waller, A. Burrows, K. Gothard and S. Vick. 2010. Brief communication: MaqFACS: A muscle-based facial movement coding system for the rhesus macaque. *American Journal of Physical Anthropology*, **143(4)**:625–630.

Partan, S.R. 2002. Single and multichannel signal composition: Facial expressions and vocalizations of rhesus macaques (Macaca mu-latta). *Behaviour*, **139**:993–1028.

Partan, S.R. and P. Marler. 1999. Communication goes multimodal. *Science*, **283(5406)**:1272–1273.

Partan, S.R. and P. Marler. 2005. Issues in the classification of multisensory communication signals. *The American Naturalist*, **166**:231–245.

Pylyshyn, Z.W. 1984. Computation and Cognition: Toward a Foundation for Cognitive Science. Cambridge, MA: MIT Press.

Rausher, H., R.M. Krauss and Y. Chen. 1996. Gesture, speech and lexical access: The role of lexical movements in the processing of speech. *Psychological Science*, **7**:226–231.

Rimé, B. and L. Schiaratura. 1992. Gesture and Speech. In Feldman R.S. and Rimé B. (eds.). Fundamentals of Nonverbal Behavior. Cambridge University Press, Cambridge (UK), pp. 239–284.

Rimé, B. 1982. The elimination of visible behaviour from social interactions: Effects of verbal, nonverbal, and interpersonal variables. *European Journal of Social Psychology*, **12**:113–129.

Rogers, W.T. 1978. The contribution of kinesic illustrators towards the comprehension of verbal behaviour within utterances. *Human Communication Research*, **5**:54–62.

Rosenfield, B. 1987. Pauses in Oral and Written Narratives. Boston: Boston University.

Rowe, C. 2002. Sound improves visual discrimination learning in avian predators. Proceedings of the Royal Society of London B, **269**:1353–1357.

Russell, J.A. 1980. A circumplex model of affect. *Journal of Personality and Social Psychology*, **39**:1161–1178.

Russell, J.A. 2003. Core affect and the psychological construction of emotion. *Psychological Review*, **110**:145–172.

Schubert, T.W. 2004. The power in your hand: Gender differences in bodily feedback from making a fist. *Personality and Social Psychology Bulletin*, **30**:757–769.

Short, J., E. Williams and B. Christie. 1976. The Social Psychology of Telecommunications. London: John Wiley & Sons.

Shattuck-Hufnagel, S., Y. Yasinnik, N. Veilleux and M. Renwick. 2007. A method for studying the time alignment of gestures and prosody in American English: 'Hits' and pitch accents in academic-lecture-style speech. In Esposito A. et al (eds.). Fundamentals of Verbal and Nonverbal Communication and the Biometric Issue. Amsterdam: IOS Press, NATO Publishing Series, pp. 34–44.

Schneider, S.S. and L.A. Lewis. 2004. The vibration signal, modulatory communication and the organization of labor in honey bees, Apis mellifera. *Apidologie*, **35**:117–131.

Slocombe, K., B. Waller and K. Liebal. 2011. The language void: The need for multimodality in primate communication research. *Animal Behaviour*, **81(5)**:919–924.

Smith, E.R. and G.R. Semin. 2004. Socially situated cognition: Cognition in its social context. *Advances in Experimental Social Psychology*, **36**:53–117.

Stahl, G. 2006. Group Cognition: Computer Support for Building Collaborative Knowledge (Acting With Technology). The MIT Press, Cambridge (MA) USA.

Tiwari, S.P. 2009. Some Fuzzy Computing Models: A Topological and Categorical Approach. Lap Lambert Academic Publishing, Germany.

Thompson, L.A. and D.W. Massaro. 1986. Evaluation and integration of speech and pointing gestures during referential understanding. *Journal of Experimental Child Psychology*, **42**:144–168.

Wheeler, B.C. 2009. Monkeys crying wolf? Tufted capuchin monkeys use anti-predator calls to usurp resources from conspecifics. Proceedings of the Royal Society of London Series B: Biological Sciences, **276**:3013–3018.

Wheeler, B.C. 2010. Decrease in alarm call response among tufted capuchin monkeys in competitive feeding contexts: Possible evidence for counterdeception. *International Journal of Primatology*, **31**:665–675.

Williams, E. 1977. Experimental comparisons of face-to-face and mediated communication: A review. *Psychological Bulletin*, **84**:963–976.

Yasinnik, Y., M. Renwick and S. Shattuck-Hufnagel. 2004. The timing of speech-accompanying gestures with respect to prosody. Proceedings of the International Conference: From Sound to Sense: +50 years of Discoveries in Speech Communication, MIT, Cambridge, USA, 10–13 June, C97–C102.

CHAPTER 8

From Annotation to Multimodal Behavior

Kristiina Jokinen and Catherine Pelachaud

1. Introduction

The usage of multimodal corpora is crucial for studying human behavior as well as modeling embodied conversational agents (Cowie et al., 2010a). Gathering data and creating corpora is complex. Interested readers can read work by Cowie and Douglas-Cowie on the topic (Cowie et al., 2010b) wherein different methodologies for creating corpora are described.

Once the corpora are gathered, there are the issues of annotating them. Defining an annotation schema is very much in relation to the phenomenon to be studied. When defining, it is necessary to decide which information should be marked and in what form, whether it should be continuous values or discrete ones, on which theoretical grounds should the annotation schemes rely, and how to mark dependencies. A lot of effort is put into deriving annotation schemes that can be used reliably by researchers working on a wide-variety of projects. Several schemas have been designed (including some standards such as ISO-Dit++ (Iso-dit)).

Embodied Conversational Agents are autonomous entities able to communicate verbally and non-verbally. They are part of an interaction system. They can be controlled at high levels where their communicative intentions and emotional states are specified or at low levels when describing their behavior. Representation languages have been designed to control them at various levels. Different factors need to be considered when elaborating on a representation language: as to which theoretical grounds the representation language is built; does it depend of a specific technology; which values for its elements does it consider; how much flexibility does it accept, etc.

In the remainder of this chapter, we first turn our attention to the presentation of annotation schemes and then to representation languages. We end the chapter by presenting some future trends.

2. Annotation Schemes

When studying human multimodal behavior, the starting point is empirical data analysis which provides the necessary generalizations and characteristics of the phenomena under study. The analysis is usually conducted by annotating the data according to an appropriate annotation scheme, and the annotations allow the data to be compared with similar annotated data as well as being re-used in further studies. Focusing on the interaction between human users and automated smart agents (Embodied Conversational Agents, virtual world characters, robot agents), this section discusses annotations from the point of view of natural and intuitive communication, and aims to give an overview of those aspects related to multimodal corpus annotation, with the purpose of building models that can be used in a variety of applications.

An annotation scheme is a collection of labels (tags) to be assigned to data elements such that they describe the material with respect to various physical, perceptual and functional dimensions. It concerns the analysis of a phenomenon with the help of analytical categories, and is to be distinguished from the representation or mark-up language which refers to the format in which the annotation categories are represented (see next section). It may include different analysis levels on which the data labeling is performed, or contain only one level with in-depth analysis categories. Metadata, such as the collection date and setting, participants' demographics, etc., are often included within the annotation scheme as well, and become important in studies related to social and personality issues.

The main purpose of the annotation is to capture the phenomena under study. Annotation categories are thus dependent on the diverse theoretical assumptions made in the description of the phenomena; for instance, linguistic annotations of the sentence structure may rely on phrase-structure, dependency, or some other type of grammar formalisms, all of which furnish the annotators with a different annotation scheme. The theory describing which categories to use and how to classify the data into these elements is also called a 'data model', although this term is often used in a more restricted sense as a formalized description of the data objects and their relations. In practical work, annotation categories are usually defined based on extensive literature, in which the phenomena are treated by extracting the essential features for the purpose of the study. Annotation schemes can thus vary due to different theoretical assumptions (i.e. they have different data models), but they may also have different granularity and completeness levels, e.g. gesture annotation may include hand-gestures but not finger movement (granularity) or encode the function rather than the form of the gesture (completeness). Whilst the different data models are often incompatible with each other and, consequently, cause incompatibilities in the annotation scheme, different granularity or completeness levels do not necessarily render annotations incompatible. Annotation schemes aim to be extensible, i.e. more annotation levels and categories can be added, and the existing categories can be further specified or combined under a more general feature, and also incremental in that the extensions can be done at different stages of the annotation process. Annotations also need to be consistent, and to this end, annotation categories have particular semantics and are applicable for the annotations they have been designed for. The annotation results have to be validated by applying the scheme to practical coding tasks and by calculating inter-coder agreement by several coders.

In general, multimodal annotation has to address the following issues:

- What to annotate: identification of an element as communicatively relevant
- How to segment the data: defining the boundaries of the elements
- How to deal with interdependence: relations between elements and annotation levels.

The first two are relevant to all annotations, whilst the third is especially characteristic of multimodal annotations. In the subsections below, we discuss these issues in more detail.

2.1 What to annotate?

In multimodal annotation, different levels are identified and various annotation features selected depending on the annotation level. In general, annotations deal with the agent's identity, physiological state, emotions and attitudes, own communication and partner interaction management, formation exchange, etc. In speech and language-related annotations, the levels correspond to different modalities and communicative phenomena. The verbal linguistic level categories concern phonemes, morphemes, words, and utterances, accompanied by phonetic and prosodic features, as well as paralinguistic vocalizations such as laughs and coughs. Pragmatic and discourse level aspects usually form their own level and include categories such as topic, focus, new information, discourse structure, and rhetorical relations, which are linked to the correlated words and sentences, analogously to interaction categories such as the speakers' communicative intentions, dialogue acts, feedback, turn-taking, and sequencing. Multimodal annotation levels concern hand-gestures, head movements, body posture, facial expressions and eye gazing, encoding, to various details, movements of the fingers, eyes, mouth, and legs. Various affective displays and social behavior are also part of multimodal annotation schemes, comprised, in particular, of emotions and personality traits, engagement, dominance, cooperation, etc. The attributes concerning the properties of the observed phenomena need not be fine-grained, but they need to capture the significant and relevant aspects of the phenomena (expressive adequacy), be explicit and consistent.

Each annotation scheme has guidelines that explain and exemplify the meanings of the categories, based on theoretical assumptions, and also practical needs of research projects. For instance, the MUMIN annotation scheme (Allwood et al., 2007) focuses on the general form and function features of multimodal elements, and it is used in the Nordic NOMCO project (Navarretta et al., 2012), which aims to create comparable annotated resources for the languages involved in the project, in order to investigate specific communicative functions of hand and head gesturing in the neighboring countries. Other frameworks have aimed at the registration of facial movements (Ekman and Friesen, 1978), hand gestures (Duncan, 2004); or studying emotions as expressed by facial movements (Ekman and Friesen, 2003). Some of these schemes can be used to annotate gestures within different scientific settings as in the construction of virtual agents (see below) or within different scientific domains (psychopathology, education).

Dialogue act annotation has been mainly derived at for the practical needs of tasks in hand, and thus it has been difficult to compare annotations or the systems which have been built on these annotations. Although, in general, the Speech Act theory by Austin and Searle has been the basis for the schemes, the particular application and interaction task has influenced the set of dialogue acts. Efforts on standardization, based on a range of dialogue act annotation schemes

(TRAINS, HCRC Map Task, Verbmobil, DIT, MUMIN, AMI, etc.) and various initiatives towards domain-independent, inter-operative and standardized sets of dialogue acts (DAMSL 1997; MATE 1999; DIT++ 2005; LIRICS 2007), have produced the ISO DIS 24617-2 (January 2010) standard for dialogue act annotation. This standard is accompanied by the annotation language DiAML which supports abstract and concrete syntax: semantics is defined for the abstract syntax in terms of information-state update operators, whilst concrete syntax defines XML representations. Similar standardization efforts are on-going concerning purely linguistic annotations, and in the area of multimodal annotations, the D-META Grand Challenge at ICMI 2012 (http://d-meta.inrialpes.fr/) aimed at setting up a basis for comparison, analysis, and further improvement of multimodal data annotations and multimodal interactive systems.

2.2 Segmentation

Once an annotation element is defined, its temporal identification is to be determined. This means that the annotator finds the element they want to annotate, and establishes where the element starts and ends. The exact start and end time may sometimes be difficult to determine due to inaccuracy of the recording device, and there can also be differences between the annotators' annotation accuracy. These kinds of inaccuracies are usually taken care of by allowing a small time-window within which the start and end times are considered equivalent.

However, other segmentation incompatibilities are more serious in that they may prevent direct comparison of the annotated data. One such issue is the relation between the element's form and function. For instance, gestures can be coded with features that describe their function besides their shape and dynamics. Following Kendon (2005), a gesture unit is identified with the help of the so-called gesture phases: preparation, stroke, and retraction, but it may also contain more specific phases if a detailed description is needed, or of a single stroke span, if the occurrence of the gesture is important rather than its internal structure.

Another issue, raised by the empirical studies, is the connection of form and function: are they tightly connected so that a particular gesture shape always corresponds to a specific function and vice versa, or do they stand in a many-to-many relation. If the annotator is expected to select gestures to be annotated only if the gesture has a communicative function, not all gestures are annotated. For example, a hand gesture related to lifting of a coffee cup may not be annotated, because it does not seem to have an obvious communicative function, but it may also be a signal of the agent being bored or having nothing to say, in which case it carries communicative meaning. Annotations thus differ depending on whether the gestures are interpreted as being intentionally communicative by the communicator (displayed or signaled) (Allwood, 2001), or the gestures are judged (by the annotator) to have a noticeable effect on the recipient.

Since emerging technology allows recognition of gestures and faces via cameras and sensors, it is possible to extract gestures and face expressions from the data. The form-features of gestures can then be automatically linked to appropriate communicative functions. Combining the top-down approach, i.e. manual annotation and analysis of the data, with the bottom-up analysis of the multimodal signals, we can visualize the speakers' communicative behavior, and also show how synchrony of conversation is constructed through the interlocutors' activity (Jokinen and Scherer, 2012). The top-down approach uses human observation, e.g. takes video recordings which are manually tagged according to some scheme, to mark communicatively important events. The bottom-up approach, on the other hand, uses automatic technological means to recognize, cluster, and interpret the signals that the communicating agents emit. These two approaches look at the communicative situations from two opposite viewpoints: they use different methods and techniques, but the object of study is the same. Communication models can thus be built and incorporated into smart applications through top-down human observations and bottom-up automatic analysis of the interaction, and the approach is beneficial for both interaction technology and human communication studies. New algorithms and models will be required for the detection and processing of speech information along with gestural and facial expressions, and existing technologies will need to be adapted to accommodate these advances. Simulations based on manual analysis of corpora on gestures and face expressions are already incorporated within the development of Embodied Conversational Agents (e.g., André and Pelachaud, 2010), and a motion capture tool to gather more precise data about the user's behavior is described in Csapo et al. (2012).

2.3 Interdependence: Modal and multimodal annotation

Two kinds of annotation of interaction data can be considered. The first is uni-modal annotation that is specific to a particular modality, e.g. dialogue act annotation or gesture annotation, and the second one is multimodal annotation proper, which takes the relation between the different communication modes into consideration. To define a multimodal relation, a basic distinction can be made between two signs being dependent or independent. Dependent signs can be compatible or incompatible. If the signs are compatible, they must either complement or reinforce each other, while incompatible signs express different contents, as for example in irony.

The correspondences between different modality levels can also be within-speaker or across-speakers, depending on whether the speech and gestures are produced by the same or different speakers. The interactive nature of communication may require a separate level of coding, since joint actions are qualitatively different from those exhibited by a single agent alone. For instance, cooperation on the construction of shared knowledge cannot be attached as a feature of the individual agent's behavior, since it is not an action that the agent just happens to perform simultaneously with a partner but a genuine joint activity that the agents produce in coordination.

Multimodal expressions can have different time spans (cf. speech segments and gesturing). An important issue here is the anchoring between various modality tracks. This can take place at different levels: at the phoneme, word, phrase or an utterance level. Most often in multimodal interaction annotations, the smallest speech segment is the word.

2.4 Annotation tools

Annotation schemes are based on the theories of communication, and they are realized in concrete data annotations. Multimodal annotation includes various independent data streams which are annotated and linked together. Usually multimodal annotation schemes agree on the modality levels (input streams) and the general descriptive dimensions on them, but differ on the number and detailed interpretation of categories. Such annotations pose technical challenges for the annotation tools and workbenches, and also for the synchronization of cross-modality phenomena. An annotation tool must thus be capable of processing multimodal information and of supporting the fusion of multimodal input streams (speech and gestures, facial expressions, etc.).

Manual annotation is notorious for being a time- and resource-consuming task, and different tools and workbenches have been developed for helping annotators. The two more common tools for speech transcription and analysis are Praat (Boersma and Weenink, 2009) and Wafesurfer (Sjölander and Beskow, 2000), whilst in the multimodal video analysis either ELAN (Elan) or Anvil (Kipp, 2001) annotation tools are commonly used. The increase in the larger corpora of annotated multimodal data has raised questions about creating, coding, processing, and managing more extensive multimodal resources, notably in the context of European collaboration projects. For instance, the European Telematics project MATE (Multilevel Annotation Tools Engineering) aimed to facilitate the use and reuse of spoken language resources, coding schemes and tools, and produced the NITE workbench (Bernsen et al., 2002) that addresses theoretical issues. Dybkjaer et al. (2002) provided an overview of the tools and standards for multimodal annotation.

2.5 Inter-coder agreement

A number of methodological recommendations have been put forward for validating the data and ensuring coherent and reliable annotations. The annotators' mutual agreement of the categories is one of the standard measures, and much attention has been devoted to it (Cavicchio and Poesio, 2009; Rietveld and van Hout, 1993). It is important to distinguish the percentage agreement (how many times the annotators are observed to assign the same category to the annotation elements), from the agreement that takes into account expected agreement (the probability that the annotators agree by chance). Agreement beyond chance can be measured by Cohen's kappa, coefficient κ, calculated as follows:

$$\kappa = (P(A) - P(E))/(1 - P(E))$$

where $P(A)$ is the proportion of times the coders agree and $P(E)$ is the proportion of times they can be expected to agree by chance. The value of κ is 1 in the case of total agreement and zero in the case of total disagreement. According to Rietveld and van Hout (1993), κ-values above 0.8 show almost perfect agreement, those between 0.6 and 0.8 show substantial agreement, those between 0.4 and 0.6 moderate agreement, those between 0.2 and 0.4 fair agreement, and those below 0.2 show slight agreement beyond chance. Generally, a value above 0.6 is considered satisfactory.

However, κ can often be very low, whilst percentage agreement is still quite high, and it has been argued that κ may not be a suitable statistic for this (Cavicchio and Poesio, 2009 for discussion). For instance, if one of the coders has strong preference for a particular category, the likelihood of the coders agreeing on that category by chance is increased and, consequently, the overall agreement measured by κ-scores decreases. Krippendorff's alpha (Krippendorff, 2004) has been suggested as an alternative to κ, as it takes care of the biased category distributions.

3. Representation Languages

Representation languages (or other forms such as script or mark-up languages (Krenn et al., 2011)) have been designed over many years to drive virtual agents' behavior. They provide information on the type of signals to be displayed; when they should appear and for how long; which other signals are also present, etc. Examples of existing languages are VHML, MPML, APML, GESTYLE, etc. The various existing languages may encode information as various as activities, culture, voice quality, signal description, emotion, etc. The readers are referred to Prendinger and Ishizuka (2004) and Krenn et al. (2011) for a broader view of existing works. However, in this section we will focus on three examples which are representatives of encoding: communicative functions, affective states, and multimodal behavior. The representation languages are respectively called FML (Heylen et al., 2008), EmotionML (Schröder et al., 2011) and BML (Kopp et al., 2006). Whilst, apart from EmotionML, the other two are not part of any standard (such as W3C or ISO), but are widely used throughout the agent community.

Before starting this description, we will introduce SAIBA, an international initiative for defining a common framework when modeling embodied conversational agents (see Figure 1). SAIBA stands for Situation, Agent, Intention, Behavior, and Animation (Kopp et al., 2006; Vilhjálmsson et al., 2007). SAIBA arises from the observation that more current agent systems have followed a similar architecture, namely to be composed of three main modules: the first one, called Intent Planner, takes as input information from the context (e.g. the environment, the conversation state). It outputs a list of communicative intentions and emotional states which are encoded through Function Mark-up Language—FML (Heylen et al., 2008) and EmotionML (Schröder et al., 2011) (see next sub-section for further information). The next module called Behavior Planner instantiates

Figure 1. SAIBA framework.

the intentions and emotional states to be conveyed in synchronized sequences of multimodal behaviors. These behaviors are encoded into the Behavior Mark-up Language—BML (Kopp et al., 2006; Vilhjálmsson et al., 2007). The third module, Behavior Realizer, receives this list of signals and computes the corresponding animation. In the remainder of this section, we will detail these representation languages.

3.1 Emotion Mark-up Language—EmotionML

Emotions are crucial for simulating liveliness in virtual agents. Emotion is a complex phenomenon involving cognitive processes and physiological and behavioral changes. Emotion is defined as "an episode of interrelated, synchronized changes in several components in response to an event of major significance..." (Scherer, 2000). It is largely accepted that five major components are part in the emotion process (Scherer, 2000):

- Neurophysiological and autonomous nervous patterns (in central and nervous systems)
- Motor expression (in face, gesture, gaze)
- Feelings (subjective experience)
- Action tendencies (action readiness)
- Cognitive processing (mental processes)

Different theories of emotion have been defined. We report here the three main theories:

- Discrete theory which speculates there is an innate neural program that leads towards specific bodily response patterns (Ekman and Friesen, 2003); there exist prototypical facial expressions for emotions.
- Dimensional theory that describes emotion along several dimensions. The circumplex of Russell (Russell, 1980) and PAD (Merhabian, 1996) are two of the main examples. The first one considers two dimensions: Pleasure and Arousal, whilst the second adds a third dimension, namely Dominance.
- Appraisal theory that views emotions as arising from the subjective evaluation of the significance of an event, of an object, of a person (Arnold, 1960; Scherer, 2000).

The first attempts (e.g. VHML (Beard and Reid, 2002)) to control virtual agents considered the six basic emotions (Ekman and Friesen, 2003). Recently EmotionML (Schröder et al., 2011), a W3C standard, was designed to allow for describing emotions using any of the three main theoretical representations. It also includes the representation of Action Tendencies (Frijda, 1986). As no common vocabulary has been agreed upon by emotion theoreticians, the choice of vocabulary for an emotion within each of these representations is left open. It is possible to use an existing one from the literature or a specific one tailored for a particular application. EmotionML encompasses tags for specifying information such as the intensity of the felt emotion, its temporal course, to which events it refers to. Coping strategies linked to appraisals is considered within EMA (Marsella and Gratch, 2009).

3.2 *Communicative intentions*

When communicating we convey not only our emotions but also our epistemic states, our attitudes, information about the world, discursive and semantic information... FML encodes any factors that are relevant when planning verbal and non-verbal behavior (NVB). So far the scope of FML has been discussed (Heylen et al., 2008) but no attempt of formalization has been made so far. This is partly due to the extensive variety of theories regarding communication. Different examples exist such as APML (De Carolis et al., 2004) which have been built from Isabella Poggi's theory of communicative acts (Poggi, 2007). Bickmore (2008) proposed the inclusion of meta-information for characterizing interaction types (e.g. encouragement, empathy, task-oriented) which he called Frames. He also advocated the inclusion of information on the interpersonal relations between speakers and those with whom they interact. Indeed communication is a process involving several partners who sense, plan and adapt to each other continuously. Interaction involves several exchanges of speaking turns. A turn-taking description should also be included (Heylen et al., 2008; Kopp et al., 2006).

3.3 *Behavior Mark-up Language—BML*

The idea behind BML is to describe multimodal behavior independently of the specific body and animation parameters characterizing a given virtual character. Its description should also be independent of the animation player to be used during an application. The choice is to define behavior within BML at a symbolic level. Several modalities are considered: face, gaze, arm-gesture, posture, etc. All the behavior encoded contains two types of information: a description of its shape and of its temporal course. Whenever possible, these descriptions rely on theoretical works. For example, facial expressions are described as a list of Action Units from the FACS—Facial Action Coding System (Ekman and Friesen, 1978). Arm gestures are decomposed into hand shape, wrist position and orientation. Their temporal structures follow the gesture phases proposed by McNeill (1992) and Kendon (2004). Similar temporal structure is applied for all modalities. Synchronization schema has been elaborated to ensure synchronization between modalities, with speech and with external events. Feedback mechanism is to be used to let the system know when a behavior has finished or had to be aborted for lack of time or infeasibility.

3.4 Expressivity

NVB can be characterized by its shape (e.g. a facial expression, a hand shape), its trajectories (of the wrist, head) and its manner of execution (speed of movement, acceleration). This manner may reflect the emotional state and personality traits of the executants. In the field of dance, describing behavior expressivity to annotate ballet has long been a challenge. Several annotation schemes exist, in particular those developed by Benesh (Benesh and Benesh, 1983) and Laban (van Laban and Ullmann, 1988). Benesh follows the metaphor of musical score. The five lines correspond to body parts. Additional signs are used to specify movement quality, including rhythm and phrasing. Laban is an important figure of modern dance. The principal components of his dance annotation schema are: body parts, movement direction and height, its duration and its dynamic quality. In order to characterize this last element, he introduces four elements: time, weight, space and flow.

From another perspective, Wallbott and Scherer (1986) studied bodily expressions of emotion. From a video corpus of actors portraying various emotional states, the authors analyzed body posture and movement. As results they found movement and posture patterns for specific emotions. A set of six expressivity parameters, namely spatial extent, temporal extent, fluidity, power, repetition and overall activation, have been implemented to control the dynamic quality of virtual agents' behavior (Hartmann et al., 2005). An extension has been proposed by Huang and Pelachaud (2012).

3.5 Lexicon

NVB is as important as the verbal behavior during the communication process. It helps speakers to plan what to say and how to say it; it is also useful for the recipients of messages in order to decode what is being said. Several taxonomies have been proposed relying on the types of meaning it carries, and its function during communication (McNeill, 1992; Poggi, 2007; Ekman and Friesen, 2003). An NVB is characterized by the shape of the signals that composed it and their associated meaning. Such link shape-meaning highly depends on the discursive context. Poggi called the pairs (signals, meaning) a communicative act. A communicative act may have several pairs (of signals, meanings) attached to it. A lexicon is like a dictionary of communicative acts that makes explicit the mapping between signals and meanings. Most agent systems have lexicon built in their system (Cassell et al., 1994; Cassell et al., 2001; Pelachaud, 2005). Lee and Marsella (2006) proposed non-verbal behavior generator (NVBG) which adds NVB based on a semantic analysis of the text to be said by the agent. Lately Bergmann and Kopp developed a computational model of multimodal behavior generator that outputs sentences with the associated hand gestures. This model is based on a statistical analysis of an annotated corpus of humans' dialogs on a specific topic (here to give spatial direction). The virtual agent is able to create on the fly complex iconic gestures relating to the route direction it describes.

4. Summary and Future Trends

This article focused on the annotation and representation of multimodal behavior for the purpose of designing and developing virtual characters and embodied agents. It has surveyed issues relevant to the annotation task itself, and given detailed examples of mark-up languages for emotion and behavior representation. We have distinguished an annotation scheme from its representation, and emphasized the dependence of the annotation scheme on the theory that described how the communicative phenomena were categorized in the study. We have also discussed general requirements for a multimodal annotation framework, including extensibility, incrementality, and uniformity, as well as presented different mark-up languages which go beyond application-specific representations, by offering theoretically consistent approaches ready for use when comparing and evaluating annotations.

On-going work on multimodal annotation focuses on developing and extending annotation schemes further, by enhancing the existing ones with more accurate and detailed feature specifications, and by broadening the set of annotation categories to cover new phenomena. For instance, active research to take place on topics such as emotion and affective mark-up languages (Schröder et al., 2011), laughter (Truong and Trouvain, 2012), analysis of audiovisual and paralinguistic phenomena (an overview is given in Schuller et al., 2013), as well as eye-gaze, turn-taking, and attention (Levitski et al., 2012; Bednarik et al., 2012; Jokinen et al., in print). Research into these areas not only increases our understanding of the particular phenomena, but also improves algorithms and techniques for their automatic recognition and analysis. This work concerns the development of principles and methods for use when creating and coding multimodal resources, as well as for processing and managing resources within new technological situations.

Recently several challenges have been introduced so as to foster research and development on multimodality, e.g. the ICMI-2012 Grand Challenges (http://www.acm.org/icmi/2012/index. php?id=challenges) and special challenge sessions at the forthcoming speech and multimodal conferences on topics such as paralinguistics, emotions, autism, engagements, gestures and data-sets. The popularity of such events is apparently related to the shared tasks the performances of which can be objectively measured, and to the shared annotated data which allow comparison and evaluation of the algorithms whilst acknowledging the complexities of multimodal problems.

Manual annotation is expensive and time-consuming, and new technology has also boosted automatic analysis of the data. Speech technology can be used to annotate spoken dialogs, whilst image processing techniques can be used for face and hand gesture recognition on video files (Jongejan, 2012; Toivio and Jokinen, 2012). Exciting new possibilities are available with the help of motion capture devices such as Kinect, cf. experiments described in Csapo et al. (2012).

An interesting question concerns the defining of cross-modality annotation categories and the minimal units suitable for anchoring correlations. A commonly used spoken correlate for gestures is the word, but when studying, e.g. speech and gesture synchrony, this seems to be too big a unit: the emphasis by a gesture stroke (Kendon, 2004) seems to co-occur with vocal stress which lands on an intonation phrase corresponding to a syllable or a mora, rather than a whole word. When designing natural communication for intelligent agents, such cross-modal timing phenomena become relevant as the delay in the expected synchrony may lead to confusion or total misunderstanding of the intended message.

REFERENCES

Abdel Hady, M. and F. Schwenker. 2010. Combining Committee-Based Semi-Supervised Learning and Active Learning. *Journal of Computer Science and Technology,* **25(4)**:681–698.

Allwood, J. 2001. Dialog Coding—Function and Grammar. Gothenburg Papers. Theoretical Linguistics, 85. Department of Linguistics, Gothenburg University.

Allwood, J., L. Cerrato, K. Jokinen, C. Navarretta and P. Paggio. 2007. The MUMIN Coding Scheme for the Annotation of Feedback, Turn Management and Sequencing. In Martin, J.C. et al. (eds.), *Multimodal Corpora for Modelling Human Multimodal Behaviour*. Special issue of the *International Journal of Language*.

André, E. and C. Pelachaud. 2010. Interacting with Embodied Conversational Agents. In Jokinen, K. and Cheng, F. (eds.), New Trends in Speech-Based Interactive Systems. New York: Springer Verlag.

Argyle, M. 1988. Bodily Communication. London: Methuen.

Arnold, M. 1960. Emotion and Personality: Vol. 1. Psychological Aspects. New York: Columbia University Press.

Bayerl, P. and H. Neumann. 2007. A Fast Biologically Inspired Algorithm for Recurrent Motion Estimation. *IEEE Transactions on Pattern Analysis and Machine Intelligence*, **29(2)**:246–260.

Beard, S. and D. Reid. 2002. MetaFace and VHML: A First Implementation of the Virtual Human Markup Language. In Marriott, A., Pelachaud,C., Rist, T., Ruttkay, Z. and Vilhjálmsson, H. (eds.), Embodied Conversational Agents: Let's Specify and Compare Them!, Workshop Notes, Autonomous Agents and Multiagent Sytems, Italy.

Bednarik, R., S. Eivazi and M. Radis. 2012. Gaze and Conversational Engagement in Multiparty Video Conversation: An annotation scheme and classification of high and low levels of engagement. In Proceedings of The 4th Workshop on Eye Gaze in Intelligent Human Machine Interaction: Eye Gaze and Multimodality, at the 14th ACM International Conference on Multimodal Interaction (ICMI-2012), October 26, 2012, Santa Monica, California, U.S.

Bergmann, K. and S. Kopp. 2009. Increasing Expressiveness for Virtual Agents—Autonomous Generation of Speech and Gesture for Spatial Description Tasks. In Decker, K., Sichman, J., Sierra, C. and Castelfranchi, C. (eds.), Proceedings of the 8th International Conference on Autonomous Agents and Multiagent Systems, pp. 361–368.

Benesh, J. and R. Benesh. 1983. Reading Dance: The Birth of Choreography, McGraw-Hill Book Company Ltd.

Bernsen, N.O., L. Dybkjær and M. Kolodnytsky. 2002. The NITE workbench—A Tool for Annotation of Natural Interactivity and Multimodal Data. In Proceedings of LREC 2002, pp. 43–49.

Bickmore, T. 2008. Framing and Interpersonal Stance in Relational Agents. The Seventh International Conference on Autonomous Agents and Multiagent Systems, Functional Markup Language Workshop, Portugal.

Bishop, C.M. 2006. Pattern Recognition and Machine Learning. (M. Jordan, J. Kleinberg and B. Schölkopf, eds.). New York: Springer Verlag.

Black, M. and Y. Yacoob. 1997. Recognizing Facial Expressions in Image Sequences Using Local Parameterized Models of Image Motion. *International Journal of Computer Vision,* **25(1)**:23–48.

Blake, A., B. Bascle, M. Isard and J. MacCormick. 1998. Statistical models of visual shape and motion. *Phil. Trans. Royal Society of London, Series A,* **386:**1283–1302.

Boersma, P. and D. Weenink. 2009. Praat: doing phonetics by computer. Version 5.1.05. Retrieved May 1, 2009, from HYPERLINK "http://www. praat.org/"

Boiten, F., N. Frijda and C. Wientjes. 1994. Emotions and Respiratory Patterns: Review and Critical Analysis. *International Journal of Psychophysiology,* **17(2):**103–128.

Breiman, L. 1996. Bagging Predictors. *Machine Learning,* **24(2):**123–140.

Cannon, W. 1927. The James-Lange Theory of Emotions: A Critical Examination and an Alternative Theory. *The American Journal of Psychology,* **39(1/4):**106–124.

Cassell, J. 2000. Nudge NudgeWinkWink: Elements of Face-to-Face Conversation for Embodied Conversational Agents. In Cassell, J., Sullivan, J., Prevost, S. and Churchill, E.F. (eds.), Embodied Conversational Agents, pp. 1–27, Cambridge, MA: MIT Press.

Cassell, J., M. Steedman, N.I. Badler, C. Pelachaud, M. Stone, B. Douville, S. Prevost and B. Achorn. 1994. Modeling the Interaction between Speech and Gesture. Proceedings of the 16th Annual Conference of the Cognitive Science Society, Georgia Institute of Technology, Atlanta, USA, August 1994.

Cassell, J., H. Vilhjálmsson and T. Bickmore, 2001. BEAT: The Behavior Expression Animation Toolkit. Proceedings of SIGGRAPH 2001, pp. 477–486.

Cavicchio, F. and M. Poesio. 2009. Multimodal Corpora Annotation: Validation Methods to Assess Coding Scheme Reliability. In Multimodal Corpora: From Models of Natural Interaction to Systems and Applications. Springer Verlag, Lecture Notes in Computer Science/Lecture Notes in Artificial Intelligence.

Core, M. and J. Allen. 1997 Coding Dialogs with the DAMSL Annotation Scheme. Presented at AAAI Fall Symposium on Communicative Action in Humans and Machines, Boston, MA, November 1997. HYPERLINK "ftp://ftp.cs.rochester.edu/pub/papers/ai/97.Core-Allen.AAAI2.ps.gz"

Cowie, R., E. Douglas-Cowie, J.-C. Martin and L. Devillers. 2010a. The Essential Role of Human Databases for Learning in and Validation of Affectively Competent Agents. In Klaus R. Scherer, Tanja Bänziger, Etienne Roesch (eds.), A Blueprint for an Affectively Competent Agent: Cross-Fertilization Between Emotion Psychology, Affective Neuroscience, and Affective Computing. Oxford: Oxford University Press, pp. 151–165.

Cowie, R., E. Douglas-Cowie, I. Sneddon, M. McRorie, J. Hanratty, E. McMahon and G. McKeown. 2010b. Induction techniques developed to illuminate relationships between signs of emotion and their context, physical and social. In K.R. Scherer, T. Banziger and E. Roesch (eds.), A Blueprint for Affective Computing: A Sourcebook and Manual, Oxford: Oxford University Press, pp. 295–307.

Craggs, R. and M. McGee Wood. 2004. A Categorical Annotation Scheme for Emotion in the Linguistic Content of Dialogue. In Affective Dialogue Systems. Proceedings of Tutorial and Research Workshop, Kloster Irsee, Germany, June 14–16. Lecture Notes in Computer Science. Springer Verlag Berlin/Heidelberg, pp. 89–100.

Csapo, A., E. Gilmartin, J. Grizou, J. Han, R. Meena, D. Anastasiou, K. Jokinen and G. Wilcock. 2012. Multimodal Conversational Interaction with a Humanoid Robot. In Proceedings of the 3rd IEEE International Conference on Cognitive Infocommunications (CogInfoCom) 2012, Kosice, Slovakia.

De Carolis, B., C. Pelachaud, I. Poggi and M. Steedman, (2004). APML—A Mark-up Language for Believable Behavior Generation. In Prendinger, H. and Ishizuka, M. (eds.), Life-like Characters. Tools, Affective Functions and Applications. Springer Verlag, Series: Cognitive Technologies, Springer-Verlag Berlin Heidelberg GmbH, pp. 65–85.

Duncan, S. 2004. Coding Manual. Technical Report available from Hyperlink "http://mcneilllab.uchicago.edu/".

Dybkjaer, L., S. Berman, M., Kipp, M.W. Olsen, V. Pirrelli, N. Reithinger and C. Soria. 2002. Survey of Existing Tools, Standards and User Needs for Annotation of Natural Interaction and Multimodal Data. ISLE Deliverable D11.1.

Ekman, P. and W.V. Friesen. 1978. Facial Action Coding System. Palo Alto: Consulting Psychologist Press.

Ekman, P. and W.V. Friesen. 2003. Unmasking the Face: A Guide to Recognizing Emotions From Facial Cues. Cambridge, Massachusetts: Malor Books.

Elan, Hyperlink "http://www.lat-mpi.eu/tools/elan/".

Frijda, N.H. 1986. The Emotions. Cambridge, UK: Cambridge University Press.

Hartmann, B., M. Mancini and C. Pelachaud. 2005. Implementing Expressive Gesture Synthesis for Embodied Conversational Agents, Gesture Workshop, LNAI, Springer Verlag, May 2005.

Heylen, D., S. Kopp, S. Marsella, C. Pelachaud and H. Vilhjalmsson. 2008. The Next Step Towards a Functional Markup Language. In: Proc. of Intelligent Virtual Agents (IVA 2008). LNAI, 5208. Berlin, Heidelberg: Springer Verlag. pp. 270–280.

Huang, J. and C. Pelachaud. 2012. Expressive Body Animation Pipeline for Virtual Agent, 12th International Conference on Intelligent Virtual Agents, USA, 2012.

ISO-DIT, http://dit.uvt.nl (last verified 01/14/2013).

Jokinen, K. 2009. Constructive Dialogue Modelling—Speech Interaction and Rational Agents. Chichester, UK: John Wiley and Sons.

Jokinen, K. 2010. Pointing Gestures and Synchronous Communication Management. In Esposito, A., Campbell, N., Vogel, C., Hussain, A. and Nijholt, A. (eds.), Development of Multimodal Interfaces: Active Listening and Synchrony. Berlin Heidelberg: Springer Verlag Publishers, LNCS 5967, pp. 33–49.

Jokinen, K., H. Furukawa, M. Nishida and S. Yamamoto. 2013. Gaze and Turn-Taking Behavior in Casual Conversational Interactions. ACM Trans.

Interactive Intelligent Systems (TiiS) Journal, Special Section on Eye Gaze and Conversational Engagement, Guest editors: Elisabeth André and Joyce Chai. Volume 3, Issue 2.

Jokinen, K. and S. Scherer. 2012. Embodied Communicative Activity in Cooperative Conversational Interactions—Studies in Visual Interaction Management. *Acta Polytechnica Hungarica. Journal of Applied Sciences*, **9(1):**19–40.

Jongejan, B. 2012 Automatic Annotation of Face Velocity and Acceleration in Anvil. Proceedings of the Language Resources and Evaluation Conference (LREC-2012). Istanbul, Turkey.

Kendon, A. 2004. Gesture: Visible Action as Utterance. Cambridge: Cambridge University Press.

Kipp, M. 2001. Anvil—A Generic Annotation Tool for Multimodal Dialogue. Proceedings of Eurospeech, pp. 1367–1370.

Kopp, S., B. Krenn, S. Marsella, A. Marshall, C. Pelachaud, H. Pirker, K. Thorisson and H. Vilhjálmsson. 2006. Towards a Common Framework for Multimodal Generation: The Behavior Markup Language. Proceedings of the 6th International Conference on Intelligent Virtual Agents (IVA), Marina Del Rey, LNCS 4133, Springer Verlag, pp. 205–217.

Krenn, B., C. Pelachaud, H. Pirker and H. Peters. 2011. Markup Languages for ECAs. In Petta, P., Pelachaud, C. and Cowie, R. (eds.), Emotion-Oriented Systems: The Humaine Handbook. Springer-Verlag Berlin Heidelberg, pp. 389–415.

Krippendorff, K. (2004). Content Analysis: An Introduction to its Methodology, 2nd ed. Beverly Hills, CA: Sage Publications.

van Laban, R. and L. Ullmann. 1988. The Mastery of Movement. Plymouth: Northcote House.

Lee, J. and S. Marsela. 2006. Nonverbal Behavior Generator for Embodied Conversational Agents. Proceedings of the 6th International Conference on Intelligent Virtual Agents, pp. 243–255.

Levitski, A., J. Radun and K. Jokinen. 2012. Visual Interaction and Conversational Activity. In Proceedings of The 4th Workshop on Eye Gaze in Intelligent Human Machine Interaction: Eye Gaze and Multimodality, at the 14th ACM International Conference on Multimodal Interaction (ICMI-2012), October 26, 2012, Santa Monica, California, U.S.

Marsella, S. and J. Gratch. 2009. EMA: A Model of Emotional Dynamics. *Journal of Cognitive Systems Research*, **10(1):**70–90.

McNeill, D. 1992. Hand and Mind: What Gestures Reveal About Thought. Chicago: University of Chicago Press.

Mehrabian, A. 1996. Pleasure-Arousal-Dominance: A General Framework for Describing and Measuring Individual Differences in Temperament. *Current Psychology*, **14(4):**261–292.

Navarretta, C., E. Ahlsén, J. Allwood, K. Jokinen and P. Paggio. 2012. Feedback in Nordic First-Encounters: A Comparative Study. Proceedings of the Language Resources and Evaluation Conference (LREC-2012). Istanbul, Turkey.

Pelachaud, C. 2005. Multimodal expressive embodied conversational agents. Proceedings of the 13th annual ACM international conference on Multimedia. SESSION: Brave New Topics 2: Affective Multimodal Human-Computer Interaction. November 6–11, Singapore, pp. 683–689.

Poggi, I. 2007. Mind, Hands, Face and Body. A Goal and Belief View of Multimodal Communication. Berlin: Weidler Buchverlag.

Prendinger, H. and M. Ishizuka. 2004. Life-Like Characters. Tools, Affective Functions, and Applications, Cognitive Technologies Series, Springer Verlag, Berlin Heidelberg.

Rietveld, T. and R. van Hout. 1993. Statistical techniques for the study of language and language behaviour. Berlin and New York: Mouton de Gruyter.

Russell, J. 1980. A Circumplex Model of Affect. *Journal of Personality and Social Psychology*, **39(6):**1161–1178.

Scherer, K. 2000. Emotion. In Hewstone, M. and Stroebe, W. (eds.). Introduction to Social Psychology: A European perspective. Blackwell, pp. 151–191.

Schuller, B., S. Steidl, A. Batliner, F. Burkhardt, L. Devillers, C. Müller and S. Narayanan. 2013. Paralinguistics in Speech and Language—State-of-the-Art and the Challenge. *Computer Speech & Language*, **27(1):**4–39.

Schröder, M., P. Baggia, F. Burkhardt, C. Pelachaud, C. Peter and E. Zovato. 2011. EmotionML—An Upcoming Standard for Representing Emotions and Related States. Proceedings of Affective Computing and Intelligent Interaction. Memphis, TN, USA.

Settles, B. 2009. Active Learning Literature Survey. Department of Computer Sciences, University of Wisconsin-Madison.

Siegel, S. and N.J. jr Castellan. 1988. Nonparametric Statistics for the Behavioral Sciences, 2nd ed. New York: McGraw-Hill.

Sjölander, K. and J. Beskow. 2000. WaveSurfer—An Open Source Speech Tool. In Proceedings of ICSLP 2000 Conference, Beijing, China.

Spooren, W. 2004. On The Use of Discourse Data in Language Use Research. In Aertsen, H., Hannay, M. and Lyall, R. (eds.). Words in Their Places: A Festschrift for J. Lachlan Mackenzie, Amsterdam: Faculty of Arts, VU, pp. 381–393.

Thiel, C., F. Schwenker and G. Palm. 2005. Using Dempster-Shafer Theory in MCF Systems to Reject Samples. Proceedings of the 6th International Workshop on Multiple Classifier Systems (MCS'05). LNCS 3541, pp. 118–127. Springer Verlag.

Toivio, J. and K. Jokinen. 2012. A comparison of automatic and manual annotation techniques in first encounter dialogues. Proceedings of the 4th Nordic Symposium on Multimodal Communication, Gothenburgh.

Truong, K.P. and J. Trouvain. 2012. Laughter annotations in conversational speech corpora—possibilities and limitations for phonetic analysis. Proceedings of the 4th International Workshop on Corpora for Research on Emotion Sentiment and Social Signals, pp. 20–24. European Language Resources Association (ELRA).

Vilhjálmsson, H., N. Cantelmo, J. Cassell, N.E. Chafai, M. Kipp, S. Kopp, M. Mancini, S. Marsella, A.N. Marshall, C. Pelachaud, Z. Ruttkay, K.R. Thorisson, H. van Welbergen and R. van der Werf. 2007. The Behavior Markup Language: Recent Developments and Challenges, Intelligent Virtual Agents, IVA'07, Paris, September.

Wallbott, H.G. and K.R. Scherer. 1986. Cues and Channels in Emotion Recognition. *Journal of Personality and Social Psychology*, **51(4)**:690–699.

Walter, S., S. Scherer, M. Schels, M. Glodek, D. Hrabal, M. Schmidt, R. Böck, K. Limbrecht, H.C. Traue and F. Schwenker. 2011. Multimodal Emotion Classification in Naturalistic User Behavior. In Proceedings of the 14th International Conference on Human-Computer Interaction, Part III, Towards Mobile and Intelligent Interaction Environments. Lecture Notes in Computer Science 6763, pp. 603–611. Springer-Verlag, Berlin Heidelberg.

Welch, P. 1967. The Use of Fast Fourier Transform for the Estimation of Power Spectra: A Method Based on Time Averaging Over Short, Modified Periodograms. *IEEE Transactions Audio and Electroacoustics*, **15(2)**:70–73.

Zheng, F., G. Zhang and Z. Song. 2001. Comparison of Different Implementations of {MFCC}. *Journal of Computer Science and Technology*, **16(6)**:582–589.

CHAPTER 9

Co-speech Gesture Generation for Embodied Agents and its Effects on User Evaluation

Kirsten Bergmann

1. Introduction

When we are face to face with others, we use not only speech, but also a multitude of nonverbal behaviors to communicate with each other. A head nod expresses accordance with what someone else said before. A facial expression like a frown indicates doubts or misgivings about what one is hearing or seeing. A pointing gesture is used to refer to something. More complex movements or configurations of the hands depict the shape or size of an object. Of all these nonverbal behaviors, gestures, the spontaneous and meaningful hand motions that accompany speech, stand out as they are very closely linked to the semantic content of the speech they accompany, in both form and timing. Speech and gesture together comprise an utterance and externalize thought; they are believed to emerge from the same underlying cognitive representation and to be governed, at least in part, by the same cognitive processes (Kendon, 2004; McNeill, 2005).

Gestures are an integral part of human communication, as Goldin-Meadow (2003, p. 4) so aptly put it: "whenever there is talk, there is gesture". There is, actually, a growing body of evidence substantiating the significant role of gestures. The importance becomes apparent in the fact that gestures already develop early in our linguistic development (Goldin-Meadow and Butcher, 2003). Even congenitally blind people, who have never seen anybody gesturing, spontaneously produce gestures while talking (Iverson and Goldin-Meadow, 1998). The important role of gestures in communication is further supported by the fact that "to date there is no report of a culture that lacks co-speech gestures" (Kita, 2009, p. 146). The role of gestures is of such significance that speech-accompanying gestures do not disappear when visual contact between speaker and listener is absent, e.g. on the telephone (Cohen, 1977; Bavelas et al., 2008).

As the use of speech-accompanying iconic gestures is such a ubiquitous characteristic of human-human communication, it is therefore desirable to endow embodied agents with similar gestural expressiveness and flexibility to improve the interaction between humans and machines with regard to naturalness and intuitiveness. Thus, research is faced with two major problems: first, how to master the technical challenge to generate flexible conversational behavior automatically in embodied conversational agents and, second, how to ensure that the produced synthetic behavior improves the human-agent conversation valued by human users.

This chapter aims to cover these two issues. First, it will be addressed how gestural behavior can be automatically generated and put in combination with speech to be realized in embodied agents, and second, it will be explored if and how automatically generated communicative behavior for embodied agents might be beneficial for human-machine interaction. Section 2 covers a phenomenological view on communicative gestures and collects evidence for a number of factors modulating gesture use which should be considered in computational models. Section 3 provides an overview of existing computational simulation accounts and Section 4 presents the *Generation Network for Iconic Gestures* (GNetIc; Bergmann and Kopp, 2009a). Finally, Section 5 deals with the question how humans judge automatically generated gestural behavior.

2. What Shapes Iconic Gestures?—Individual and Common Patterns in Gesture Use

In Figure 1, examples are given from three speakers who are describing the same stimulus, a round church window. The speaker on the left-hand side uses a two-handed gesture in which the shape of the hands statically depicts the shape of the window. The gesture of the speaker in the middle is similar such that the hand is shaped in a way that bears a resemblance with the shape of the window. The difference is that the one in the middle is performed with only one hand. Finally, the speaker on the right-hand side depicts the window by drawing its shape in the air.

So the same stimulus is depicted in quite different ways, showing that there is no one-to-one mapping between a gesture and the object it depicts. McNeill and Duncan (2000) termed this phenomenon 'idiosyncrasy' implying that gestures are not held to standards of good form, but are rather created locally by speakers while speaking. This view receives support by Gullberg (1998, p. 51) who stated: "One of the most salient aspects of gesture is that people differ in their use of it". This becomes particularly obvious in gesture frequency: while some speakers rarely move their hands at all, other use gestures all the time (Krauss et al., 1996; Jacobs and Garnham, 2007). Marsh and Watt (1998) found that individuals vary in their preferences for particular gestural techniques of representation. Also, regarding handedness in the use of gestures, a right- or left-hand preference was reported, and there exist speaker-specific handshape preferences (Bergmann, 2012). Widely unexplained is, however, the question of how these differences among individuals arise. Although a number of correlates are possible, such as cultural background, personality traits, or cognitive skills, only very few studies have addressed this question so far. Regarding gesture frequency, Hostetter and Alibali (2007) found that individual differences in gesture rate are associated with the speakers' verbal and spatial skills. Individuals with low verbal fluency and high spatial visualization skill were found to gesture the most.

The fact that people are able to recover and interpret the meaning of iconic gestures, as shown by Cassell and Thórisson (1999), suggests that there is at least some systematicity in the way speakers encode meaning in gestures. The examples above clearly indicate that there is a certain degree of iconicity in the gestures since the circular shape of the window becomes apparent in all of them. There are, however, differences in how the physical form of the gestures corresponds to the object they depict: in the gestures of the first two speakers, the hand(s) adopt the shape of the window to be depicted, that is, there is a resemblance

Figure 1. Gestures from three different speakers all of which depicting the same object, a round church window.

between the hand configuration and the object shape. Likewise, in the third speaker's gesture, the round shape of the reference object is depicted by the movement of the speaker's drawing hand. That is, there obviously exist several 'techniques' to represent the same kind of meaning in gestures: whereas in the left hand and the middle examples, the hands adopt a static posture as a model for the circular shape, the same shape is depicted by a circular movement of the drawing hand in the right-hand example. For an adequate account of how meaning is transformed into gesture form, these representation techniques certainly have to be taken into account. Concrete mapping rules are more likely to be found within a set of gestures belonging to the same representation technique than across all instances of iconic gesture use. Empirical studies, however, reveal that similarity to the referent cannot fully account for all occurrences of iconic gesture use (Kopp et al., 2007). Rather, recent findings actually indicate that a gesture's form is also influenced by specific contextual constraints such as the discourse context (Holler and Stevens, 2007; Gerwing and Bavelas, 2004), the linguistic context (Kita and Özyürek, 2003; Gullberg, 2010; Bavelas et al., 2002), and gesture history (McNeill, 2005; Bergmann, 2012).

In sum, gesture use in humans is subject to both individual and common patterns of how meaning is mapped onto gesture form. Thus, with regard to computational modeling of gesture production, these complex influences of referent characteristics, contextual factors, and inter-individual differences should be taken into account. The following section reviews the state of the art in computational generation models and Section 4 presents one model in detail—the GNetIc model—that aims to cover the above-mentioned issues.

3. Generating Co-Speech Gestures for Embodied Agents

Different modeling approaches have been tested in an effort to translate systematic characteristics of coverbal gestures, shared among speakers, into generative models. In line with the two major concerns in gesture, simulation as suggested by the empirical literature can be broken down into two categories. First, *model-based* accounts that focus on how meaning is mapped onto physical gesture forms as well as on influences of contextual factors on gesture use. Second, *customized* attempts emphasizing inter-individual differences in communicative behavior trying to model individual gesture style. Finally, there is another trend in behavior generation, namely *data-based* attempts to simulate (individual) speakers' gesturing behavior based on corpora of gesture use.

3.1 Model-based gesture generation

The first systems investigating the challenge of iconic gesture generation were lexicon-based approaches. In general, these systems were characterized by a straightforward mapping of meaning onto gesture form. The *Behavior Expression Animation Toolkit* (BEAT) was among the first of a new generation of toolkits to allow the generation of synthetic speech alongside synchronized nonverbal behaviors, such as hand gestures and facial displays, to be realized with an animated human figure (Cassell et al., 2001). This approach of mapping text onto multimodal behavior was characterized by representing linguistic and social context and applying behavior generation rules based on empirical results. A similar approach was taken with the *Nonverbal Behavior Generator* (NVBG), proposed by Lee and Marsella (2006). The system analyzes the syntactic and semantic structure of surface texts and takes the affective state of the embodied agent into account to generate appropriate nonverbal behaviors. Based on a study from the literature and a video analysis of emotional dialogues, the authors developed a list of nonverbal behavior generation rules. The *Real Estate Agent* (REA) is a more elaborate system as it aims to model the bi-directional process of communication (Cassell, 2000). That is, in addition to the generation of nonverbal behaviors, the system also seeks to understand aspects of these same modalities' use by a human interlocutor. The focus of gesture generation in the REA system is the context-dependent coordination of (lexicalized) gestures with speech, accounting for the fact that gestures do not always carry the same meaning as speech.

Relying on empirical results, the systems mentioned so far focus on the context-dependent coordination of gestures with concurrent speech, whereby gestures are drawn from a lexicon. Flexibility and generative power of gestures to express new content, therefore, is obviously very limited. A different attempt that is closely related to the generation of speech-accompanying gestures in a spatial domain is Huenerfauth's (2008) system which translates English texts into American Sign Language (ASL) focusing on classifier predicates which are complex and descriptive types of ASL sentences. These classifier predicates have several similarities with iconic gestures accompanying speech. The system also relies on a library of prototypical templates for each type of classifier predicates in which missing parameters are filled in adaptation to the particular context.

The NUMACK system (Kopp et al., 2007) tries to overcome the limitations of lexicon-based gesture generation by considering patterns of human gesture composition. Based on empirical results, referent features are linked to morphological gesture features by an intermediate level of image description features. These explicate the imagistic content of iconic gestures, consisting of separable, qualitative features describing the meaningful geometric and spatial properties of entities.

3.2 Customized gesture generation

The research reviewed so far is devoted to build general models of gesture use, i.e. systematic inter-personal patterns of gesture use are incorporated exclusively. What is not considered in these systems is individual variation which is investigated by another line of research. Hartmann et al. (2006) identified six *expressivity parameters* of gesture quality as an intermediate level of behavior parameterization between holistic, qualitative communicative functions such as mood, personality, and emotion on the one hand, and low-level animation parameters like joint angles on the other hand. These parameters were applied to the gesture engine of the embodied agent GRETA, whose library of known prototype gestures are tagged for communicative function. Recently, Mancini and Pelachaud (2010) implemented the concept of expressivity parameters to create distinctive behavior patterns for embodied conversational agents. Their proposed algorithm generates nonverbal behavior for a given communicative intention and emotional state, driven by the agent's general behavior tendency ('Baseline') and modulated by dynamic factors such as emotional states, relation with interlocutor, physical constraints, social roles, etc.

Rehm et al. (2008) presented another variant that makes use of the gestural expressivity parameters to generate culture-specific gestures. Differences between the cultures were identified and integrated in a probabilistic model for generating agent behaviors. In a Bayesian network, the culture to be simulated is connected with dimensions of culture as a middle layer. These culture-specific dimensions were then connected with gesture parameters.

3.3 Data-based gesture generation

In another line of research, data-driven methods are employed to simulate individual speakers' gesturing behavior. Stone et al. (2004) recombine motion-captured pieces with new speech samples to recreate coherent multimodal utterances. Units of communicative performance are re-arranged while retaining temporal synchrony and communicative coordination that characterizes peoples' spontaneous delivery. The range of possible utterances is naturally limited to what can be assembled out of the pre-recorded behavior. Neff et al. (2008) aim at generating character-specific gesture style capturing the individual differences of human speakers. Based on statistical gesture profiles learned from annotated multimodal behavior, the system takes arbitrary texts as input and produces synchronized conversational gestures in the style of a particular speaker. The resulting gesture animations succeed in making an embodied character look more lively and natural and have empirically been shown to be consistent with a given performer's style. The approach does not need to account for the meaning-carrying functions of gestures, since the gestures focused on are discourse gestures and beats.

In summary, research on automatic gesture production has either emphasized general patterns in the formation of iconic gestures, or concentrated on individual gesturing patterns. In the next section, a modeling approach will be presented which goes beyond by accounting for both, systematic commonalities across speakers and idiosyncratic patterns of individual speakers.

4. The GNetIc Generation Approach for Iconic Gesture

A recent approach to generate iconic gesture forms from an underlying imagistic of contents is the *Generation Network for Iconic Gestures* (GNetIc; Bergmann and Kopp, 2009a). A formalism called Bayesian decision networks (BDNs)—also termed Influence Diagrams (Howard and Mattheson, 2005)—is employed that supplement standard Bayesian networks by decision nodes. This formalism provides a representation of a finite sequential decision problem, combining probabilistic and rule-based decision making. GNetIc is a feature-based account of gesture generation, i.e. gestures are represented in terms of characterizing features as their representation technique and physical form features. These make up the outcome variables in the model that divide into *chance variables* quantified by conditional probability distributions in dependence on other variables ('gesture occurrence' (G), 'representation technique' (RT), 'handedness' (H), and 'handshape' (HS)), and *decision variables* for which sufficient data is lacking or a sound theoretical account is available. These can be modeled by way of explicit rules in the respective decision nodes (for the gesture form features 'palm orientation' (PO), 'back-of-hand orientation' (BoH), 'movement type' (MT), and 'movement direction' (MD)). Factors which potentially contribute to either of these choices are considered as input variables. So far, three different factors have been incorporated into this model: discourse context, the previously performed gesture, and features of the referent. The probabilistic part of the network is learned from data of the SaGA corpus (Lücking et al., 2013) by applying machine learning techniques. The definition of appropriate rules in the decision nodes is based on theoretical considerations of the meaning-form relation via gestural representation techniques and a data-based empirical analysis of these techniques (for details see Bergmann, 2012). See Figure 2 for a schema of the generation network.

4.1 Generating gestures with GNetIc models

The generation network described above can be used directly for gesture formulation. To make it run in an autonomous embodied agent, however, a few pre- and post-processing additional steps are necessary to complete the mapping from representations of visuo-spatial information and discourse context to an adequate speech-accompanying iconic gesture. The gesture formulator has access to a structured blackboard since it is part of a greater speech and gesture generation architecture in which all modules operate concurrently and proactively on this blackboard. Details of this architecture are described in Bergmann and Kopp (2009). Information is accessed from that blackboard and results are written back to it. The initial situation for gesture formulation is a so-called *Imagistic Description Tree* (IDT; Sowa and Wachsmuth, 2009) representation (kind of a structured mental imagery) of the object to be referred to. In a pre-processing step, this representation is analyzed to extract all features that are

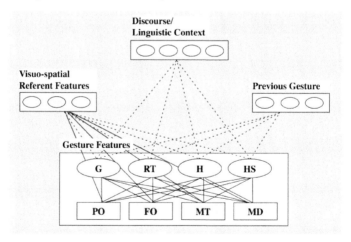

Figure 2. Schema of a gesture generation network in which gesture production choices are consideredeither probabilistically (chance nodes drawn as ovals) or rule based (decision nodes drawn as rectangles). Each choice is depending on a number of contextual variables. The links are either learned from speaker-specific corpus data (dotted lines) or defined in a set of if-then rules (solid lines).

required as initial evidence for the network: (1) whether an object can be decomposed into subparts, (2) whether it has any symmetrical axes, (3) its main axis, and (4) its position in the world. Further information drawn upon by the decision network concerns the discourse context. It is provided by other modules in the overall generation process and can be accessed directly from the blackboard. All evidence available is then propagated through the network resulting in a posterior distribution of probabilities for the values in each chance node. By applying a winner-takes-all rule to the posterior probability distribution we decide which value to fill in a feature matrix specifying the gesture morphology. Alternatively, maximum a-posteriori sampling could be applied at this stage of decision making to result in non-deterministic gesture specifications.

4.2 Modeling results

The previously described generation model has been embedded in a larger production architecture realized using a multi-agent system toolkit, a natural language sentence planner (SPUD; Stone et al., 2003), the Hugin toolkit for Bayesian inference (Madsen et al., 2005), and the ACE realization engine (Kopp and Wachsmuth, 2004). With this prototype implementation, an embodied agent is enabled to explain the same virtual reality buildings that were also used in the SaGA corpus study. Being equipped with proper knowledge sources, i.e. communicative plans, lexicon, grammar, propositional and imagistic knowledge about the world, the agent can randomly pick a landmark and a certain spatial perspective towards it, and then creates his explanations including speech and gestures autonomously. By simply switching between the GNetIc networks learned from different speakers, the embodied agent's gesturing can immediately be individualized differently. Accordingly, the gesturing behavior for a particular referent in a respective discourse context varies. Currently the system has five different individual networks (speakers P1, P5, P7, P8 and P15 from the SaGA corpus) as well as the average network learn from the combined data at its disposal.

In Figure 3, examples are given from five different simulations each of which based on exactly the same initial situation, i.e. all gestures are referring to the same referent (a round window of a church) and are generated in exactly the same discourse situation. The resulting nonverbal behavior varies significantly depending on the decision network underlying the simulation: For P7, no gesture is produced at all, whereas for P5 and P8, static posturing gestures are produced which, however, differ in their low-level morphology. For P5, the

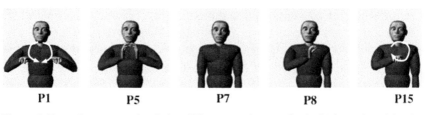

P1   P5   P7   P8   P15

Figure 3. Example gestures simulating different speakers, each of which produced for the same referent (a round window of a church) in the same initial situation.

hands adopt a C-shape while in the simulation for P8, an O-shape is used with the right hand only. P1 and P15 both use drawing gestures which, however, differentiate in their handedness.

4.3 Data-based evaluation

In the following, we will look at whether and how GNetIc models afford an adequate approximation of empirically observed behavior. For this purpose, the gestures determined by GNetIc were compared with those found in the SaGA corpus (Bergmann and Kopp, 2009a; Bergmann and Kopp, 2010). To this end, the data corpus the GNetIc models were built from (SaGA data from five different speakers: 473 noun phrases, 288 gestures) has been divided into training data (80%) and test data (20%). The training set was used for structure learning and parameter estimation of the decision networks. For each speaker's test set, it was tested whether the gestures generated with an aggregated decision network (learned from all of the five speakers) as well as the gestures generated with the individual decision networks (learned from only this speaker's data).

In total, individual networks achieved a mean accuracy of 62.4% (SD = 11.0), while the mean for the aggregated network was 57.8% (SD = 22.0). By trend, all individual networks performed better than networks learned from non-speaker specific data. Particularly for the case of P5, the accuracy of both, technique and handshape decisions, is remarkably better with the individual network than with the general one.

In contrast to chance nodes, which are individualized both in their connections to the input nodes and the local conditional probabilities, the decision nodes employ identical rules in each network. Yet, decisions are made in a particular order and decision nodes follow to operate upon previously determined chance node values. In consequence, a decision node may sometimes not match the test data values, because earlier nodes have already produced mismatching values. Therefore, the rule-based decision nodes were evaluated locally, i.e. taking the test case data for previous decisions as a basis. This way, the quality of decision-making rules was evaluated directly to avoid dealing with the above-mentioned problem. The mean accuracy for rule-based choices made in all network's decision nodes is 57.8% (SD = 15.5). Altogether, given the large potential variability for each of the variables, the results are quite satisfying. For example, the mean deviation of the predicted BoH orientation (the direction of the vector running along the back of hand) is 37.4 degrees, with the worst, opposite rating corresponding to a deviation of 180 degrees. Note also that we are using a very strict measure for accuracy as every decision made with GNetIc is assessed against what has been exactly annotated. Even minor variation (e.g. 'right' instead of 'right/up') leads to a mismatch classification. However, even gestures whose features are partly classified as mismatches may very well communicate adequate semantic features.

5. How Humans Judge Embodied Agents Using Co-Speech Gestures

There is increasing evidence that endowing embodied agents with human-like, nonverbal behavior may lead to enhancements of the likeability of the agent, trust in the agent, satisfaction with the interaction, naturalness of interaction, ease of use, and efficiency of task completion (Bickmore and Cassell, 2005; Heylen et al., 2002). With regard to effects of co-speech gestures, Krämer et al. (2003) found no effect on agent perception when comparing a gesturing agent with a non-gesturing one. The agent displaying gestures was perceived just as likable, competent, and relaxed as the agent that did not produce gestures. In contrast, Cassell and Thórisson (1999) reported that nonverbal behavior (including beat gestures) resulted in an increase of perceived language ability and life-likeness of the agent, as well as smoothness of interaction. A study by Rehm and André (2007) revealed that the perception of an agent's politeness depended on the graphical quality of the employed gestures. Moreover, Buisine and Martin (2007) found effects of different types of speech-gesture cooperation in agent's behavior. They found that redundant gestures increased ratings of explanation quality, expressiveness of the agent, likeability and a more positive perception of the agent's personality. In an evaluation of speaker-specific gesture style simulation, Neff et al. (2008) reported that the proportion of subjects who correctly recognized a speaker from generated gestures was significantly above chance. Recently, Niewiadomski et al. (2010) considered the role of verbal vs. nonverbal vs. multimodal emotional displays on warmth, competence and believability, whereby nonverbal behavior consisted of facial expressions accompanied by emotional gestures. It was found that all dependent measures (warmth, competence and believability) increased with the number of modalities used by the agent.

5.1 Effects of gesture generation with different GNetIc models

The GNetIc account has been evaluated with regard to whether the generated gestures can be beneficial for human-agent interaction (Bergmann et al., 2010). Two aspects were addressed in this study. First, the quality of the produced iconic gestures as rated by human users, and second, whether an agent's gesturing behavior could systematically alter a user's perception of the agent's warmth and competence as the two fundamental dimensions of social cognition. The warmth dimension captures whether we judge someone as are friendly and well-intentioned, and the competence dimension captures whether someone has the ability to deliver on those intentions. A number of studies have shown that warmth and competence assessments determine whether and how we intend to interact with others (Cuddy et al., 2011): We seek the company of people who are assumed to be warm and avoid those who appear less sociable (i.e. cold). With regard to competence, we prefer to cooperate with people we judge as competent, while incompetent people are disregarded.

To investigate both questions, the flexibility afforded by GNetIc to generate speech-accompanying gestures was exploited in a study comparing five different conditions: two individual conditions (ind-1 and ind-2 with GNetIc networks learned from the data of individual speakers: subject P5 in ind-1, subject P7 in ind-2), a combined condition with a network generated from the aggregated data of five different speakers (including P5 and P7), two control conditions (no gestures at all and simple random choices at the chance nodes in the network). Note that in all conditions, the gestures were produced from identical input values and accompanied identical verbal utterances. The random condition resulted still in appropriate iconic gestures, as the decision nodes still applied their sound mappings, but these gestures could occur at atypical positions in the sentence or employ random handedness.

In a between-subject design, 110 participants received a description of a church by the embodied agent Max, produced fully autonomously with a speech and gesture production architecture containing GNetIc. Immediately after receiving the descriptions, participants filled out a questionnaire in which two types of dependent variables had to be assessed. First, participants were asked for their evaluation of the presentation quality with respect to Max's verbal eloquence, gestures' helpfulness, the comprehensibility of Max's explanations, and the vividness of the agent's mental image of the church. Second, participants were asked to state their person perception of the embodied agent in terms of items like 'polite', 'authentic', or 'cooperative'. In the analysis, these items were grouped into scales for warmth and competence.

It turned out that both individual GNetIc conditions outperformed the other conditions in that gestures were perceived as more helpful, overall comprehension was rated higher, and the agent's mental image was judged as more vivid (see Figure 4). Similarly, the two individual GNetIc networks outperformed the control conditions regarding agent perception in terms of warmth and competence. The combined GNetIc condition was rated worse than the individual GNetIc conditions in all of these categories. Further, the no gesture condition was rated more positively than the random condition, in particular for the subjective measures of overall comprehension, the gesture's role for comprehension, and vividness of the agent's mental image. That is, with regard to these aspects it seems better to make no gestures than to randomly generate gestural behavior even though it is still reasonably iconic.

Overall, individualized gesturing was strikingly beneficial with regard to how embodied agents and their communicative skills are judged by human users, suggesting that modeling individual speakers with proper abilities for the target behavior results in even better

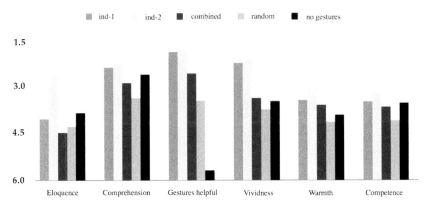

Figure 4. Participants' ratings of the embodied agent Max regarding the quality of the produced iconic gestures and the agent's personality.

behavior judged from the perspective of human interaction partners. This may be due to the fact that individual networks ensure a greater coherence of the produced behavior. As a consequence, the agent may appear more coherent and self-consistent which, in turn, may make its behavior more predictable and easier to interpret for the user.

5.2 Interaction effects of coverbal gestures and agent appearance over time

The previous section has clearly shown that co-speech gestures are able to play a beneficial role for how embodied agents are judged by human recipients. This finding is, however, still limited to one particular embodied agent and measurements were only taken at one particular moment of time. So, to elucidate the role of *gesture use* in embodied agents in relation to the variables *agent appearance* and *point of measurement*, a follow-up study has been conducted particularly addressing interaction effects of these variables with regard to the two major dimensions of social cognition: warmth and competence (Bergmann et al., 2012).

This question has been investigated in a study with repeated measures manipulating the agent behavior (gestures absent vs. gestures present; Bergmann et al., 2012). For the gesture-present condition, the GNetIc model built from speaker P5 of the SaGA corpus (the model to be most successful model according to the previous section) was employed again. This time, two different embodied agents were used: the child-like agent 'Billie' and the robot-like agent 'Vince' (see Figure 5). The study was conducted in two consecutive phases. In the first phase, participants were provided with a short introduction by the embodied agent which took approximately 15

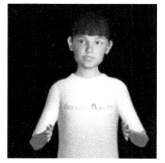

Figure 5. The two embodied agents employed in the study both showing a gesture for a square as it occurs in the stimuli participants were provided with: the robot-like agent 'Vince' (left) and the human-like agent 'Billie' (right).

seconds: "Hello, my name is Billie/Vince. In a moment you will have the possibility to getting to know me closer. But first you will be provided with a questionnaire concerning your first impression of myself". Subsequently, participants were asked to state their first impression regarding their perception of the embodied agent's personality in a questionnaire (T1, took approximately five minutes). In the second phase, the embodied agent described a building with six sentences which took approximately 45 seconds. Each sentence was followed by a pause of three seconds. Participants were instructed to carefully watch the presentation given by the embodied agent in order to be able to answer questions regarding content and subjective evaluation of the presentation afterwards. Immediately upon receiving the descriptions by the agent, participants filled out a second questionnaire (T2) stating their perception of the embodied agent's personality at this point of time.

It turned out that with regard to perceived *warmth*, the agents were overall perceived as warmer than at measuring point T2. In addition, there was a significant interaction effect of point of measurement and agent appearance: ratings of warmth decreased for the robot-like agent between the two points of measurement, while ratings remained constant for the human-like agent. In other words, participants rated the robot-like agent as being warmer than the human-like agent at measuring point T1, whereas ratings of warmth did not differ between robot-like and human-like agents at T2.

Regarding perceived *competence* of the embodied agents, there was a significant effect for the interaction of the point of measurement and agent behavior. While gesture use was found to result in an increase of perceived competence of the agents between the two points of measurement, ratings of agent competence slightly decreased when the agents did not use any gestures.

What can we learn from these findings for the design of (interactions with) virtual agents? First of all, the results clearly show that there is a second chance to make first impressions. In particular, with regard to competence, employing virtual agents with gestures helps to increase participants' ratings—independent of the agents' appearance. As a consequence, we can advise to endow virtual agents with gestural behavior to improve their perceived competence.

6. Conclusions

This chapter discussed the possibilities, approaches and effects of modeling co-speech gestures for embodied agents. The GNetIc model, described in detail, combines data-driven and rule-based decision making to account for both inter-individual differences in gesture use, as well as common patterns of how meaning is mapped onto gesture form. In addition, it accounts for the fact that the physical appearance of generated gestures is influenced by multiple factors such as characteristic features of the referent accounting for iconicity, contextual linguistic/discourse factors, or a speaker's previous gesture use. Evaluations show that the gestures producible with this model can reasonably approximate a certain range of human iconic gesturing and are acceptable to human users.

The results demonstrate that automatically generated gestural behavior is actually beneficial with regard to the impact of embodied agents on human addressees. Notably, different models were found to result in different behavior, with consistently differing perception and evaluation by human recipients. As a consequence, it seems reasonable to detect particularly appropriate speakers and to individualize an embodied agent's communicative behavior accordingly. This does, of course, raise the question concerning the characteristics of 'successful' gesture style. At this point, the potential of embodied agents comes to the fore as they provide the flexibility to turn on and off aspects of the behavior model to observe how human addressees respond. Individualization of communicative behavior, however, bears the danger of narrowing acceptance down to a certain population of users, since gesture perception, like production, may be subject to inter-individual differences. For instance, Martin et al. (2007) found the rating of gestural expressivity parameters to be influenced by a human addressee's personality traits. Accordingly, an important lesson to be learned, therefore, concerns the role of evaluation studies as an integral part of the communicative behavior modeling process. While prediction accuracy is highly prized in many evaluations of behavior simulation, the impact of how humans perceive an embodied agent's expressive behavior should always be a major criterion to help producing adequate behavior and increase the acceptance of the agent.

Acknowledgements

This research is partially supported by the Deutsche Forschungsgemeinschaft (DFG) in the Collaborative Research Center 673 "Alignment in Communication" and the Center of Excellence 277 in "Cognitive Interaction Technology" (CITEC).

REFERENCES

Bavelas, J., C. Kenwood, T. Johnson and B. Philips. 2002. An experimental study of when and how speakers use gestures to communicate. *Gesture*, **2(1)**:1–17.

Bavelas, J., J. Gerwing, C. Sutton and D. Prevost. 2008. Gesturing on the telephone: Independent effects of dialogue and visibility. *Journal of Memory and Language*, **58**:495–520.

Bergmann, K. 2012. The Production of Co-Speech Iconic Gestures: Empirical Study and Computational Simulation With Virtual Agents. Bielefeld, Germany: Bielefeld University.

Bergmann, K. and S. Kopp. 2009. Increasing expressiveness for virtual agents–Autonomous generation of speech and gesture for spatial description tasks. In Decker, K., Sichman, J., Sierra, C., and Castelfranchi, C. (eds.), Proceedings of the 8th International Conference on Autonomous Agents and Multiagent Systems (AAMAS 2009), pp. 361–368. Ann Arbor, MI: IFAAMAS.

Bergmann, K. and S. Kopp. 2009a. GNetIc—Using Bayesian decision networks for iconic gesture generation. In Ruttkay, Z., Kipp, M., Nijholt, A., and Vilhjalmsson, H. (eds.), Proceedings of the 9th International Conference on Intelligent Virtual Agents, pp. 76–89. Springer Verlag, Berlin/Heidelberg.

Bergmann, K. and S. Kopp. 2010. Modelling the production of co-verbal iconic gestures by learning Bayesian Decision Networks. *Applied Artificial Intelligence*, **24**:530–551.

Bergmann, K., F. Eyssel and S. Kopp. 2012. A second chance to make a first impression? Hoe appearance and nonverbal behavior affect perceived warmth and competence of virtual humans over time. In Walker, M., Neff, M., Paiva., A., and Nakano, Y.I. (eds.), Proceedings of the 12th Conference on Intelligent Virtual Agents. Springer Verlag, Berlin/Heidelberg.

Bergmann, K., S. Kopp and F. Eyssel. 2010. Individualized gesturing outperforms average gesturing—evaluating gesture production in virtual humans. In Allbeck, J., Badler, N., Bickmore, T., Pelachaud, C., and Safonova, A. (eds.), Proceedings of the 10th Conference on Intelligent Virtual Agents. pp. 104–117. Springer Verlag, Berlin/Heidelberg.

Bickmore, T. and J. Cassell. 2005. Social dialogue with embodied conversational agents. In van Kuppevelt, J., Dybkjaer, L., and Bernsen, N. (eds.), Advances in Natural, Multimodal Dialogue Systems. Kluwer Academic Publishers, New York.

Buisine, S. and J.C. Martin. 2007. The effects of speech-gesture cooperation in animated agents' behavior in multimedia presentations. *Interacting with Computers*, **19**:484–493.

Cassell, J. 2000. More than just another pretty face: Embodied conversational interface agents. *Communications of the ACM*, **43**:70–78.

Cassell, J. and K. Thórisson. 1999. The power of a nod and a glance: Envelope vs. emotional feedback in animated conversational agents. *Applied Artificial Intelligence*, **13**:519–538.

Cassell, J., H. Vilhjálmsson and T. Bickmore. 2001. BEAT: The behavior expression animation toolkit. In Proceedings of SIGGRAPH '01, pp. 477–486, New York, NY.

Cohen, A. 1977. The communicative functions of hand illustrators. *Journal of Communication,* **27**:54–63.

Cuddy, A.J., P. Glick and A. Beninger. 2011. The dynamics of warmth and competence judgments, and their outcomes in organizations. *Research in Organizational Behavior,* **31**:73–98.

Gerwing, J. and J. Bavelas. 2004. Linguistic influences on gesture's form. *Gesture,* **4**:157–195.

Goldin-Meadow, S. 2003. Hearing Gesture—How Our Hands Help Us Think. Harvard University Press, Cambridge, MA.

Goldin-Meadow, S. and C. Butcher. 2003. Pointing toward two-word speech in young children. In Kita, S., (ed.), *Pointing*, pp. 85–107. Lawrence Erlbaum Association, Mahwah, NJ.

Gullberg, M. 1998. Gesture as a Communication Strategy in Second Language Discourse: A Study of Learners of French and Swedish. Lund University Press, Lund.

Gullberg, M. 2010. Language-specific encoding of placement events in gestures. In Bohnemeyer, J. and Pederson, E. (eds.), Event Representation in Language and Cognition. Cambridge University Press, Cambridge, UK.

Hartmann, B., M. Mancini and C. Pelachaud. 2006. Implementing expressive gesture synthesis for embodied conversational agents. In Gibet, S., Courty, N., and Kamp, J.-F. (eds.), Gesture in Human-Computer Interaction and Simulation, pp. 45–55. Springer Verlag, Berlin/Heidelberg.

Heylen, D., I. van Es, A. Nijholt and B. van Dijk. 2002. Experimenting with the gaze of a conversational agent. In Proceedings International CLASS Workshop on Natural, Intelligent and Effective Interaction in Multimodal Dialogue Systems, pp. 93–100. Copenhagen, Denmark.

Holler, J. and R. Stevens. 2007. An experimental investigation into the effect of common ground on how speakers use gesture and speech to represent size information in referential communication. *Journal of Language and Social Psychology,* **26**:4–27.

Hostetter, A. and M. Alibali. 2007. Raise your hand if you're spatial—relations between verbal and spatial skills and gesture production. *Gesture,* **7**:73–95.

Howard, R. and J. Matheson. 2005. Influence diagrams. *Decision Analysis,* **2**:127–143.

Huenerfauth, M. 2008. Spatial, temporal and semantic models for American Sign Language generation: Implications for gesture generation. *International Journal of Semantic Computing,* **2(1)**:21–45.

Iverson, J. and S. Goldin-Meadow. 1998. Why people gesture when they speak. *Nature,* **396**:228.

Jacobs, N. and A. Garnham. 2007. The role of conversational hand gestures in a narrative task. *Journal of Memory and Language,* **56**:291–303.

Kendon, A. 2004. Gesture—Visible Action as Utterance. Cambridge, UK: Cambridge University Press.

Kita, S. 2009. Cross-cultural variation of speech-accompanying gesture: A review. *Language and Cognitive Processes*, **24(2):**145–167.

Kita, S. and A. Özyürek. 2003. What does cross-linguistic variation in semantic coordination of speech and gesture reveal? Evidence for an interface representation of spatial thinking and speaking. *Journal of Memory and Language*, **48:**16–32.

Kopp, S. and I. Wachsmuth. 2004. Synthesizing multimodal utterances for conversational agents. *Computer Animation and Virtual Worlds*, **15:**39–52.

Kopp, S., P. Tepper, K. Ferriman, K. Striegnitz and J. Cassell. 2007. Trading spaces: How humans and humanoids use speech and gesture to give directions. In Nishida, T. (ed.), *Conversational Informatics*, pp. 133–160. John Wiley, New York.

Krämer, N., B. Tietz and G. Bente. 2003. Effects of embodied interface agents and their gestural activity. In Rist, T., Aylett, R., Ballin, D., and Rickel, J. (eds.), Proceedings of the 4th International Workshop on Intelligent Virtual Agents, pp. 292–300. Springer Verlag, Berlin/Heidelberg.

Krauss, R., Y. Chen and P. Chawla. 1996. Nonverbal behavior and nonverbal communication: What do conversational hand gestures tell us? *Advances in Experimental Social Psychology*, **28:**389–450.

Lee, J. and S. Marsella. 2006. Nonverbal behavior generator for embodied conversational agents. In Gratch, J., Young, M., Aylett, R., Ballin, D., and Olivier, P. (eds.), Proceedings of the 6th International Conference on Intelligent Virtual Agents, pp. 243–255, Springer Verlag, Berlin/Heidelberg.

Lücking, A., K. Bergmann, F. Hahn, S. Kopp and H. Rieser. 2013. Data-based analysis of speech and gesture: the Bielefeld Speech and Gesture Alignment corpus (SaGA) and its applications. *Journal on Multimodal User Interfaces*, **7(1-2):**5–18.

Madsen, A., F. Jensen, U. Kjærulff and M. Lang. 2005. HUGIN—The tool for Bayesian networks and influence diagrams. *International Journal of Artificial Intelligence Tools*, **14:**507–543.

Mancini, M. and C. Pelachaud. 2010. Generating distinctive behavior for embodied conversational agents. *Journal of Multimodal User Interfaces*, **3:**249–261.

Marsh, T. and A. Watt. 1998. Shape your imagination: Iconic gestural-based interaction. In Proceedings of the IEEE Virtual Reality Annual International Symposium. Los Alamitos, CA, USA: IEEE Computer Society.

Martin, J.-C., S. Abrilian and L. Devillers. 2007. Individual differences in the perception of spontaneous gesture expressivity. Integrating Gestures, 71 p.

McNeill, D. 2005. Gesture and Thought. University of Chicago Press, Chicago, IL.

McNeill, D. and S. Duncan. 2000. Growth points in thinking-for-speaking. In McNeill, D. (ed.), Language and Gesture, pp. 141–161. Cambridge University Press, Cambridge, UK.

Neff, M., M. Kipp, I. Albrecht and H.-P. Seidel. 2008. Gesture modeling and animation based on a probabilistic re-creation of speaker style. *ACM Transactions on Graphics*, **27(1)**:1–24.

Niewiadomski, R., V. Demeure and C. Pelachaud. 2010. Warmth, competence, believability and virtual agent. In Allbeck, J., Badler, N., Bickmore, T., Pelachaud, C., and Safonova, A. (eds.), Proceedings of the 10th Conference on Intelligent Virtual Agents, pp. 272–285. Springer Verlag, Berlin/Heidelberg.

Rehm, M. and E. André. 2007. Informing the design of agents by corpus analysis. In Nishida, T. and Nakano, Y. (eds.), Conversational Informatics. John Wiley and Sons, Chichester, UK.

Rehm, M., Y. Nakano, E.A.T. Nishida, N. Bee, B. Endrass, M. Wissner, A. Lipi and H.-H. Huang. 2008. From observation to simulation: Generating culture-specific behavior for interactive systems. *AI & Society*, **24**:267–280.

Sowa, T. and I. Wachsmuth. 2009. A computational model for the representation and processing of shape in coverbal iconic gestures. In Coventry, K., Tenbrink, T., and Bateman, J. (eds.), Spatial Language and Dialogue, pp. 132–146. Oxford University Press, Oxford, UK.

Stone, M., C. Doran, B. Webber, T. Bleam and M. Palmer. 2003. Microplanning with communicative intentions: The SPUD system. *Computational Intelligence*, **19**:311–381.

Stone, M., D. DeCarlo, I. Oh, C. Rodriguez, A. Stere, A. Lees and C. Bregler. 2004. Speaking with hands: Creating animated conversational characters from recordings of human performance. ACM Transactions on Graphics—Proceedings of ACM SIGGRAPH 2004, **23(3)**:506–513.

CHAPTER 10

A Survey of Listener Behavior and Listener Models for Embodied Conversational Agents

Elisabetta Bevacqua

1. Introduction

One of the major issues that must be pursued when developing Embodied Conversational Agents (ECAs) is to create virtual humanoids able to sustain natural and satisfying interactions with users. For this, ECAs must be endowed with human-like capabilities not only while speaking but also while listening. In human-human communications, interlocutors show their participation in the dialogue, to push it forward and make the speaker go on. Similarly, in user-ECA interactions, agents must not freeze while the user is speaking since the absence of the appropriate behavior would deteriorate the quality of the interaction. In this chapter, the behavior of listeners is investigated as well as the efforts made by researchers on the one hand to gain insight into the signals performed by interlocutors and on the other hand to develop computational models to generate the listener behavior for ECAs while interacting with a user.

Communication takes place when a response is elicited (Gumperz, 1982). Therefore, human-human communication is not a one-way phenomenon, and participants do not alternate their roles like in a volley match, just uttering sentences while speaking and freezing while listening. Communication is much more like a dance, in which people move together, sometime they lead and sometime they follow, waltzing toward a common goal: exchange information, coordinate actions, display emotional state, establish a relationship and fulfill their needs (Castelfranchi and Parisi, 1980; Poggi, 2007). So, like the speaker, listeners play an important role in the conversation, and they have to actively participate in the interaction in order to push it forward and make the speaker go on. In fact, whenever people listen to somebody, they do not assimilate passively all her/his words, but they actively participate in the interaction providing information about how they feel and what they think of the speaker's speech. Listeners emit expressions like "a-ah", "mmh" or "yes" to assure *grounding*[1], nod their head in agreement, smile to show liking, frown if they do not understand and shake their head to make clear that they refuse the speaker's words. In short, while listening, people assume an active role in the interaction showing above all that they are attending the exchange of communication. According to the listener's behavior, the speaker can estimate how her/his interlocutor is reacting and can decide how to carry on the interaction: for example, by interrupting the conversation if the listener is not interested or rephrasing a sentence if the listener showed signs of incomprehension.

Due to the undeniable importance of the behavior showed by listeners, ECAs must be able to generate this type of behavior in order to interact successfully with a user. To pursue such a goal, the listener's behavior must be investigated thoroughly as well as the signals performed by interlocutors. This chapter aims to provide such a proper survey, reporting also the progress made by researchers in the human-machine interface field to create ECAs able to show an appropriate listening behavior. The chapter is structured in three parts. In the first one, we investigate the behavior shown by listeners during an interaction. The second part is devoted to studies that use virtual agents to determine listener signals. Finally, in the third part, we focus on virtual agent systems able to generate the listener behavior while the agent interacts with a user.

2. Listener Behavior

Many studies have been conducted on the behavioral signals displayed by the listener during conversations in order to understand and define

---

[1] Grounding: the process by which partners in a dialogue gain common ground, that is mutual belief and shared conception (Clark and Schaefer, 1989).

the typical dynamics that people apply. These studies have been conducted by Yngve (1970), Allwood et al. (1993), Poggi (2007), Welji and Duncan (2004), Clark and Schaefer (1989), Schegloff and Sacks (1982) and many more. One of the first studies about the expressive behaviors shown by people engaged in an interaction has been presented by Yngve (1970). His work focused mainly on those signals used to manage turn-change, both by the speaker and the listener. To describe this type of signals, Yngve introduced the term "backchannel". In this conception, backchannels are seen as non-intrusive acoustic and non-verbal signals provided during the speaker's turn. Backchannels are used to set the common ground without bringing much new information.

Allwood et al. (1993) extended Yngve's theory. They chose the term of *feedback* (originally introduced by Wiener (1948)) to encompass a wider set of verbal and non-verbal signals that the listener can provide. Feedbacks include not only non-intrusive acoustic and gestural signals, but also single words or short phrases like "I see", "Oh, that's great!" that can need a full speaking turn to be expressed. In particular, Allwood et al. analyzed the linguistic feedbacks from a semantic and pragmatic point of view, considering them as mechanisms which allow interlocutors to exchange information about four basic communicative functions (Allwood et al., 1993):

- *Contact*: the interlocutor wants to and can continue the interaction,
- *Perception*: the interlocutor is willing and able to perceive the message,
- *Understanding*: the interlocutor is willing and able to understand the message,
- *Attitude*: the interlocutor wants to and can show her/his attitudinal reactions to the communicated content, that is if she/he believes or disbelieves, likes or not, accepts or refuses what the speaker is saying.

Within their studies, Allwood and colleagues showed that feedback behavior can convey more than one communicative function at a time, for example an interlocutor can show, through a feedback signal, that she/he is both understanding and agreeing.

Through feedback signals, a listener can transmit both positive and negative information (Poggi, 2007). A speaker does not need to know only when the listener can perceive her/his words or understand the content of her/his speech, but she/he needs also to know when the perception decreases and when what she/he is saying is not understandable any more. These negative signals help her/him to apply strategies to make the conversation successful again. Like Allwood et al., Poggi sees feedback as a means used by listeners to provide information about some conditions of the communicative process (Poggi, 2007): the functions of attention, comprehension, believability, interest and agreement. In addition, she defined feedback signals as a mechanism used by interlocutors to fulfill the other party's "control goals" during an interaction (Castelfranchi and Parisi, 1980; Poggi, 2007). While talking, people aim to obtain or give information, have the interlocutors do something, in short they have some "central goals" they want to see fulfilled. But, at the same moment, people have also some side goals, called "control goals", that, even if not explicitly mentioned, are quite relevant: the goals of knowing if their interlocutor is listening, understanding, agreeing to do what they ask for, and so on. In this idea, feedback signals provide information about the fulfillment of these control goals, showing the interlocutor's level of comprehension and personal reaction.

2.1 Verbal and non-verbal backchannels

Communication between humans takes place through several modalities. Voice, head movements, facial expressions, gaze shifts, gestures and changes in posture can be performed to convey a communicative function both while speaking and listening. A backchannel can be constituted by either a single verbal or non-verbal signal or it can be a combination of several signals on different modalities. In this chapter, we focus on backchannels performed on the face, head, gaze and voice modalities.

2.1.1 Verbal and acoustic signals

Backchannels that received a great deal of attention are the listener's linguistic signals that include both verbal and vocal signals. For verbal signals, we mean single words like "yeah", "ok", "no" or short sentences as "I see", while vocal signals, also called paraverbals, consist of brief sounds like "mm mh", "aha", "uhhh". Allwood et al. studied them from a semantic point of view in Allwood et al. (1993) and, besides defining the communicative functions they can convey, they discovered that the meaning of a linguistic backchannel depends strongly on the polarity of the previous utterance, for example a negative answer that follows a negative phrase shows acceptance, whereas a negative answer after a positive utterance shows disagreement. Poggi and Gardner noticed that the canonical use and value of backchannels like "yeah", "mm hm" and "mm" change by varying the intonation of the voice (Poggi, 1981; Gardner, 1998). Campbell (2007) studied the importance of non-verbal speech utterances emitted both by the speaker and by the listener during an interaction. He found out that components, like laughs, grunts or sound as "umms" and "ahhs", can be easily recognized and that they help to identify the status of each member of the conversation. This constant stream allows participants to determine the progress of the interaction. Especially, it helps the speaker to be informed of the successfulness of the communication and, consequently, to transmit the content more efficiently. For example, laughs, grunts, etc. can indicate the degree of attentiveness of the listener. Campbell's work shows how this information can be automatically annotated and can provide an estimation of the participants' status. Moreover, the prosodic variation of these utterances can be quite relevant. This variation may indicate the level of the relationship between the speaker and the listener (Campbell, 2006).

2.1.2 Head movements

Head movements are the communicative signals most employed by the listeners to provide backchannels, in particular head nods and shakes (Rosenfeld, 1978). In fact, during conversation participants use their head a lot; Heylen (2005) studied several communicative functions transmitted by participants through head movements. In particular, listeners perform head signals to provide continuous backchannels to the speaker. They respond almost instantly to speaker's non-verbal request for backchannel in the form of head movements. Head nod and head shake are produced to silently say "yes" and "no"; that is why they are called symbolic gestures. Allwood and Cerrato (2003) and Cerrato and Skhiri (2003) studied how people use specific head movements as backchannel signals, in particular when they co-occur with vocal expressions. Among all possible movements they decided to focus on nods, jerks (a sudden abrupt motion), shakes, waggles (a back and forth, left to right movement) and side-way turns (a single turn on the left or on the right). They saw, for example, that head nods and jerks produced with short facial expressions showed continuation of contact, perception and understanding of the message. Shakes were performed with short negative answers to show disagreement and waggles where often accompanied by statements of doubt and hesitation. While Cerrato showed that listener's head movements are synchronized with her/his vocal expressions, Hadar and colleagues (Hadar et al., 1985) showed that about a quarter of the head movements performed by listeners are also in synchrony with the speaker's speech.

2.1.3 Facial expressions

The face is an important channel of non-verbal communication. During interaction, both while speaking and listening, facial expressions are used to accomplish several fundamental tasks. They can help to regulate the flow of the conversation (Duncan, 1972), for example by smiling to the speaker, the listener can encourage her/him to go on in the conversation (Poggi and Pelachaud, 2000). Facial expressions have been found to convey one's opinion about events, people or objects (Chovil, 1992), through this kind of facial displays, called *personal reactions*, a person shows what she feels or thinks, in particular a listener can show disgust and disliking by wrinkling her/his nose, or liking and approval by smiling and raising her/his eyebrows. Facial expressions provide both consciously and non-consciously information about one's moods and emotional states (Argyle, 1988; Ekman and Friesen, 1969). Not displaying the correct facial expression at the right moment can be a source of misunderstanding and convey the wrong message (Poggi and Pelachaud, 2000). From the listener's point of view, through facial expressions, a person can show, in the first place, if she is following the conversation or not. Lowering the eyelids and yawning could be a clear sign of boredom and lack of attention, while keeping the eyes wide open and raising the eyebrows show that the listener is keeping up with the conversation and that she/he is interested. One of the most frequent and studied listener's facial displays is the smile. Brunner (1979) conducted several tests to understand the role of smiles produced by a listener during conversation. He found out that placement of smiles in conversations is very similar to that of other typical backchannel signals like paraverbals and head nods, so smiles can be considered as backchannels.

2.1.4 Gaze behavior

Gaze behavior in human-human interaction is one of the most studied aspects of non-verbal activity during conversation. It has been found to serve many functions as, for example, helping to regulate the flow of conversation, ensuring smooth turn-taking behavior and defining (through duration, direction and type of gaze) the relationship between participants (Kendon, 1967). In listener's behavior, gaze fulfills a quite important role. At a perception level, turning the gaze toward the speaker allows the listener to see better and so to perceive visual input better; such perception is even more optimized if also the head turns in the direction of the speaker, enhancing the eyes' view and allowing the listener to hear better (Heylen, 2005). So, gaze also becomes the principal mean used to show attention and interest. Directing gaze toward the speaker is usually associated with focus of attention (Heylen, 2005; Kendon, 1967; Argyle and Cook, 1976). Turning even the head and the body in the direction of the speaker signals an even higher level of interest and attention (Peters, 2000). To make a conversation successful, creating a bond with one's interlocutor is quite relevant. Building this type of relationship is linked to the notion of *engagement*. According to Conte and Castelfranchi (1995) and Poggi (2007), engagement is defined as "the value that a participant in an interaction attributes to the goal of being together with the other participant(s) and of continuing the interaction". Another definition of engagement is provided by Sidner et al. (2004); they defined the engagement as "the process by which two (or more) participants establish, maintain and end their perceived connection during interactions they jointly undertake". In terms of engagement, gaze behavior represents a useful way in which the listener may let the speaker know her/his level of intention to maintain engagement without the need to be explicit and interrupt the flow of conversation. An example of gaze behavior that communicates a high level of engagement is "mutual gaze", that is looking at each other or looking at a common object (Sidner et al., 2004).

3. Evaluating Backchannel Signals

The previous section has provided an overview on the behavioral signals that people can display while listening to a speaker. A similar behavior must be generated for an ECA when it interacts with a user. In fact, providing the agent with the capability of displaying backchannel signals, that users can interpret, will make the agent more realistic and human-like. The user will feel more at ease since the ECA shows the appropriate level of understanding and actively participates in the conversation. Although a lot has been written about human behavior expressed while listening, the existing literature does not cover all possible signals and all the meanings that these signals can convey, particularly when several different signals are displayed at the same time. For such a reason, several perceptual studies have been performed. We will describe some of them in detail in the following two sections. The main aim of these studies was two-fold: on the one hand, they wanted to investigate the relation between the form of a signal (or combination of signals) and the possible transmitted meanings and on the other hand, they aimed to understand how people interpret backchannels performed by virtual agents. Several tests to study the listener's behavior have already been performed before (Gratch et al., 2002; Cassell and Thórisson, 1999; Granström et al., 2002); however, nobody tried to associate one or more meanings to backchannel signals. Moreover, most of the previous experiments consisted in evaluating the listener's signals during human-human interactions rather than in human-agent interaction.

3.1 Non-verbal signals

In two perceptual studies (Bevacqua et al., 2007; Heylen et al., 2007), Bevacqua, Heylen and colleagues asked people to watch a set of backchannel signals performed by a virtual agent and to assign a meaning to each of them. The meanings were chosen among the communicative functions suggested in the literature (Allwood et al., 1993; Poggi, 2003). All backchannel signals were displayed *context-free*, that is, the discursive context of the speaker's speech was not defined. The first experiment presented in Bevacqua et al. (2007) served as a pilot test to understand how users perceived backchannel signals displayed by a virtual agent. Were participants able to assign a meaning to these signals? Could they easily distinguish between positive and negative signals? Did backchannels shown by an ECA allow polysemy and synonymy? For polysemy and synonymy, the authors meant that the same signal could have different meanings and a single meaning could be expressed with different signals or combination of signals.

Fourteen video clips were created using signals chosen among those proposed in the literature (Allwood and Cerrato, 2003; Poggi, 2007), for example *head nod, head nod+smile, head nod+raise of the eyebrows*. Some signals were simple, containing just a single action (like *head nod* or *head shake*), while others were obtained by combining several actions (like *head shake+frown+tension of the lips*). To display the signals, the 3D agent Greta (Niewiadomski et al., 2009) was used. Twelve French participants took part in the evaluation. They watched one video at a time and then selected for each of them none, one or more meanings among the 12 proposed: agreement, disagreement, acceptance, refusal, interest, not interest, belief, disbelief, understanding, not understanding, liking, not liking.

To analyze the data, the authors coded the answers given by the subjects as positive or negative, according to the following principles: agreement, acceptance, liking, belief, understanding and interest were considered as positive answers and disagreement, refusal, not liking, not understanding and not interest were considered as negative answers. Results confirmed that participants were able to assign a meaning to backchannel signals even when displayed by a virtual agent. Most backchannel signals conveyed either a positive or a negative connotation. All signals containing the actions head nod and smile were associated to positive meanings, whereas negative meanings were mostly attributed to signals containing head shake and frown. Only three signals (*raise of the left eyebrow*, *head tilt on the right+raise eyebrows* and *eyes wide open*) were hard to interpret for the users in terms of positive and negative. The authors supposed that some signals are particularly hard to interpret when showed context-free since their meaning can strongly depend on the context and to the listener's personality. With regard to the second hypothesis, backchannel signals polysemy and synonymy, results confirmed that participants associated different meanings to the same signals and that they attributed the same meaning to different signals or a combination of signals. For example, *head nod* can convey the meanings of agreement, acceptance and understanding. The meaning of refusal can be associated to the signals *head shake, head shake+frown, frown+tension of the lips, head shake+frown+tension of the lips*.

In order to generalize these findings, the pilot experiment needed to be performed with more subjects. Moreover, as the authors have tested combinations of signals, they decided to prepare a second version of the experiment in order to assess the meaning of each single action. In particular, the authors aimed to get a better understanding about how different actions contribute to the interpretation of a facial expression. A first question to explore was: is it possible to identify a signal (or a combination of signals) for each meaning? For example, is there a signal more relevant than others for a specific meaning or can a single meaning be expressed through different signals or a combination of signals? The researchers hypothesized that for each meaning they could find a prototypical signal. A second question was: does a combination of signals alter the meaning of backchannel single signals? The hypothesis was that in some cases, adding a signal to another could significantly change the perceived meaning. In that case, the independent variable was the combination of signals and the dependent variable was the meaning attributed to each signal by the subjects. Sixty French subjects were involved in that experiment. The 3D agent Greta displayed 21 video clips (see Table 1) showing signals chosen among those proposed by Allwood and Cerrato (2003) and Poggi (2007). After each video clip and before moving on, participants could select one meaning according to their opinion about which meaning fitted that particular backchannel signal best.

The same list of meanings proposed in the pilot test was used in this evaluation too. It was possible to select several meanings for one signal and when none of the meanings seems to fit, participants could

Table 1.   Backchannel signals performed by the 3D agent Greta. The action *tension* means tension of the lips.

1. nod	8. raise eyebrows	15. nod and raise eyebrows
2. smile	9. shake and frown	16. shake, frown and tension
3. shake	10. tilt and frown	17. tilt and raise eyebrows
4. frown	11. sad eyebrows	18. tilt and gaze right down
5. tension	12. frown and tension	19. eyes wide open
6. tilt	13. gaze right down	20. raise left eyebrows
7. nod and smile	14. eyes roll up	21. tilt and sad eyebrows

just select either "I don't know" or "none" (if they thought that there was a meaning but different from the ones proposed). Through this test, prototypical signals for most of our meanings were determined.

For the positive meanings, agreement is meant by a nod, as well as acceptance. To mean liking a *smile* appears as the most appropriate signal. A *nod* associated to a *raise of the eyebrows* seems to convey understanding; however, only 17 subjects out of 30 thought so. As for interest and belief, no significant results were found. A combination of *smile+raise eyebrows* could be a possibility for interest. For the negative meanings, disagreement and refusal are meant by a *shake*, whereas not liking is represented by *frown* and *tension of the lips*. A *tilt+frown* as well as a *raise of the left eyebrow* mean disbelief for most of the subjects. The best signal to mean not understanding seems to be a *frown*. And *tilt* and *gaze right down* as well as *eyes roll up* are more relevant for the meaning not interest. It also appeared that a combination of signals could significantly alter the perceived meaning. For instance, *tension* alone and *frown* alone do not mean not liking, but the combination *frown+tension* does. The combination *tilt+frown* means disbelief, whereas *tilt* alone and *frown* alone do not convey this meaning. *Tilt* alone and *gaze right down* alone do not mean not interest as significantly as the combination *tilt+gaze*. Conversely the signal *frown* means not understanding but when the signal *shake* is added, *frown+shake* significantly loses this meaning. The resulting set of interpretable signals has been used to define a library of prototypical backchannel signals, called *backchannel lexicon*. Such a library of behaviors has been implemented for the virtual agent Greta and it has been used to determine the appropriate animation the agent should display while listening. For example, if the agent intends to communicate its agreement toward what the user is saying, the system will find in the backchannel lexicon the appropriate behavior to display, for instance a head nod.

3.2 Multimodal signals

As described in Section 2.1.1, backchannels are provided not only through the visual modality, but also through voice by uttering paraverbals, words or short sentences (Gardner, 1998; Allwood et al., 1993). For such a reason to create credible virtual agents, this type of signals must be taken into account. Bevacqua et al. (2010) proposed to improve user-agent interaction by introducing multimodal signals in the backchannels performed by the ECA Greta. Moreover, they presented a perceptual study with the aim of getting a better understanding about how multimodal backchannels are interpreted by users. Like in their previous studies (Bevacqua et al., 2007; Heylen et al., 2007), video clips of a virtual agent performing context-free multimodal backchannel signals were shown. The participants were asked to assign none, one or several meanings to each signal. Again, the meanings proposed were: agreement, disagreement, acceptance, refusal, interest, not interest, belief, disbelief, understanding, not understanding, liking, not liking.

To create videos, seven visual cues (raise eyebrows, nod, smile, frown, raise left eyebrow, shake and tilt+frown) and eight acoustic cues (seven vocalizations plus silence: ok, ooh, gosh, really, yeah, no, m-mh and (silence)) were selected. The visual cues were chosen among those studied in previous evaluations (Bevacqua et al., 2007; Heylen et al., 2007), whereas the vocalizations were selected using an informal listening test (Bevacqua et al. (2010) for more details). The authors hypothesized that (i) a multimodal signal created by the combination of visual and acoustic cues representative of a meaning would obtain the strongest attribution of the given meaning; (ii) sometimes the meaning conveyed by each acoustic and visual cues is different by the meaning transmitted by their combination; and (iii) multimodal signals obtained by the combination of visual and acoustic cues that have strongly opposite meanings are rated as nonsense (as for instance *nod+no, shake+ok, shake+yeah*).

The evaluation was performed in English and 55 participants accessed anonymously to it through a web browser where the multimodal signals were played one at a time. Participants used a bipolar 7-points Likert scale for each meaning: from -3 (extremely negative attribution) to +3 (extremely positive attribution). Assigning 0 to all dimensions meant that participants could not find a meaning among those proposed for the given signal. They could also judge the signal as completely nonsense.

The 95% confidence interval was calculated for all the meanings. Table 2 reports all signals for which the mean was significantly above zero (for positive meanings) or below zero (for negative meanings). For each dimension of meaning (i.e. agreement/disagreement, acceptance/refusal, etc.), an analysis of variance (ANOVA) was performed. For all dimensions, there was an effect of different visual cues and an effect of acoustic cues. No effect of the interaction between the visual and acoustic cues was found.

Table 2.   Meanings significantly associated to the multimodal backchannels.

	ok	ooh	gosh	really	yeah	no	m-mh	(silence)
raise eyebrows	AG, AC, U	U		I		NL		
nod	B, AG, AC, U	AC, L, U, I	AG, AC, U, I	L, U, I	B, AG, I, AC, U		B, AG, AC	B, AG, AC, U
smile	B, AG, AC, U, L, I	B, AG, AC, U, I, L	AG, L	AG, AC, U, L, I	AG, AC, U, L, I	DB	B, AG, AC, L	
frown	AG, AC	NL	NL	NL, I		DA, NL		DB, N, U, NL
raise left eyebrows	AG, AC	U	DB	DB, I		DB, R		DB, NL
shake		DB, NL	DB, NL	DB, NI	DB, NL	DA, R	DA, R, NL	DA, DB, R, NL, NI
tilt+frown	AC	U		DB, I	AC, L	DA, R, NL		DB, NU

AG = agreement, AC = acceptance, DA = disagreement, R = refusal, L = liking, NL = not liking, B = belief, DB = disbelief, I = interest, NI = not interest, U = understanding, NU = not understanding.

A *t*-test was performed and it showed that the signal *nod+yeah* was more strongly judged as showing agreement than any other signal. *Nod* has the second highest attribution of agreement; however, it was not significantly above the following *nod+m-mh*. The signal *shake+no* was not more strongly judged as showing disagreement than the other signals. The highest disagreement mean is for *shake*; however, it is not significantly different from *shake+no*, *shake+m-mh*. There is a difference between *shake* and *shake+yeah*, which is the fourth highest disagreement attribution. The highest refusal was attributed to *shake* and five out of seven signals associated with the vocalization *no*: *raise eyebrows+no*, *frown+no*, *shake+no*, *raise left eyebrow+no* and *tilt+frown+no*. There were no significant differences between any of these, nor between *shake* and *shake+ooh* or *shake+m-mh*. The refusal attributed to *shake* was higher than the one for *shake+really* and others that have a higher mean. The signal *raise eyebrows+gosh* was not even significantly associated to interest. The highest meaning of interest was equally attributed to *smile+ok, nod+ok, nod+ooh, smile+ooh*. The highest acceptance was attributed to the *nod, nod+yeah, nod+ok, nod+ooh, nod+m-mh, smile+ok, tilt+frown+ok* which were not differentiated in terms of degree of attribution. *Nod+ok* was, however, rated as showing more acceptance than *nod+really* and *nod+no*. The highest meaning of disbelief was attributed to *shake+yeah, raise left eyebrow+no, raise left eyebrow+really*, and *tilt+frown+no*. No difference was observed between these. Highest attribution of understanding was observed for *raise eyebrows+ooh, nod+ooh, nod+really, nod+yeah* and *nod*. *Raise eyebrows+ooh* was not more strongly judged as showing agreement than the other signals. A significant difference was even found between *nod-ooh* and *raise eyebrows+ooh*: *nod-ooh* was more strongly associated to the understanding than *raise eyebrows+ooh*. In conclusion the first hypothesis was only partially satisfied.

Results showed that the strongest attribution for a meaning is not always conveyed by the multimodal signals obtained by the combination of visual and acoustic cues representative of the given meaning. For example, disagreement is not more strongly conveyed by the multimodal signals *shake+no*, as we hypothesized. Other signals, like *shake* and *shake+m-mh*, convey this meaning as well. That means that the meaning conveyed by a multimodal backchannel cannot be simply inferred by the meaning of each visual and acoustic cue that composes it. It must be considered and studied as a whole to determine the meaning it transmits when displayed by virtual agents. More results that go in the direction of such a conclusion were obtained. The authors found that some multimodal signals convey a meaning different from the ones associated to the particular visual and acoustic cues when presented on their own. For example, a high meaning of no interest was attributed to *frown+ooh*, although in our previous studies (Heylen, 2007) the signal *frown* was associated mainly to disbelief and, in our preliminary and informal listening test, the vocalization *ooh* was associated to understanding.

As regard to the third hypothesis, the evaluation showed that multimodal signals composed by visual and acoustic cues that have strongly opposite meanings are rated as nonsense. Four multimodal signals were significantly rated as nonsense: *nod+no, shake+yeah, shake+ok* and *shake+really*. What is more, it is interesting to notice that a high attribution of nonsense does not necessarily exclude the attribution of other meanings. Thus, the high nonsense signal of *shake+yeah* was also highly judged as showing disbelief. A possible explanation would be that these signals might be particularly context dependent.

This evaluation provided a better insight about several multimodal backchannels and the meanings they convey. The results have been used to enrich and expand the backchannel lexicon of the virtual agent Greta.

4. Listening Virtual Agents

Several researchers have approached the issue of implementing ECAs endowed with believable listener behavior. All the systems proposed so far share the same intent: they try to generate appropriate signals and to predict their timing, that is, the right time when a backchannel signal should be triggered. Several studies have proposed models for the selection and the generation of backchannel signals and they can be grouped into three different classes according to the cues they look for to trigger a listener's signal:

- *Content based*: these systems analyze the content of the speech or the actions performed by the user to decide when a signal must be provided.

- *Acoustic cues based*: this type of systems is based on the detection of several kinds of acoustic information in the speaker's speech, like variation in pitch or pauses between words or utterances, that can be used to determine when a backchannel signal could be triggered.

- *Multimodal cues based*: recent studies have shown that not only prosodic information can be a valid cue for backchannel timing, but also visual signals can be quite useful. In fact, often listener's backchannels are elicited by speaker's visual actions. This class of models takes into account user's verbal and non-verbal behavior to trigger a backchannel signal, thus it can be considered as an extension of the previous class.

The following sections show in detail each class of models and the corresponding systems, implemented so far, that apply the proposed strategies.

4.1 Content-based models

Beun and van Eijk (2004) proposed a model to generate feedback utterances according to the estimated user's mental state. The system used dialogue rules to determine discrepancies between its own ontology and the user's mental state. For example, the system looks for words in the wrong order, or an incorrect combination of words in the user's speech or user's incorrect actions onto a particular object.

For instance, if the user asks to "edit a process", the system could suppose that the user has a wrong belief, that is that a process can be edited. When a discrepancy is found, a feedback utterance is generated.

Nakano et al. (1999) developed a spoken dialogue system, called WIT, that enables robust utterance understanding and real-time listener's responses. The content of the responses generated by this system is highly appropriate, but since it lacks control on the timing of response emission, users can feel uncomfortable with the resulting delayed interaction.

Heylen et al. (2004) introduced affective feedback in the tutoring system INES. Such a system was used to let students practice on the computer the task of giving an injection in the arm of a virtual patient. INES, graphically represented by a talking head, provides emotional feedback according to the level of correctness of the student's actions. Four emotions have been considered: joy, distress, happy-for and sorry-for; for example, INES shows joy when the student succeeds. The aim of this type of feedback consists in making the learning process more effective by taking into account the user's emotional state. The system is not able to recognize the student's emotional state, but it makes some assumptions looking at the student's achievements and failures in doing injections.

According to Kopp et al. (2008), a conversational agent should be able to respond in a pertinent and reasonable way to the statements and the questions asked by a user. Backchannels provided by this type of agent should derive from its internal state that describes how it feels and what it thinks about the speaker's speech. The model proposed by Kopp reflects this idea and it has been tested with Max, a virtual human developed at the A.I. Group at Bielefeld University. Max appears in an information desk in the Heinz-Nixdorf-Museums Forum and provides visitors with information about the museum and the exhibitions. To communicate with Max, visitors type what they want to say on a keyboard and Max responds through both verbal and non-verbal behaviors. The backchannel model implemented for Max is based on a reasoning and deliberative processing that plans how and when the agent must react according to its intentions, beliefs and desires. Max is able to display multimodal backchannels (like head nod, shake, tilt and protrusion with various repetitions and different movement quality) that are triggered solely according to the written input that the user types on a keyboard. The system evaluates the input in an incremental fashion, parsing the words typed by the user to update constantly its knowledge about the topic of the conversation. To determine backchannel timing, the system applies the *end-of-utterance* detection, since listener's signals are often emitted on phrases boundaries. So far, the end of an utterance is signaled when the key "enter" is typed; in the future an incremental parsing of the user's typed speech will be used to predict this relevant moment. Other appropriate moments, in which a backchannel could be provided, are found by considering the type of words pronounced by the user: a backchannel is more appropriate after a relevant word, like a noun or a verb, than after an article. A rule-based approach helps the system to determine which type of backchannel should be provided after a specific event, for example, after a pause the system can select a slight nod or an acoustic response like "mhm". After every new word, a probabilistic component computes the probability of emission for each single backchannel. Such a probability depends on three other probabilities: the probability of having successfully understood the new word, the probability to provide a certain backchannel B and the conditional probability that Max performs the backchannel B with a level x of understanding. Thanks to this system, Max is able to have long and coherent conversations with users.

4.2 Acoustic cues-based models

Other researchers have based their backchannel systems on the recognition of the user's signals. In particular, the systems belonging to this class look for audio cues. They are based on the concept that often backchannel signals are provided when the speaker is able to perceive them more easily. This moment in an interaction corresponds to a grammatical completion, usually accompanied by a region of low pitch (Ward, 1996).

A first approach was proposed by Cassell et al. (1994). They implemented a system able to automatically generate and animate interactions between two or more virtual agents. The system produces appropriate and synchronized speech, intonation, facial expressions, and gestures. The speech is computed by a dialogue planner that generates both the text and the prosody of the sentences. To drive lip movement, head motion, shift of the gaze, facial expressions and gestures, the system uses the speaker/listener relationship, the text, and the turn conveying intonation. In particular, listener behavior is generated at utterances boundaries where silence usually occurs. The system is able to determine if there is enough time for producing a signal by looking at the timing of phonemes and pauses computed by the text-to-speech synthesizer.

Pelachaud (2005) presented Greta, an ECA that exhibits non-verbal behavior synchronized and consistent with speech. The system can automatically generate the animation of two agents while interacting in the same graphic window. The system computes the behavior of the agent who is playing the listener inserting a non-verbal backchannel signal (like a head nod, a raise of the eyebrows or a smile) according to some probabilistic rules in correspondence with pauses, a particular pitch accent and so on.

Ward and Tsukahara (2000) provided evidences for the assumption that often backchannel signals are provided when the speaker is able to perceive it more easily. They proposed a model based on acoustic cues to determine the right moment to provide a backchannel. Backchannel signals are provided when the speaker talked with a low pitch lasting 110 ms after 700 ms of speech and provided that backchannel has not been displayed within the preceding 800 ms. To evaluate their system, they tested it on a corpus of pre-recorded dyad conversations. Results showed that the system (based on the low pitch rule) predicted the occurrence of backchannel signals better than random: the accuracy was 18% versus 13% for English and 34% versus 24% for Japanese.

Fujie et al. (2004) presented the humanoid robot ROBISUKE, a conversational robot able to provide appropriate feedbacks to the user before the end of an utterance. The robot uses the spoken dialogue system developed by Nakano et al. (1999) to determine the content of the feedback according to the content of the user's speech. The system employs a network of finite state transducers to link recognized words to content. Then, to determine the right timing of the feedback, it extracts prosody information.

Thórisson (1997) developed a virtual agent, called Gandalf, capable of interacting with users using verbal and non-verbal signals. Gandalf is provided with a face and a hand. It has knowledge about the solar system. Its interaction with users consists in providing information about the universe. The solar system is displayed in the same screen where Gandalf stands and the agent can travel from planet to planet telling user's facts about each one. During the interaction with the user, Gandalf is able not only to display facial expressions, *attentional* cues (like gazing at the user or at the object it is talking about) and appropriate behaviors for managing turn-taking, but it is also capable of producing real-time backchannel signals. To generate backchannels, the system evaluates the duration of the pauses in the speaker's speech. A backchannel (a short utterance or a head nod) is displayed when a pause, longer than 110 ms, is detected. Gandalf is based on a multi-layer multimodal architecture that endows the agent with multimodal perception and action generation skills. Each layer has sufficient information to decide which action to perform at a specific time scale. Besides information about the user's voice intonation (obtained through an automatic intonation analysis), the system collects, by a human observer in a Wizard of Oz setting, data about the user's hand gestures and gaze direction. Such data is used to automatically control the virtual agent's gaze and its turn-taking behavior.

Another backchannel system has been implemented by Cassell and colleagues for REA, the Real Estate Agent (Cassell and Thórisson, 1999). REA is a virtual humanoid whose task consists in showing users the characteristics of houses displayed behind her. She interacts with users through verbal and non-verbal behaviors; REA's text to speech synthesizer allows her to vary the intonation of her voice. Like Gandalf, REA's responses are generated on the basis of a pause duration model. She provides a backchannel signal at each pause that lasts more than 500 ms (Cassell and Thórisson, 1999). Still like Gandalf, REA can emit backchannels as short utterances and head nods, but in addition she can show puzzlement (for example by raising its eyebrows) asking for repair when she does not understand what the speaker says.

Cathcart et al. (2003) proposed a model based on pause duration. They presented a shallow model that uses human dialogue data for predicting where backchannel signals should appear. This dialogue data was extracted from the HCRC Map Task Corpus, a set of 128 task-oriented dialogues between English speakers. From the analysis of this corpus, Carthcart and colleagues found out that backchannel can be expected at phrase boundaries and that these boundaries occurred every five to fifteen syllables (Knowles et al., 1996). On the basis of this result, the system inserted a backchannel every n words (they approximated syllables by words), where n was determined by the frequency of backchannels that occurred in the data. They evaluated their model in three different situations: (1) the model simply inserted a backchannel signal every n words; (2) the model provided backchannel only when a pause, longer than a certain duration, was detected; (3) the model integrated both methods. Results showed that the integration of the two approaches increased noticeably the accuracy of predicting a backchannel.

4.3 Multimodal cues-based models

Several studies have shown that speaker's non-verbal behavior can be helpful in defining when a backchannel signal could be provided by the listener during a conversation. In certain conditions, for example when the conversation is progressing smoothly and successfully, people tend to synchronize themselves like in a dance. Based on this fact, Maatman et al. (2005) derived from the literature a list of useful rules to predict when a backchannel can occur according to user's verbal and non-verbal behavior. They concluded that backchannel signals (like head nods or short verbal responses that invite the speaker to go on) appear at a pitch variation in speaker's voice; listener's frowns, body movements and shifts of gaze are produced when the speaker shows uncertainty. Mimicry behavior is often displayed by the listener during the interaction; for example, listener mimics shift postures, gaze shift, head movements and facial expressions.

Later on, Gratch et al. (2007) developed the "Rapport Agent", an agent that provides solely non-verbal backchannels when listening. This agent was implemented to study the level of rapport that users feel while interacting with a virtual agent capable of providing backchannel signals. The system analyzes the user's non-verbal behavior (head nods, head shakes, head movements, mimicry) and some features of the user's voice to decide when a backchannel must be displayed. Tests performed with the Rapport Agent showed that the system can elicit a feeling of rapport in users. In the evaluation, a participant, the speaker, had to watch a cartoon movie and then tell the story to another subject, the listener. The speaker and the listener were separated; the listener could hear the speaker and see her/him on a screen, while the speaker could see an avatar on a screen (he was told that the avatar reproduced perfectly the listener's behaviors). The subjects were randomly assigned to one of the following conditions:

- *Responsive*: the avatar is controlled by the Rapport Agent system.
- *Unresponsive*: the avatar is driven by a script that generates random backchannel signals; as a consequence its behavior is not related to the behavior that the speaker is actually displaying.

The results of the test showed that, in responsive condition, participants spoke more than in the other condition. Moreover, speakers' speech was more fluent in responsive condition than in unresponsive condition. In fact, in the latter condition they produced more disfluencies, that is, filled pauses and stutters.

Recently Morency et al. (2008) proposed an enhancement of this type of system introducing a machine learning method to find the speaker's multimodal features that are important and can affect timing of the agent's backchannel. The system uses a sequential probabilistic model for learning how to predict and generate real-time backchannel signals. The model is designed to work with two sequential probabilistic models: the Hidden Markov Model and the Conditional Random Field. To train the predicted model, a corpus of 50 human-to-human conversations has been used. From the video and audio recordings they extracted several prosodic features like pitch variations, vowel volume, energy speech variation; lexical features, as individual words, incomplete words, emphasized words and so on; speaker's features, like gaze behavior; listener's features, as backchannel emission. Speaker's features taken into account can have an individual influence (a single feature is involved in the triggering of a backchannel, like a long pause) or a joint influence (more than one feature influence the listener's backchannel, like a short pause and a look at the listener). The model takes as input the speaker's multimodal features and returns a sequence of probabilities of listener backchannel. In such a sequence, the peak probabilities crossing a fixed threshold are selected as good opportunities to provide the backchannel. Morency and colleagues also approached the problem of agent's expressiveness. By varying the threshold they obtained virtual agents with different level of expressiveness, that is, that provides more or less backchannels during an interaction.

Within the European Project SEMAINE[2], autonomous talking agents have been created. These agents, called Sensitive Artificial Listening (SAL) agents, operate as chatbots inviting the human interlocutor to chat and to bring him or her in a particular mood. Despite having limited verbal skills, the SAL agents are designed to sustain realistic interaction with human users (an example of interaction is shown in Figure 1). A particular concern of the project was to have the agents produce appropriate listening behaviors. The listener module integrated in the SEMAINE architecture (Schröder et al., 2011) has been proposed by Bevacqua et al. (2012). This module, called Listener Intent Planner (LIP), decides *when* the agent must provide a backchannel and *which* signal must be displayed. A set of probabilistic rules based on the literature (for example Bertrand et al. (2006); Ward and Tsukahara (2000)) is used to produce a backchannel signal when certain speaker's visual and/or acoustic behaviors are recognized. To identify those user's behaviors, she/he is continuously tracked through a video camera and a microphone.

When a user's behavior satisfies one of the rules, a backchannel is triggered. The LIP differs from previous listener models mostly in the type of backchannels that it can generate. Earlier models have mainly considered *reactive* backchannels, an automatic behavior used to show contact and perception, whereas the LIP also generates *responsive* backchannels, that is, attitudinal signals are able to transmit the agent's communicative functions. Responsive signals are used to show, for example, that the agent agrees or disagrees with the user, or

---

[2] http://www.semaine-project.eu/

Figure 1. The user interacts with a SAL agent. The corresponding video can be watched here http://www.youtube.com/watch?v=vrXhl0ismMo.

that it believes but at the same time refuses the speaker's message. In order to display an appropriate backchannel signal according to the agent's communicative functions, a lexicon has been defined. Such a lexicon has been created through the perceptual studies described earlier in this paper (see Section 3). The LIP module can also generate backchannels of mimicry of some user's non-verbal behaviors, such as head shake, nod and tilt, smile, raise of the eyebrows.

The authors are interested in this type of signals since several studies have shown that mimicry, when not exaggerated to the point of mocking, has several positive influences on interactions: it can make the interaction an easier and more pleasant experience, improving the feeling of engagement (Chartrand and Bargh, 1999; Cassell et al., 2001; Chartrand et al., 2005).

5. Conclusion

A survey of the behavior performed by the listener engaged in a dialogue with a speaker has been proposed in this chapter. The investigation of such behavior is fundamental for developing ECA systems able to sustain natural and successful interaction with users. Several perceptual studies on backchannels displayed by virtual agents have been presented. The aim of such evaluations is two-fold: on the one hand, the authors wanted to get a better understanding about the relation between the form of this type of signals and their functional meaning and on the other hand, they aimed to gain insight into the way people interpret backchannels performed by virtual humanoids. The chapter is concluded by a description of the progress made in the human-machine interface field to create ECAs able to show an appropriate listening behavior. A classification of listener models has been proposed. As we have seen, most of the systems generate the listening agent's behavior only according to the user's behavior. Even if such a strategy allows for creating virtual agents that look like if they are listening, interacting with them is still limited and frustrating: a user perceives quite easily that the agent does not really understand what she/he is saying. Very few systems take into account the content of the user's speech and they are still restricted to specific domains. An agent able to show a listening behavior that is coherent both with the user's behavior and with the content of the speech remains a big challenge.

Acknowledgements

The perceptual studies presented in this chapter have been conducted in collaboration with other colleagues. We wish to acknowledge the precious work done by Catherine Pelachaud, Dirk Heylen, Marc Schröder, Sylwia Hyniewska, Sathish Pammi and Marion Tellier.

REFERENCES

Allwood, J. and L. Cerrato. 2003. A study of gestural feedback expressions. In Paggio, P., Jokinen, K., and Jonsson, A. (eds.), First Nordic Symposium on Multimodal Communication, pp. 7–22, Copenaghen.

Allwood, J., J. Nivre and E. Ahlsén. 1993. On the semantics and pragmatics of linguistic feedback. *Journal of Semantics*, **9(1)**:1–26.

Argyle, M. 1988. Bodily Communication. London, Methuen & Co.

Argyle, M. and M. Cook. 1976. Gaze and Mutual gaze. Cambridge University Press, Cambridge, England.

Bertrand, R., P. Blache, R. Espesser, G. Ferré, C. Meunier, B. Priego-Valverde, and S. Rauzy. 2006. Le CID (corpus of interactional data): protocoles, conventions, annotations. Travaux Interdisciplinaires du Laboratoire Parole et Langage d'Aix-en-Provence (TIPA), **25**:25–55.

Beun, R. and R. van Eijk. 2004. Conceptual discrepancies and feedback in human-computer interaction. Proceedings of the conference on Dutch directions in HCI. ACM New York, NY, USA.

Bevacqua, E., E. de Sevin, S. Hyniewska and C. Pelachaud. 2012. A listener model: introducing personality traits. *Journal on Multimodal User Interfaces, special issue Interacting with Embodied Conversational Agents*, **6(1)**:27–38.

Bevacqua, E., D. Heylen, M. Tellier and C. Pelachaud. 2007. Facial feedback signals for ECAs. Proceedings of AISB'07 Annual convention, workshop "Mindful Environments", Newcastle upon Tyne, UK, pp. 147–153.

Bevacqua, E., S. Pammi, S.J. Hyniewska, M. Schröder and C. Pelachaud. 2010. Multimodal backchannels for embodied conversational agents. Proceedings of the 10th International Conference on Intelligent Virtual Agents. LNCS 6356. Philadelphia (PA), USA.

Brunner, L. 1979. Smiles can be back channels. *Journal of Personality and Social Psychology*, **37(5)**:728–734.

Campbell, N. 2006. Speech Synthesis and Discourse Information. Proceedings of the fifth Slovenian and first International Language Technologies Conference (IS-LTC 2006), October 9-10, Ljubljana, Slovenia 2006, pp. 11–16.

Campbell, N. 2007. The role and use of speech gestures in discourse. *Archives of Acoustics*. **32(4)**:803–814.

Cassell, J., Y. Nakano, T. Bickmore, C. Sidner and C. Rich. 2001. Non-verbal cues for discourse structure. Proceedings of the 39th Annual Meeting on Association for Computational Linguistics, pp. 114–123. Association for Computational Linguistics Morristown, NJ, USA.

Cassell, J., C. Pelachaud, N. Badler, M. Steedman, B. Achorn, B. Douville, S. Prevost and M. Stone. 1994. Animated conversation: Rule-based generation of facial expression, gesture and spoken intonation for multiple conversational agents. SIGGRAPH 94 21st International ACM Conference on Computer Graphics and Interactive Techniques, Orlando, FL, USA, July 24–29, 1994.

Cassell, J. and K.R. Thórisson. 1999. The power of a nod and a glance: Envelope vs. emotional feedback in animated conversational agents. *Applied Artificial Intelligence*, **13**:519–538.

Castelfranchi, C. and D. Parisi. 1980. Linguaggio, conoscenze e scopi. Il Mulino, Bologna.

Cathcart, N., J. Carletta and E. Klein. 2003. A shallow model of backchannel continuers in spoken dialogue. Proceedings of the tenth Conference on European chapter of the Association for Computational Linguistics, Budapest, Hungary, pp. 51–58.

Cerrato, L. and M. Skhiri. 2003. Analysis and measurement of communicative gestures in human dialogues. Proceedings of AVSP, St. Jorioz, France, pp. 251–256.

Chartrand, T. and J. Bargh. 1999. The Chameleon Effect: The Perception-Behavior Link and Social Interaction. *Journal of Personality and Social Psychology*, **76**:893–910.

Chartrand, T., W. Maddux and J. Lakin. 2005. Beyond the perception-behavior link: The ubiquitous utility and motivational moderators of nonconscious mimicry. In Hassin, R.R., Uleman, J.S. and Bargh, J.A. (eds.), *The Newunconscious*, pp. 334–361. New York: Oxford University Press.

Chovil, N. 1992. Discourse-oriented facial displays in conversation. *Research on Language and Social Interaction*, **25**:163–194.

Clark, H. and E. Schaefer. 1989. Contributing to discourse. *Cognitive Science*, **13**:259–294.

Conte, R. and C. Castelfranchi. 1995. Cognitive and Social Action. London: University College.

Duncan, S. 1972. Some signals and rules for taking speaking turns in conversations. *Journal of Personality and Social Psychology*, **23**:283–292.

Ekman, P. and W. Friesen. 1969. The repertoire of nonverbal Behavior: Categories, origins, usage and coding. *Semiotica*, **1(1)**:49–98.

Fujie, S., K. Fukushima and T. Kobayashi. 2004. A Conversation Robot with Back-channel Feedback Function based on Linguistic and Nonlinguistic Information. Proceedings ICARA International Conference on Autonomous Robots and Agents, Palmerston North, New Zealand, pp. 379–384.

Gardner, R. 1998. Between speaking and listening: The vocalisation of understandings. *Applied Linguistics*, **19(2)**:204–224.

Granström, B., D. House and M. Swerts. 2002. Multimodal feedback cues in human-machine interactions. In Bel, B. and Marlien, I., (eds.), Speech Prosody Conference, pp. 347–350, Aix-en-Provence.

Gratch, J., J. Rickel, E. Andre, J. Cassell, E. Petajan and N. Badler. 2002. Creating interactive virtual humans: some assembly required. Intelligent Systems, *IEEE (see also IEEE Intelligent Systems and Their Applications)*, **17(4)**:54–63.

Gratch, J., N. Wang, J. Gerten, E. Fast and R. Dufy. 2007. Creating rapport with virtual agents. Proceedings of the 7th International Conference on Intelligent Virtual Agents, LNCS 4722, pp. 125–138. Paris, France.

Gumperz, J. 1982. Discourse strategies. Cambridge, England: Cambridge University Press.

Hadar, U., T. Steiner and F. Cliford Rose.1985. Head movement during listening turns in conversation. *Journal of Non-verbal Behavior*, **9(4)**:214–228.

Heylen, D. 2005. A closer look at gaze. In Pelachaud, C., Andr, E., Kopp, S. and Ruttkay, Z. (eds.), Creating Bonds with Embodied Conversational Agents, AAMAS Workshop, pp. 3–9. Utrecht: University of Utrecht.

Heylen, D. 2007. Multimodal Backchannel Generation for Conversational Agents. Proceedings of the workshop on Multimodal Output Generation (MOG 2007), Aberdeen, Scotland.

Heylen, D., E. Bevacqua, M. Tellier and C. Pelachaud. 2007. Searching for prototypical facial feedback signals. Proceedings of 7th International Conference on Intelligent Virtual Agents IVA, LNCS 4722, pp. 147–153, Paris, France.

Heylen, D., M. Vissers, R. Akker and A. Nijholt. 2004. Affective Feedback in a Tutoring System for Procedural Tasks. ISCA Workshop on Affective Dialogue Systems, LNCS 3068, pp. 244–253, Kloster Irsee, Germany.

Kendon, A. 1967. Some functions of gaze-direction in social interaction. *Acta Psychologica*, **26**:22–63.

Knowles, G., A. Wichmann and P. Alderson. 1996. Working with Speech: Perspectives on Research into the Lancaster/IBM Spoken English Corpus. London: Longman.

Kopp, S., J. Allwood, K. Grammer, E. Ahlsen and T. Stocksmeier. 2008. Modeling Embodied Feedback with Virtual Humans. In Wachsmuth, I. and Knoblich, G. (eds.), Modeling Communication with Robots and Virtual Humans. LNAI 4930, pp. 18–37. Berlin, Heidelberg: Springer.

Maatman, R.M., J. Gratch and S. Marsella. 2005. Natural behavior of a listening agent. Proceedings of the 5th International Conference on Interactive Virtual Agents, pp. 25–36, Kos, Greece.

Morency, L.-P., I. de Kok and J. Gratch. 2008. Predicting listener backchannels: A probabilistic multimodal approach. Proceedings of the 8th International Conference on Intelligent Virtual Agents, LNCS 5208, Tokyo, Japan. Springer.

Nakano, M., K. Dohsaka, N. Miyazaki, J. Hirasawa, M. Tamoto, M. Kawamori, A. Sugiyama and T. Kawabata. 1999. Handling rich turn-taking in spoken dialogue systems. Proceedings of the 6th European Conference on Speech Communication and Technology, Budapest, Hungary. EUROSPEECH.

Niewiadomski, R., E. Bevacqua, M. Mancini and C. Pelachaud. 2009. Greta: An interactive expressive ECA system. Proceedings of Conference on Autonomous Agents and Multi-Agent Systems (AAMAS09), Budapest, Hungary.

Pelachaud, C. 2005. Multimodal expressive embodied conversational agents. MULTIMEDIA '05: Proceedings of the 13th Annual ACM International Conference on Multimedia, pp. 683–689, New York, NY, USA. ACM.

Peters, C. 2000. Foundations of an agent theory of mind model for conversation initiation in virtual environment. Special Issue on Behavior Planning for Life-Like Characters and Avatars. AI Communication, **13**:169–181.

Poggi, I. 1981. Le interiezioni. Studio del linguaggio e analisi della mente. Torino: Boringhieri.

Poggi, I. 2003. Mind markers. In Trigo, N., Rector, M. and Poggi, I., (eds.), Gestures. Meaning and Use. Oporto, Portugal: University Fernando Pessoa Press.

Poggi, I. 2007. Mind, Hands, Face and Body. a Goal and Belief View of Multimodal Communication. Berlin: Weidler Buchverlag.

Poggi, I. and C. Pelachaud. 2000. Performative facial expressions in animated faces. *Embodied Conversational Agents*, Cambridge, MA: MIT Press, pp. 154–188.

Rosenfeld, H. 1978. Conversational control functions of nonverbal behavior. In Siegmann, A.W. and Feldstein, S. (eds.), *Nonverbal behavior and Communication*, pp. 291–328. Hillsdale, New Jersey: Lawrence Erlbaum.

Schegloff, E. and H. Sacks. 1982. Discourse as an interactional achievement: Some uses of "uh huh" and other things that come between sentences. In Tannen, D. (ed.), Analyzing Discourse: Text and Talk, pp. 71–93. Georgetown: Georgetown University Press.

Schröder, M., E. Bevacqua, R. Cowie, F. Eyben, H. Gunes, D. Heylen, M. ter Maat, G. McKeown, S., Pammi, M. Pantic, C. Pelachaud, B. Schuller, E. de Sevin, M. Valstar, and M. Wollmer. 2011. Building autonomous sensitive artificial listeners. *IEEE Transactions on Affective Computing*, **99(1)**:134–146.

Sidner, C.L., C.D. Kidd, C. Lee and N. Lesh. 2004. Where to look: A study of human-robot interaction. Proceedings of Intelligent User Interfaces Conference, pp. 78–84. ACM Press.

Thórisson, K.R. 1997. Gandalf: an embodied humanoid capable of real-time multimodal dialogue with people. Proceedings of the First International Conference on Autonomous Agents, pp. 536–537, ACM New York, NY, USA.

Ward, N. 1996. Using prosodic clues to decide when to produce back-channel utterances. Proceedings of the Fourth International Conference on Spoken Language Processing (ICSLP'96), **3**:1728–1731.

Ward, N. and W. Tsukahara. 2000. Prosodic features which cue back-channel responses in English and Japanese. *Journal of Pragmatics*, **23**:1177–1207.

Welji, H. and S. Duncan. 2004. Characteristics of face-to-face interactions, with and without rapport: Friends vs. strangers. In Symposium on Cognitive Processing Effects of 'Social Resonance' in Interaction, 26th Annual Meeting of the Cognitive Science Society.

Wiener, N. 1948. Cybernetics and Control and Communication in the Animal and the Machine. MIT Press, Cambridge, MA.

Yngve, V. 1970. On getting a word in edgewise. Papers from the Sixth Regional Meeting of the Chicago Linguistic Society, pp. 567–577.

CHAPTER 11

Human and Virtual Agent Expressive Gesture Quality Analysis and Synthesis

Radoslaw Niewiadomski, Maurizio Mancini
and Stefano Piana

1. Introduction

Nonverbal communication is an essential element of human-human communication, equally important as the verbal message. It consists of bodily *nonverbal signals*, including *facial expressions*, *hand/ arm gestures*, *posture shifts*, and so on (Argyle, 1998). In particular, the communicative meaning of gestures usually depends on two components: *shape* and *expressive quality*. While the role of the former, for instance the configurations of the hand in time, is well known (McNeill, 1996), studies about the latter, i.e., how a particular mental intention is communicated through a gesture's expressive quality, are quite recent. Nevertheless it has been experimentally shown that a gesture's expressive quality may communicate social relations and communicative intentions, such as: emotional states (Castellano et al., 2007), affiliation (Lakens and Stel, 2011), cultural background (Rehm, 2010), dominance (Varni et al., 2009; Jayagopi et al., 2009), agreement (Bousmalis et al., 2009) or group cohesion (Hung and Gatica-Perez,

2010). Some authors also suggest that gesture's *expressivity* could be important in the communication of social states such as empathy (Varni et al., 2009) or even sexual interest (Grammer et al., 2000).

In this chapter, the expression *"expressive gesture quality"* refers to those features of nonverbal behaviors that describe how a specific gesture is performed, for example its temporal dynamics, fluidity or energy. In the domain of *Human-Computer Interaction (HCI)*, expressive gesture quality is important at least for two reasons. First of all, researchers try to detect the expressive qualities in human nonverbal behavior and to infer their communicative meaning. In this case, the final goal is the recognition of, for example, the user's emotional state or mood. Second, expressive gesture synthesis is studied in the design and implementation of virtual agents, i.e., anthropomorphic autonomous characters displayed, for instance, on the computer screen that use various verbal and nonverbal forms of communication (Cassell et al., 2000). Virtual agents may modulate their expressive gesture quality to better transmit their communicative intentions, e.g., their emotional state or mood, to the user.

This chapter presents an overview of studies that enhance the communicative capabilities of human-computer interfaces by taking into account the expressive qualities of nonverbal behavior. It also presents a detailed description of two systems for expressive gesture quality analysis and synthesis in HCI. In more details, the next section is split into three parts: in the first and second parts, we review some of the methods described in the literature for expressive gesture quality analysis and synthesis by illustrating various algorithms; in the third part, we present systems in which a continuous expressive gesture quality analysis and synthesis *loop* is performed. In Section 3, we present a case study—the analysis and synthesis of some expressive gesture features in the EyesWeb XMI platform for the creation of multimodal applications (Camurri et al., 2007) and the virtual agent called Greta (Niewiadomski et al., 2011). Finally, in Section 4, we provide a conclusive overview by comparing existing algorithms for gesture quality analysis and synthesis in HCI.

2. State of the Art

2.1 Expressive gesture quality analysis

In HCI, a central role is played by automated gesture analysis techniques, aiming to extract and describe physical features of human behavior and use them to infer information related to, for example, their emotional state, personality, or social role. The ability for systems to understand users' behavior and to respond to them with appropriate feedback is an important requirement for generating socially tuned machines (Schröder et al., 2011; Urbain et al., 2010). Indeed, the expressive gesture quality of movement is a key element in both understanding and responding to users' behavior.

Many researchers (Johansson, 1973; Wallbott and Scherer, 1986; Gallaher, 1992; Ball and Breese, 2000; Pollick, 2004) investigated human motion features and encoded them into categories. Some authors refer to body motion using dual qualifiers such as slow/fast, small/large, weak/energetic, and unpleasant/pleasant. Behavior expressivity has been correlated to energy in communication, to the relation between temporal/spatial features of gestures, and/or to personality/emotion. Harald G. Wallbott (Wallbott, 1998) deems that behavior expressivity is related to the notion of quality of the mental, emotional, and/or physical state, and the intensity of this state. Behaviors do not only encode content information, that is, *"What is communicated"* through a gesture shape, but also expressive information, that is, *"How it is communicated"* through the manner of execution of the gesture.

There exist at least two important aspects of expressive gesture quality analysis, that is, the low level feature detection and its high level interpretation in terms of its eventual communicative meaning. Both of them have received important contributions in the last years. In next two subsections, we present both these aspects.

2.1.1 Expressive gesture features detection

Several low-level features were proposed to describe the expressivity of the movement. Theories from arts and humanities, such as for example Laban's Effort theory (Laban and Lawrence, 1947) are some of the sources analysis techniques are grounded on. Several algorithms have been proposed to measure the features that can be extracted from a movement. Interestingly, the same features can be computed in many different ways.

The *Spatial Extent* and the *Fluidity* of movement are two such features that are often analyzed. Among others, Cardakis et al. (2007) analyze the user's gesture extent by measuring the distance between two hands, whereas hands' fluidity is computed as the sum of the variance of the norms of the hands' motion vectors. Similarly, Camurri et al. (2004a) compute the *Contraction Index*, which is the ratio between the area of the minimum rectangle surrounding the actor body and the body silhouette. In Bernhardt and Robinson's (2007) work, the maximum distance of hand and elbow from body is taken. Camurri et al. (2004a) approximate movement fluidity with the *Directness Index*, revealing whether a movement follows a straight line or a sinuous trajectory. Mazzarino et al. (2007) propose two different methods to measure fluidity. First, they estimate hand's fluency by finding gesture start and ending time, then they determine the amount of movement phases in a given time window: the lower the number of phases, the higher the fluency. Second, they analyze discrepancies between different body parts' movements: fluency is then evaluated by comparing the quantity of motion of upper and lower body. More recently, Mazzarino and Mancini (2009) propose to estimate human movement *Smoothness* by computing the correlation between trajectory curvature and velocity in a given time window. Smoothness is computed for each frame of a video and for a particular point on the body. Thus, it is possible to establish several points (e.g., the two hands) and compute smoothness separately for each of them.

The other group of expressive characteristics of a gesture focuses on different temporal aspects of its realization. For instance, temporal aspects of gesture in Bernhardt and Robinson (2007) are measured by the average hand (and elbow) speed. A slightly different method to estimate the temporal quality of the gesture is used in Mancini and Castellano (2007). For this purpose, the authors compute the velocity of the barycenter of the hand.

Movement *Power* and *Impulsivity* were also addressed by different computational methods. In Cardakis et al. (2007), power is the first derivative of the motion vectors, whereas Mancini and Castellano (2007) operationalize power by the acceleration of the hand. More complex algorithms are used to compute impulsivity of the movement. In Mazzarino and Mancini (2009), it is characterized as a local peak in the time series of quantity of motion. For this purpose, the authors detect any significant rise of quantity of motion in a given time window.

Finally, the *Overall Body Activation* is analyzed by Camurri et al. (2004a) through the computation of the *Quantity of Motion (QoM)*. It is measured as the difference of the person's body silhouettes area computed on consecutive video frames. In Cardakis et al. (2007), the user's movement is estimated as the sum of the motion vectors of color-tracked user's hands.

2.1.2 Communicative meaning of expressive gesture qualities

The high-level meaning of expressive gesture features has been recently investigated, to extract the communicative high-level message of gestures and body movements. Camurri et al. (2003, 2006) and

Castellano (2006) classified expressive gesture in human full-body movement (music and dance performances) and in motor responses of participants exposed to music stimuli: they identified parameters deemed important for emotion recognition and showed how these parameters could be tracked by automated recognition techniques.

Other studies show that expressive gesture analysis and classification can be obtained by means of automatic image processing (Drosopoulos et al., 2003; Balomenos et al., 2005) and that the integration of multiple modalities (facial expressions and body movements) is successful for multimodal emotion recognition (Gunes and Piccardi, 2005).

Several systems have been proposed in which visual feedback/response is provided by analyzing some features of the users' behavior. In such systems, the input data can be obtained from dedicated hardware (joysticks, hand gloves, etc.), audio, and video sources. SenToy (Paiva et al., 2003) is a doll with sensors in its arms, legs and body. Several body positions of the doll are associated with emotional states. According to how the users manipulate the doll, they can influence the emotions of characters in a virtual game: depending on the expressed emotions, the synthetic characters perform different actions. Taylor et al. (2005) developed a system in which the reaction of a virtual character is driven by the way in which the user plays a music instrument. Kopp et al. (2003) designed a virtual agent able to imitate natural gestures performed by humans using motion-tracked data. When mimicking, the agent extracts and reproduces the essential form features of the gesture stroke, which is the most important gesture phase. Reidsma et al. (2006) designed a virtual rap dancer that invites users to join him in a dancing activity. Users' dancing movements are tracked by a video camera and guide the virtual rap dancer.

Castellano et al. (2007) investigate how emotional states can be communicated through speech, face, and gesture both in a separate and in a joint way. In particular, gestures are analyzed by tracking user's hands and computing some meta-movement features. At first, the user's body silhouette is extracted from the input video by performing background subtraction. Then, hands are localized using skin color tracking and their geometrical barycenter is determined. The authors extract two types of indicators: movement cues and features. Movement cues correspond to data computed directly from points (hands' barycenter) moving on a 2D plane (the video frame), like speed, acceleration, and fluidity. Movement features are meta-indicators, that is, descriptors of the movement cues variation over time, like initial and final slope, maximum value, number of peaks, and so on. These cues are provided as input to a Bayesian classifier determining which emotional state could be associated with the performed gesture. Those Bayesian classifiers are used before and after performing data fusion between modalities. In the first case, the classifiers are applied separately to speech, face, and gesture features. Separate results are then combined via a voting algorithm. In the second case, there is a single classifier that receives all the features coming from different modalities as input.

Sanghvi et al. (2011) evelute children's emotional reaction while playing chess with a robot by analyzing their upper body movements and posture. For this, computer vision algorithms are applied (e.g., CAMShift) to extract the children's body silhouette from the input video. Body (frontal/backward) lean angle and curvature of the back, Quantity of Motion, and a Contraction Index are determined. Then, the authors compute the 1st, 2nd, and 3rd derivative of each feature's time-series and the histograms of each derivative. The result is a stream of features that is classified using 63 classifiers. The best results are obtained by the ADTree and OneR classifiers and they show that Quantity of Motion and the 2nd derivative of the movement features are the most significant indicators in discriminating the user's emotional state.

Kleinsmith and Bianchi-Berthouze (2011) revise psychological and neuroscientific works demonstrating the importance of both gesture movement and form in the process of human affect recognition. By showing participants a set of, for example, emotional upright and upside-down reversed videos, researchers prove that emotion recognition is still possible but with lower rate, revealing the contribution of form in the process. The authors present a system for recognition of affective postures based on sequences of static postures. Non-acted expressions of affect are collected by motion capturing the body joints rotations of human video game players at the time when the game is won or lost. The system is composed by two separate modules: the first one for classification of static postures and the second one for classification of a sequence of postures. Each posture is described as a vector of body joints rotations. The input of the posture classification module consists in a static posture while the output is a probability distribution for the set of labels: **defeated, triumphant, neutral**. Then a decision rule is applied to sequences of static postures, by computing the cumulative probability that each label appears in the sequence.

Previous work from the same authors (Kleinsmith et al., 2011) focuses on the recognition of non-acted affective states grounded only on body postures. In particular, they present models based on low-level description of body configuration. Each body posture is represented by a vector of features: each feature represents the normalized rotation of one of the body joints around one of the three axes (e.g., rotation of the left/right shoulder/elbow around the $x/y/z$ axis). Then models for automatic recognition of four emotional states (frustrated, triumphant, concentrating, and defeated) are defined by providing 103 postures represented by their vectors of low-level features and the corresponding label.

In Bianchi-Berthouze (2012), the body movements of video game players are analyzed both by observers and from motion capture data to understand their role, e.g., movements that are functional to the game vs. movements that express affect. The motion capture data consists of the players' body joints rotation and the amount of body movements is computed by a normalized sum of all of the joints over a gaming session.

2.2 Expressive gesture quality synthesis

Two approaches for expressive gesture generation are widely used: animation based on motion capture data and procedural animation. First of all, expressive movement can be re-synthesized from motion capture data. An example of such an approach is proposed by Tsuruta et al. (2010) to generate emotional dance motions. In this work the authors parameterize "standard" captured motions by modifying the original speed of motion or altering its joint angles. Emotional dance motions are parameterized by a small number of parameters obtained empirically. Five emotional attitudes are considered: neutral, passionate, cheerful, calm, and dark. The parameters influence directly the joint values of a very simple body model consisting of 6 degrees of freedom (DOF), namely knees, waists, and elbows.

Several models were proposed for procedural animation. In Allbeck and Badler (2003), the choice of nonverbal behavior and the movement quality depends on the agent's personality and emotional state. The way in which the agent performs its movements is influenced by a set of high-level parameters derived from Laban Movement Analysis (Laban and Lawrence, 1947), and implemented in the Expressive Motion Engine, EMOTE (Chi et al., 2000). The authors use two of the four categories of the Laban's annotation scheme: Effort and Shape. Effort corresponds to the dynamics of the movement and it is defined by four parameters: space (relation to the surrounding space: direct/indirect), weight (impact of movement: strong/light), time (urgency of movement: sudden/sustained), and flow (control of movement: bound/free). The Shape component describes body movement in relation to the environment. In Allbeck and Badler's model, this component is described using three dimensions: horizontal (spreading/enclosing), vertical (rising/sinking) and sagittal dimension (advancing/retreating). EMOTE acts on the animation as a filter. The model adds expressivity to the final animation. It can also be used to express some properties of the virtual agent or its emotional state. For this purpose, the EMOTE parameters were mapped to the emotional states (OCC model; Ortony et al., 1988) and personality traits (OCEAN model; Goldberg, 1993).

Neff and Fiume (2002, 2003) propose a pose control model that takes into account several features of nonverbal behavior such as the timing of movement, the fluent transition between different poses and its expressive qualities. For each body posture, different properties can be defined like its tension, amplitude, or extent. The model allows a human animator to vary, for example, how much space a character occupies during a movement or to define whether the posture should be relaxed or tensed. In more detail, Neff and Fiume (2002) propose a model of tensed vs. relaxed nonverbal behaviors in a physically based animation framework. This model is based on the observed relation between expressive features of movement and forces of gravity and momentum. The system allows an animator to control explicitly the tension for each DOF in the character animation by taking into account the gravity and external forces. Tense movements are short, greater acceleration and without overshoot, while relaxed ones start slowly and with a delay (necessary to overcome inertia) and finish with a visible overshoot.

In a more recent work (Neff and Fiume, 2003), the same authors focus on three different aspects of movement, i.e., timing, amplitude, and spatial extent of gesture. First, they propose a sequential realization of the movements where different joints are no longer animated at the same time but they move in a sequence (i.e., some of the joints follow the others) to make movement more fluid (natural). The human animator can specify the offset and the type (forward or reverse successions of joints) manually; then the system automatically propagates the time shifts to all joints, which for this purpose are organized in a hierarchical structure. They also model the amplitude of the movements, i.e., they adjust the ranges over which a motion occurs. The amplitude of the movements is modeled by multiplying the distance of the joints for each pose (keyframe) of the animations. For each inter-pose, two deltas are calculated: one measuring their distance from the average to the end of the pose and the second measuring the distance from the average to the end state of the previous pose, then they are multiplied by the amplitude factor. The average values are applied to poses from the second to $n - 1$ in a sequence. A similar mechanism is used to calculate the spatial extent, i.e., the space where the action is realized.

Hartmann et al. (2005) define and implement a set of parameters that allow one to alter the way in which an agent expresses its actual communicative intention. This model is based on perceptual studies conducted by Wallbott and Scherer (1986), Wallbott (1998), and Gallaher (1992). These works define a large number of dimensions that characterize gesture expressivity. Hartmann et al. (2005) implement six of these dimensions (see Section 3 for more details). Three of them, namely spatial extent, temporal extent and power, act on the parameters defining the gestures and the facial expressions. They modify respectively the amplitude of a signal (which corresponds to the physical displacement of a facial feature or the wrist position), the movement duration (linked to the execution velocity of the movement), and the dynamic properties of movement (namely acceleration). Another dimension, fluidity, modifies the movement's trajectory as well as works over several behaviors of a given modality. In the latter case, it specifies the degree of fluidity between consecutive behaviors. The last two dimensions, overall activity and repetitivity, refer to the quantity of signals and to their repetition.

Szczuko et al. (2009) propose a fuzzy controller that can be used to modify a key-frame-based animation by applying two expressive features: fluidity and level of exaggeration. In this approach an animation is modified by adding some additional gesture phases, namely preparation, overshoot, and phase movement hold. The preparation is a slight movement in a direction opposite to the main direction, while the overshoot is an additional movement of the last bone in the chain that overpasses the target position and goes back. According to the authors, adding these phases to an animation modifies its expressive quality without changing the communicative meaning of the displayed behavior. In the paper, the authors propose an approach that allows a human animator to specify explicitly the fluidity of animation as well as its expressiveness through a set of discrete values: {fluid, middle, abrupt} and {natural, middle, exaggerated}. This approach is based on fuzzy rules defined in a perceptual study. According to the given values of expressiveness and fluidity, the fuzzy rules control the duration and the amplitude of additional phases.

A similar approach, proposed in Kostek and Szczuko (2006), can be used to generate emotionally characterized body animations. In this approach, the authors first manually create a set of animations of certain gestures. Second, the emotional content and intensity of these animated gestures are evaluated in a perceptive study. Third, in order to find the relation between perceived emotion and the low-level features of evaluated gestures' animations, the authors apply rough set exploration algorithms on the low-level features of animated gestures (i.e., the amplitude, the length and the speed of every joint's movement). The authors show that the most significant features for communicating emotional content are related to the duration of the gesture phases (stroke position, duration of hold, preparation and stroke phase). They argue that the duration of the gesture phases is the most significant feature in communicating emotional states; it is more important than the amplitude of the movement. Finally they also propose values of low-level gesture features that can be applied to any gesture animation in order to modify its emotional content.

Hsieh and Luciani (2005, 2006) use a physically based particle modeling approach to model several modern dance figures. Their models allow the user to control the expressive qualities of movement such as light vs. strong, free vs. bound, or sudden vs. sustained. In more details, for each dance figure the authors define one physically based particle model. Each figure needs its own description using different set of parameters (e.g., masses, forces) that influence the dynamics of the movement. Thus, every dance figure is defined using different set of mathematical equations that describe its forces, velocity, potential, and kinetic energy. The models describe the energy flow for each figure rather than its spatial criteria. So, instead of defining explicitly its trajectory, a movement is generated implicitly by the involved forces. The single models can be combined in more complex behaviors.

2.3 Expressive gesture quality loop between humans and machines

Applications offering real-time expressive gesture quality analysis and synthesis are becoming widely developed in multimodal interactive scenarios. Their final aim is the creation of credible and natural multimodal interaction between human and machine: that is, they aim to create an *expressive gesture quality loop*, i.e., bidirectional interaction that exploits the communicative role of gesture expressivity, as illustrated in Figure 1.

The idea of an interactive HCI system that intentionally uses the expressive qualities of the behavior was proposed by Caridakis et al. (2007). Their system allows a virtual agent to mimic the qualitative features of the user's behavior. For this purpose, recognized expressive

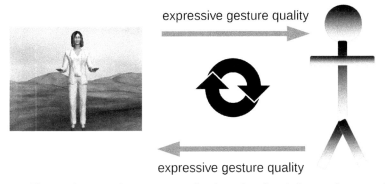

expressive gesture quality

expressive gesture quality

Figure 1. The user's expressive gesture quality is analyzed to influence the agent; in a symmetrical way, the agent's expressive gesture quality influences the user's behavior.

(Color image of this figure appears in the color plate section at the end of the book.)

features of the human behavior are mapped to the agent behaviors. Consequently, the agent does not repeat the human movements but its individual behavior is modified to fit the user's expressive behavior profile. The system was only partially implemented and it was not working in a real-time. Thus, natural human-machine interaction was not possible. Instead, the gesture expressivity parameters detected with the gesture analysis module were used to control manually a virtual agent that implements Hartmann's model of expressive behavior (Hartmann et al., 2005; see also Sections 2.2 and 3). Caridakis et al. (2007) proposed a mapping between the expressive features of human behavior and the agent's expressivity parameters: the sum of the variance of the norms of the motion vectors is associated to the agent's *Fluidity*; the first derivative of the motion vector to *Power*; the distance between hands to *Spatial Extent*; the sum of the motion vectors to *Overall Activity*.

A similar solution was proposed more recently by Mancini and Castellano (2007). Differently from Cardakis et al.'s work, Mancini and Castellano build a truly interactive system that takes as input the video data, extracts high-level behavior features using the EyesWeb XMI platform (Camurri et al., 2007), and finally synthesizes them with a virtual agent. Similarly to Caridakis et al. (2007), the agent copies only the expressive qualities of the movement of the human but realizes different gestures. The video input is treated with EyesWeb XMI to perform quantitative analysis of the human movement in real-time. The virtual agent uses the expressive model proposed by Hartmann et al. (2005) (see previous section). The following mapping between the features detected by EyesWeb XMI and the agent's expressivity quality of movement is then performed: *Contraction Index* is mapped to *Spatial Extent*, *Velocity* of movement to *Temporal Extent*, *Acceleration* to *Power*.

Finally *Directness Index*, a measure of the movement straightness, is associated with *Fluidity*.

Pugliese and Lehtonen (2011) propose the creation of an *enactive loop*: a user and a virtual agent simulate the situation in which two humans are in the same room but they are separated by a glass, so they can communicate only by body movements. But, as it happens in the above applications, the interaction is only at the expressive gesture quality level. The proposed system allows defining a mapping between the user's detected *Quantity of Motion* and *Distance* (the distance of the person from the glass) onto the same agent's movement features. However, such mapping is free, that is, the agent could simply imitate the user or it could, for example, respond with opposite behaviors (e.g., if the user moves quickly, the agent moves slowly and so on).

3. A Case Study: The EyesWeb XMI Platform and the Greta ECA

In this section, we introduce a concrete example of two existing systems for expressive gesture quality analysis and synthesis, implementing some of the algorithms described in the previous sections. The EyesWeb XMI platform is a modular system that allows both expert (e.g., researchers in computer engineering) and non-expert users (e.g., artists) to create multimodal installations in a visual way (Camurri et al., 2007). The platform provides modules, called blocks, that can be assembled intuitively (i.e., by dragging, dropping, and connecting them with the mouse) to create programs, called patches, that exploit system's resources such as multimodal files, webcams, sound cards, multiple displays and so on.

The Greta (Niewiadomski et al., 2011) is a virtual agent able to communicate verbally as well as nonverbally various communicative intentions. Concerning nonverbal communication it is able to display facial expressions, gestures, torso and head movements. Greta is controlled using two XML-like languages BML and FML-APML. It is a part of several interactive multimodal systems working in real-time, e.g., SEMAINE (Schröder et al., 2011) or AVLaughterCycle (Urbain et al., 2010).

3.1 Expressive gesture quality analysis framework

In this section, we describe a framework for multi-user nonverbal expressive gesture quality analysis. Its aim is to facilitate the construction of computational models to analyze the nonverbal behavior explaining the emotions expressed by the users.

That is, the computed features are used to analyze users' nonverbal behavior, the emotions they express, and the level of social interaction. For this purpose we use the EyesWeb XMI software platform (Camurri et al., 2004b; Camurri et al., 2007), video cameras or Kinect sensors, and we extract several low-level movement features (e.g., movement energy) that are used to compute mid- and high-level features (e.g., impulsivity or smoothness) in a multilayered approach, possibly up to labeling emotional states or social attitudes. The choice of these features is motivated by previous works on the analysis of emotion using a minimal set of features (e.g., Camurri et al., 2003; Glowinski et al., 2011).

An overview of the software architecture is sketched in Figure 2. In the following sections, we describe a subset of the proposed framework's components. The first one is the *Physical Layer*, performing measurements on the user's physical position in space as well as the user's body joints configuration. The *Low-Level Features* include those features directly describing the physical features of movements, such as its speed, amplitude and so on. The *Mid-Level Features* can be described by models and algorithms based on the low-level features; for example, the movement smoothness can be computed given its velocity and curvature.

3.1.1 Physical layer

For the purpose of the expressive gesture quality analysis described in this case study we use one or more Kinect sensors. Kinect is a motion sensing input device by Microsoft, originally conceived for the Xbox.

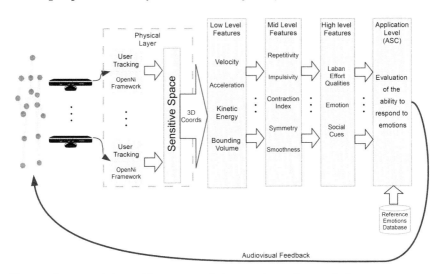

Figure 2. Framework for multi-user nonverbal expressive gesture analysis.

We implement support for multiple such devices in the EyesWeb XMI platform, enabling to track movement in a larger sensitive space, e.g., tracking a user that moves in separate rooms. This also allows a higher number of users to be tracked simultaneously: each Kinect device focuses on a different space area, next the data captured by each device are merged to obtain a single sensitive area. The motion capture measurements provided by each Kinect device are then processed to share the same absolute reference system.

Multiple users' detection and tracking is performed thanks to different EyesWeb XMI software modules (blocks). To communicate with the Kinect sensor, EyesWeb XMI supports both the OpenNi framework (version 1.5.4.0) and the Microsoft Kinect SDK (version 1.6). The two APIs support the streaming of both color images and depthmaps captured by the Kinect's optical sensors. A depthmap is a grayscale image where the color intensity represents the distance from the sensor measured in mm. Both OpenNi and Microsoft Kinect SDK support user segmentation and tracking by providing 2D and 3D measurements of multiple users joints: the two APIs are similar in terms of speed and real-time performances (the Microsoft API is slightly less influenced by occlusions), the sets of joints tracked by the two APIs are similar but the Microsoft one can track a bigger number of joints (see Figure 3).

Using the OpenNi Framework, EyesWeb supports the automatic calibration of the user tracking system, and provides functionalities to save configuration files. This feature allows to avoid the tuning phase of Kinect, which consists of the automatic calibration phase requiring from 10 to 15 seconds (during the tuning phase the tracking measurements are less precise).

Two different EyesWeb blocks, called "Kinect Extractor OpenNi" and "Kinect Extractor SDK", were developed to interface with Kinect

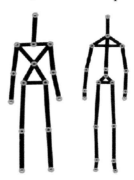

Figure 3. 2D coordinates of the tracked joints produced by the Microsoft SDK (on the left) and by the OpenNi API (on the right).

using the OpenNi or the Microsoft APIs; both blocks provide the data from the Kinect's sensors. Multiple instances of these blocks may be used in a single application in order to use several Kinect devices at the same time; for each device, the outputs provided by the blocks are: a set of tracked users, the image from the color camera or, alternatively, the image from the infrared camera (available only using OpenNi), an image representing the reconstructed depthmap, where the distance from the sensor is mapped to a gray-level in the image. The block developed to support the Microsoft SDK can also output information about face tracking and an audio stream.

3.1.2 Low-level features

Wallbott identified movement expansiveness as a relevant indicator for distinguishing between high and low arousal emotional states (Wallbott, 1998). He also observed that the degree of movement energy is an important factor in discriminating emotions. In his study, highest ratings for the energy features corresponded to hot anger and joy, while lowest values corresponded to sadness and boredom. Meijer (1989) highlighted that emotional warmth and empathy are usually expressed by open arms. Camurri et al. (2003) showed that movement activity is a relevant feature in recognizing emotion from the full-body movement of dancers. Results showed that the energy in the anger and joy performances was significantly higher than in the grief ones. From the above studies we defined and implemented two low-level user's full-body movement features: Bounding Volume and Kinetic Energy. These features can be considered a 3D extension of the two previously developed 2D low-level features—Contraction Index and Motion Index (or Quantity of Motion) (see Section 2.1 for details):

- *Bounding Volume* (BV): It is the volume of the smallest parallelepiped enclosing the user's body. Figure 4 shows an example of BV computation. The BV can be considered as an approximation of the degree of "body openness", of the user for example, if the user stretches her arms side ways or up, the BV increases.
- *Kinetic Energy* (KE): It is computed from the speed of the user's body segments, tracked by Kinect, and their percentage mass as referred by Winter (1990). In particular, the full-body kinetic energy is equal to:

$$E_{FB} = \frac{1}{2}\sum_{i=0}^{n} m_i v_i^2$$

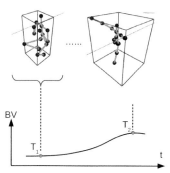

Figure 4. Bounding volume real-time computation: at time T_1 the user has a contracted posture, thus the BV value is very low; at time T_2 the user reaches an expanded posture, exhibiting a higher value for the BV.

(Color image of this figure appears in the color plate section at the end of the book.)

where m_i is the mass of the i-th user's body segment (e.g., head, right/left shoulder, right/left elbow and so on) and v_i is the velocityof the i-th segment, computed as the difference in the position of the segment at the current Kinect frame and the position at the previous frame.

3.1.3 Mid-level features: Impulsivity index

Impulsive body movements are those performed quickly, with a high energy and by suddenly moving spine/limbs in a straight direction. We adopt this definition after reviewing some literature about human movement analysis and synthesis. Wilson et al. (1996) found that stroke gestures have a lower number of execution phases compared to other conversational gestures, suggesting that their execution is shorter in time. In Laban's theory, impulsive gestures have quick time and free flow, that is, they are executed quickly with energy flowing through body in a consciously uncontrolled way. Finally, Bishko (1991) defines impulsive gestures as "an accent leading to decreasing intensity".

The measure we propose for the Impulsivity Index mid-level feature is a combination of the two low-level features: Kinetic Energy (KE) and Bounding Volume (BV). KE is firstly used to identify the gesture duration dt: for example, using an adaptive threshold, when KE becomes greater than the threshold we identify the gesture beginning time; when KE goes below the threshold we identify the ending time. Then, if KE is higher than a fixed energy threshold and the gesture length dt is lower than a fixed time threshold, then Impulsivity Index is equal to the ratio between the variation of BV and the gesture length dt:

```
let time threshold = 0.45 s;
let energy threshold = 0.02;
```

if (*KE* ≥ *energy threshold*) then evaluate *GestureTimeDuration dt*;

if (*dt* ≤ *time threshold*) then *ImpulsivityIndex* = Δ*BV* /*dt*;

3.2 Expressive gesture quality synthesis

In the virtual agent called Greta (Niewiadomski et al., 2011), expressive gesture synthesis is implemented according to the model proposed by Hartmann et al. (2005). The procedural animation generated with Greta can be modified by the use of six high-level parameters: *Overall Activity* (OAC), *Repetitivity* (REP), *Spatial Extent* (SPC), *Temporal Extent* (TMP), *Fluidity* (FLD) and *Power* (PWD).

Two of them, namely the OAC and REP, influence the general agent's activity. The first one corresponds to the general amount of activity. As this parameter increases (or decreases), the number of nonverbal behaviors increases (or decreases). Each nonverbal behavior has a value associated with it; the OAC parameter is associated to a threshold: only gestures that overpass this threshold are displayed. The second general parameter, REP, specifies the possibility that the gesture stroke will be repeated. In this manner the generation of rhythmic repetitions of the same behavior is allowed.

The four remaining parameters SPC, TMP, FLD and PWR can be specified for each gesture separately. SPC controls the amplitude of movement. It implements McNeill diagram (McNeill, 1996), where wrist positions are defined in terms of 17 sectors. SPC can be used to create expanded vs. contracted gestures. TMP controls temporal aspect of movement. It modifies the speed of execution of gesture phases. Gestures are slow if the value of the parameter has a negative value, and fast when the value is positive. It is calculated according to the Fitts's law (Fitts, 1954) that predicts that the time required to move to a target area in function of the target distance. Consequently the value of temporal extent changes the duration of each phase, and the execution time of keyframes. The third parameter, FLD, controls the smoothness and continuity of movement (e.g., smooth, graceful vs. sudden or jerky). Higher values allow smooth and continuous execution of movements, while lower values create discontinuity in the movements. Fluidity is modeled in two different manners. First of all, its value influences the interpolation of the gesture, i.e., the trajectory of the movement. For this purpose it is mapped to the continuity parameter of *Tension-Continuity-Bias* (TCB) splines. Second, it also influences the transition between two gestures. The higher is the fluidity the more likely two gestures will be fluently co-articulated without passing through a rest intermediate position.

Finally the last parameter, PWR, controls the dynamic properties of the movement. Powerful movements are expected to have higher acceleration and deceleration magnitudes. They should also exhibit less overshoot. This effect is obtained by mapping the power value to tension and bias parameters of TCB spline curves. Thus, similarly to the fluidity parameter, power modifies the trajectory of the movement. By controlling the PWR value, one may specify gestures that are weak and relaxed or, oppositely, strong and tensed.

4. Conclusion

In this chapter, we focused on the expressive quality of gesture in the HCI perspective. We discussed two aspects which are complementary: expressive gesture quality analysis and synthesis. We argued that often the same features detected in human behavior are synthesized on virtual agents. At the same time, different computational methods, both for analysis and synthesis, may focus on the same expressive feature. For this reason we conclude the chapter comparing these features and computational methods, both for analysis and synthesis. The result of this comparison is presented in Table 1.

Table 1 shows that, in particular, Fluidity and Power do not have a single universal interpretation. These features are also difficult to analyze and synthesize. Indeed, in the evaluation of the model by Hartmann et al. (2005), these two expressive features received the lowest recognition rates when synthesized with the Greta agent. It is easy to notice that sometimes different algorithms model similar features using different names, as it is in the case of Power and Tension.

To summarize this chapter, one can observe that there is a growing interest in the analysis and synthesis of the expressive quality of gesture. It also spreads over other research domains such as robotics (e.g., Le et al., 2011). However, more research is needed to specify and model expressive features of nonverbal behavior in a more realistic way. The recent developments of less invasive motion capture-based methods seem to be a promising methodology to study expressive gesture quality and to build new expressive models. The creation of efficient broad-consumer tools, such as Kinect, opens new challenges in the domain of the expressive gesture quality analysis.

Acknowledgements

We would like to thank Dr. E. Bevacqua from National Engineering School of Brest (France), Dr. G. Volpe from University of Genoa (Italy),

Table 1.   A comparison of different approaches for expressive gesture quality analysis and synthesis.

Measure	Analysis	Synthesis
Spatial aspects of movement	maximum distance of hand and elbow (Bernhardt and Robinson, 2007) distance between two hands (Cardakis et al., 2007) minimum rectangle surrounding the body (contraction index, Mancini and Castellano, 2007)	joint distance from the body (Hartmann et al., 2005), (Neff and Fiume, 2003)
Temporal aspects of movement	average hand and elbow speed (Bernhardt and Robinson, 2007) hand barycenter velocity (Mancini and Castellano, 2007)	gesture phase duration computed according to the Fitt's law (Hartmann et al., 2005)
Fluidity, Smoothness	sum of variance of the norms of the motion vectors (Caridakis et al., 2007) approximated by the Directness: ratio between the length of the shortest line and the actual trajectory (Camurri et al., 2004a) discrepancy between movements of different body parts (Mazzarino et al., 2007) relation of the sequence of motion and non-motion phases (Mazzarino et al., 2007)	sequential joint activation (Neff and Fiume, 2003) duration of and amplitude of the preparation and overshoot phase (Szczuko et al., 2009) trajectory interpolation procedure (Hartmann et al., 2005) concatenation of two gestures (Hartmann et al., 2005)
Energy, Power, Impulsivity	hand and elbow acceleration (Bernhardt and Robinson, 2007) first derivate of motion vectors (Caridakis et al., 2007) hand acceleration (Mancini and Castellano, 2007) peak in the time series of quantity of motion (Mazzarino et al., 2007) kinetic energy of user's joints (Piana et al., 2012)	tension and bias of TCB spline lines (Harmann et al., 2005)
Tension, Stiffness		timing and amplitude of gesture phases (Ness and Fiume, 2002)
Overall activity	sum of the motion vectors (Cardakis et al., 2007) Quantity of motion, velocity (Mancini and Castellano, 2007)	Threshold-based gesture selection algorithm (Hartmann et al., 2005)

and Jennifer Hofmann from University of Zurich (Switzerland) for their valuable comments and suggestions.

This research has been partially funded by the European Community Seventh Framework Programme (FP7/2007-2013) ICT, under grant agreement No. 289021 (ASCInclusion) and grant agreement No. 270780 (ILHAIRE—Incorporating Laughter into Human Avatar Interactions: Research and Experiments, http://www.ilhaire.eu).

REFERENCES

Allbeck, J. and N. Badler. 2003. Representing and Parameterizing Agent Behaviors. In H. Prendinger and M. Ishizuka (eds.), Life-like Characters: Tools, Affective Functions and Applications, Springer, Germany, pp. 19–38.

Argyle, M. 1998. Bodily Communication. Methuen and Co., London, 2nd edition.

Ball, G. and J. Breese. 2000. Emotion and Personality in a Conversational Agent. In J. Cassell, J. Sullivan, S. Prevost, and E. Churchill (eds.), Embodied Conversational Characters. MIT Press, Cambridge.

Balomenos, T., A. Raouzaiou, S. Ioannou, A. Drosopoulos, K. Karpouzis and S. Kollias. 2005. Emotion Analysis in Man-Machine Interaction Systems. In S. Bengio and H. Bourlard (eds.), Machine Learning for Multimodal Interaction, LNCS 3361, Springer Verlag, pp. 318–328.

Bernhardt, D. and P. Robinson. 2007. Detecting Affect from Non-stylised Body Motions. In A.C. Paiva, R. Prada, and R.W. Picard (eds.), Proceedings of the 2nd International Conference on Affective Computing and Intelligent Interaction (ACII '07), Springer Verlag, Berlin, Heidelberg, pp. 59–70.

Bianchi-Berthouze, N. 2012. Understanding the role of body movement in player engagement. Paper accepted to Human-Computer Interaction.

Bishko, L. 1991. The use of Laban-based analysis for the discussion of computer animation. In The 3rd Annual Conference of the Society for Animation Studies.

Bousmalis, K., M. Mehu and M. Pantic. 2009. Spotting Agreement and Disagreement: A Survey of Nonverbal Audiovisual Cues and Tools. In Proceedings of IEEE International Conference Affective Computing and Intelligent Interfaces, pp. 1–9.

Camurri, A., I. Lagerlöf and G. Volpe. 2003. Recognizing Emotion from Dance Movement: Comparison of Spectator Recognition and Automated Techniques. *International Journal of Human-Computer Studies*, **59(1-2)**:213–225.

Camurri, A., B. Mazzarino and G. Volpe. 2004. Analysis of Expressive Gesture: The EyesWeb Expressive Gesture Processing Library. In A. Camurri and G. Volpe (eds.), Gesture-based Communication in Human-Computer Interaction, LNAI 2915, Springer Verlag, pp. 460–467.

Camurri, A., B. Mazzarino, M. Ricchetti, R. Timmers and G. Volpe. 2004. Multimodal Analysis of Expressive Gesture in Music and Dance Performances. In A. Camurri and G. Volpe (eds.), Gesture-based Communication in Human-Computer Interaction, LNAI 2915, Springer Verlag, pp. 20–39.

Camurri, A., G. Castellano, M. Ricchetti and G. Volpe. 2006. Subject Interfaces: Measuring Bodily Activation during an Emotional Experience of Music. In S. Gibet, N. Courty, and J.F. Kamp (eds.), *Gesture in Human-Computer Interaction and Simulation*, volume 3881, Springer Verlag, pp. 268–279.

Camurri, A., Coletta, P., Varni, G. and Ghisio, S. (2007). Developing Multimodal Interactive Systems with EyesWeb XMI. In Proceedings of the 2007 Conference on New Interfaces for Musical Expression (NIME07), ACM, pp. 302–305.

Caridakis, G., A. Raouzaiou, E. Bevacqua, M. Mancini, K. Karpouzis, L. Malatesta and C. Pelachaud. 2007. Virtual Agent Multimodal Mimicry of Humans, In J.-C. Martin, P. Paggio, P. Kühnlein, R. Stiefelhagen, and F. Pianesi (eds.), Language Resources and Evaluation, special issue of Multimodal Corpora For Modelling Human Multimodal Behavior, Springer, Netherlands, pp. 367–388.

Cassell, J., J. Sullivan, S. Prevost and E.F. Churchill. 2000. Embodied Conversational Agents. The MIT Press, Cambridge, Massachusetts.

Castellano, G. 2006. Human full-body movement and gesture analysis for emotion recognition: A dynamic approach. Paper presented at HUMAINE Crosscurrents meeting, Athens.

Castellano, G., L. Kessous and G. Caridakis. 2007. Multimodal Emotion Recognition from Expressive Faces, Body Gestures and Speech. In F. de Rosis (ed.), Proceedings of the Doctoral Consortium of 2nd International Conference on Affective Computing and Intelligent Interaction.

Chi, D., M. Costa, L. Zhao and N. Badler. 2000. The EMOTE Model for Effort and Shape. In Proceedings of the 27th Annual Conference on Computer Graphics and Interactive Techniques (SIGGRAPH 2000). ACM Press/ Addison-Wesley Publishing Co., New York, NY, USA, pp. 173–182.

Drosopoulos, A., T. Balomenos, S. Ioannou, K. Karpouzis and S. Kollias. 2003. Emotionally-Rich Man-Machine Interaction Based on Gesture Analysis. *Human-Computer Interaction International*, in the procedings of the HCI International conference, Greece **4:**1372–1376.

Fitts, P.M. 1954. The Information Capacity of the Human Motor System in Controlling the Amplitude of Movement. *Journal of Experimental Psychology*, **47(6):**381–391.

Gallaher, P.E. 1992. Individual Differences in Nonverbal Behavior: Dimensions of Style. *Journal of Personality and Social Psychology*, **63(1):**133–145.

Glowinski, D., N. Dael, A. Camurri, G. Volpe, M. Mortillaro and K. Scherer. 2011. Toward a Minimal Representation of Affective Gestures, *IEEE Transcations on Affective Computing,* **2(2):**106–118.

Goldberg, L.R. 1993. The Structure of Phenotypic Personality Traits. *American Psychologist*, **48(1):**26–34.

Grammer, K., K. Kruck, A. Juette and B. Fink. 2000. Non-Verbal Behavior as Courtship Signals: The Role of Control and Choice in Selecting Partners. *Evolution and Human Behavior*, **21(6):**371–390.

Gunes, H. and M. Piccardi, 2005. Fusing Face and Body Display for Bi-modal Emotion Recognition: Single Frame Analysis and Multi-frame

Post Integration. In Proceedings of the First international conference on Affective Computing and Intelligent Interaction, pp. 102–111.

Hartmann, B., M. Mancini, S. Buisine and C. Pelachaud. 2005. Design and Evaluation of Expressive Gesture Synthesis for Embodied Conversational Agents. In Third International Joint Conference on Autonomous Agents and Multi-Agent Systems, Utrecht, Holland, pp. 1095–1096.

Hsieh, C.-M. and A. Luciani. 2005. Generating Dance Verbs and Assisting Computer Choreography. ACM Multimedia, Singapore, **12**:774–782.

Hsieh, C.-M. and A. Luciani 2006. Minimal Dynamic Modeling for Dance Verbs. *Journal of Visualization and Computer Animation*, **17**:359–369.

Hung, H. and D. Gatica-Perez. 2010. Estimating Cohesion in Small Groups Using Audio-Visual Nonverbal Behavior. *IEEE Transactions on Multimedia*, **12(6)**:563–575.

Jayagopi, D., H. Hung, C. Yeo and D. Gatica-Perez. 2009. Modeling Dominance in Group Conversations Using Non-verbal Activity Cues. *IEEE Trans. on Audio, Speech, and Language Processing, Special Issue on Multimodal Processing for Speech-based Interactions*, **17(3)**:501–513.

Johansson, G. 1973. Visual Perception of Biological Motion and a Model for Its Analysis. *Perception and Psychophysics*, **14**:201–211.

Kleinsmith, A. and N. Bianchi-Berthouze. 2011. Form as a Cue in the Automatic Recognition of Non-acted Affective Body Expressions. Proceedings of Affective Computing and Intelligent Interaction 2011, LNCS 6974, Springer Berlin Heidelberg, pp. 155–164.

Kleinsmith, A., N. Bianchi-Berthouze and A. Steed 2011. Automatic Recognition of Non-acted Affective Postures. *IEEE Transactions on Systems, Man, and Cybernetics,* **41(4)**:1027–1038.

Kopp, S., T. Sowa and I. Wachsmuth. 2003. Imitation Games with an Artificial Agent: From Mimicking to Understanding Shape-related Iconic Gestures. In A. Camurri and G. Volpe (eds.), Gesture-based Communication in Human-Computer Interaction, LNAI 2915, Springer Verlag, pp. 436–447.

Kostek, B. and Szczuko, P. (2006). Rough Set-Based Application to Recognition of Emotionally-Charged Animated Character's Gestures, Transactions on Rough Sets V, LNCS 4100, Springer Berlin Heidelberg, pp. 146–166. http://link.springer.com/chapter/10.1007%2F11847465_7

Laban, R. and Lawrence, F.C. (1947). *Effort.* Macdonald and Evans, USA.

Lakens, D. and M. Stel. 2011. If They Move in Sync, They Must Feel in Sync: Movement Synchrony Leads to Attributions of Rapport and Entitativity. *Social Cognition*, **29(1)**:1–14.

Le, Q.A., S. Hanoune and C. Pelachaud. 2011. Design and Implementation of an Expressive Gesture Model for a Humanoid Robot. In 11th IEEE-RAS International Conference on Humanoid Robots (Humanoids 2011), Bled, Slovenia, pp. 134–140.

Mancini, M. and G. Castellano. 2007. Real-time Analysis and Synthesis of Emotional Gesture Expressivity. In Proceedings of the Doctoral Consortium of 2nd International Conference on Affective Computing and Intelligent Interaction.

Mazzarino, B., M. Peinado, R. Boulic, G. Volpe and M.M. Wanderley. 2007. Improving the Believability of Virtual Characters Using Qualitative Gesture Analysis. In S. Gibet, N. Courty, and J.F. Kamp (eds.), Gesture in Human-Computer Interaction and Simulation, volume 3881, Springer Verlag, pp. 48–56.

Mazzarino, B. and M. Mancini. 2009. The Need for Impulsivity and Smoothness—Improving HCI by Qualitatively Measuring New High-Level Human Motion Features. Proceedings of the International Conference on Signal Processing and Multimedia Applications SIGMAP, Milan, Italy, pp. 62–67.

McNeill, D. 1996. Hand and Mind: What Gestures Reveal about Thought. University of Chicago Press, Chicago, Illinois.

Meijer, M. 1989. The Contribution of General Features of Body Movement to the Attribution of Emotions. *Journal of Nonverbal Behavior,* **13(4)**:247–268.

Neff, M. and E. Fiume. 2000. Modeling tension and relaxation for computer animation. In Proceedings of the 2002 ACM SIGGRAPH/Eurographics Symposium on Computer Animation (SCA '02), pp. 81–88.

Neff, M. and E. Fiume. 2003. Aesthetic edits for Character Animation. In Proceedings of the 2003 ACM SIGGRAPH/Eurographics Symposium on Computer Animation (SCA '03), pp. 239–244.

Niewiadomski, R., E. Bevacqua, Q.A. Le, M. Obaid, J. Looser and C. Pelachaud. 2011. Cross-media Agent Platform. In Proceedings of 2011 Web3D ACM Conference, Paris, France, pp. 11–19.

Ortony, A., G.L. Clore and A. Collins. 1988. The Cognitive Structure of Emotions. Cambridge University Press, Cambridge.

Paiva, A., R. Chaves, M. Piedade, A. Bullock, G. Andersson and K. Höök. 2003. Sentoy: A Tangible Interface to Control the Emotions of a Synthetic Character. In AAMAS '03: Proceedings of the Second International Joint Conference on Autonomous Agents and Multiagent Systems, New York, NY, ACM Press, pp. 1088–1089.

Piana, S., M. Mancini and A. Camurri. 2012. Automated Analysis of Non-verbal Expressive Gesture. Sixth International Workshop on Human Aspects in Ambient Intelligence, International Joint Conference on Ambient Intelligence, Pisa, Italy.

Pollick, F.E. 2004. The Features People Use to Recognize Human Movement Style. In A. Camurri and G. Volpe (eds.), Gesture-based Communication in Human-Computer Interaction, LNAI 2915, Springer Verlag, pp. 10–19.

Pugliese, R. and K. Lehtonen. 2011. A Framework for Motion Based Bodily Enaction with Virtual Characters, In H. Vilhjálmsson, S. Kopp, S. Marsella, and K. Thórisson (eds.), Proceedings of Intelligent Virtual Agents 2011, LNCS 6895, Springer, Berlin/Heidelberg, pp. 162–168.

Rehm, M. 2010. Non-symbolic Gestural Interaction for AmI. In H. Aghajan, R.L.-C. Delgado, and J.C. Augusto (eds.), Human-Centric Interfaces for Ambient Intelligence, ACM Press, Burlington: Elsevier Science, pp. 327–345.

Reidsma, D., A. Nijholt, R. Poppe, R. Rienks and H. Hondorp. 2006. Virtual Rap Dancer: Invitation to Dance. In Conference on Human Factors in Computing Systems, Montréal, Québec, Canada. ACM Press, pp. 263–266.

Sanghvi, J., G. Castellano, I. Leite, A. Pereira, P.W. McOwan and A. Paiva. 2011. Automatic Analysis of Affective Postures and Body Motion to Detect Engagement with a Game Companion. In Proceedings of the 6th International Conference on Human-Robot Interaction, New York, NY. ACM Press, pp. 305–312.

Schröder, M., E., Bevacqua, R. Cowie, F. Eyben, H. Gunes, D. Heylen, M. ter Maat, G. McKeown, S. Pammi, M. Pantic, C. Pelachaud, B. Schuller, E. de Sevin, M. Valstar and M. Wollmer. 2011. Building Autonomous Sensitive Artificial Listeners. *IEEE Transactions on Affective Computing*, **3(2)**:165–183.

Szczuko, P., B. Kostek and A. Czyżewski. 2009. New Method for Personalization of Avatar Animation. In K. Cyran, S. Kozielski, J. Peters, U. Stanczyk and A. Wakulicz-Deja (eds.), Man-Machine Interactions, Advances in Intelligent and Soft Computing, Springer Berlin Heidelberg, pp. 435–443.

Taylor, R., D. Torres and P. Boulanger. 2005. Using Music to Interact with a Virtual Character. In The 2005 International Conference on New Interfaces for Musical Expression, pp. 220–223.

Tsuruta, S., W. Choi and K. Hachimura. 2010. Generation of Emotional Dance Motion for Virtual Dance Collaboration System. Digital Humanities, In Procedings of DH Humanities 2010, London, UK, pp. 368–371.

Urbain, J., R. Niewiadomski, E. Bevacqua, T. Dutoit, A. Moinet, C. Pelachaud, B. Picart, J. Tilmanne and J. Wagner. 2010. AVLaughterCycle. Enabling a Virtual Agent to Join in Laughing with a Conversational Partner Using a Similarity-driven Audiovisual Laughter Animation. *Journal of Multimodal User Interfaces*, **4(1)**:47–58.

Varni, G., A. Camurri, P. Coletta and G. Volpe. 2009. Toward a Real-Time Automated Measure of Empathy and Dominance. *CSE*, **4**:843–848.

Wallbott, H.G. 1998. Bodily Expression of Emotion. *European Journal of Social Psychology*, **28**:879–896.

Wallbott, H.G. and K.R. Scherer. 1986. Cues and Channels in Emotion Recognition. *Journal of Personality and Social Psychology*, **51(4)**:690–699.

Wilson, A., A. Bobick and J. Cassell. 1996. Recovering the Temporal Structure of Natural Gesture. In Proceedings of the Second International Conference on Automatic Face and Gesture Recognition, Killington, pp. 66–71.

Winter, D. 1990. Biomechanics and Motor Control of Human Movement. John Wiley and Sons, Toronto.

CHAPTER 12

A Distributed Architecture for Real-time Dialogue and On-task Learning of Efficient Co-operative Turn-taking

Gudny Ragna Jonsdottir and Kristinn R. Thórisson

1. Introduction

Building automatic dialogue systems that match human flexibility and reactivity has proven difficult. Many factors impede the progress of such systems, useful as they may be, from the low-level of real-time audio signal analysis and noise filtering to medium-level turn-taking cues and control signals, all the way up to high-level dialogue intent and content-related interpretation. Of these, we have focussed on the dynamics of turn-taking—the real-time[1] control of who has the turn and how turns are exchanged and how to integrate these in an expandable architecture for dialogue generation and control. Manual categorization of silences, prosody and other candidate turn-giving signals, or analysis of corpora to produce static decision trees for this purpose cannot address the high within- and between-individual variability observed in natural interaction. As an alternative, we have

---

[1] By "real-time" here we mean conducting dialogue at a pace acceptable to, and in line with, human expectations, as understood and learned from real-world experience.

developed an architecture with integrated machine learning, allowing the system to automatically acquire proper turn-taking behavior. The system learns cooperative ("polite") turn-taking in real-time by talking to humans via Skype. Results show performance to be close to that of human, as found in naturally occurring dialogue, with 20% of the turn transitions taking place in under 300 milliseconds (msecs) and 50% under 500 msecs. Key contributions of this work are the methods for constructing more capable dialogue systems with an increasing number of integrated features, implementation of adaptivity for turn-taking, and a firmer theoretical ground on which to build holistic dialogue architectures.

As many have argued, turn-taking is a fundamental and necessary mechanism for real-time verbal (and extraverbal) information exchange, and should, in our opinion, be one of the key focus areas for those interested in building complete artificial dialogue systems. Turn-taking skills include minimizing overlaps, minimizing silences, giving proper back-channel feedback, barge-in techniques, and other components which most people handle fluidly and with ease. People use various multimodal behaviors including intonation and gaze, for example, to signal that they have finished speaking and are expecting a reply (Goodwin, 1981). Based on continuously streaming information from our sensory organs, most of us pick up on such signals without problems, infer the correct state of dialogue, and what the other participants intend, and then automatically produce multimodal information in real-time that achieves the goals of the dialogue. In amicable conversations, participants usually share the goal of cooperation. Turn exchange—a negotiation-based activity based on the massive historical training ("socialization"') of the participants—usually proceeds so smoothly that people do not even realize the degree of complexity inherent in the processes responsible for making it happen.

The challenge of endowing synthetic agents with such skills lies not only in the integration of perception and action in sensible planning schemes but especially in the fact that these have to be tightly coordinated while marching to a real-world clock. How easy or difficult this is is dictated by the architectural framework in which the mechanisms are being implemented, and a prime reason for the broad overview we give of our dialogue architecture here.

In spite of recent progress in speech synthesis and recognition, lack of temporal responsiveness is one of a few key components that clearly sets current dialogue systems apart from humans; speech recognition systems that have been in development for over two decades are still far from addressing the needs of real-time dynamic dialogue (Jonsdottir et al., 2007). Many researchers have pointed out the lack of implemented systems intended to manage dynamic open-microphone/full-duplex dialogue (cf. Moore, 2007; Allen et al., 2001; Raux and Eskenazi, 2007), where the system is sufficiently aware of when it is given the turn, and can be naturally interrupted at any point in time by the human, and vice versa.

Although syntax, semantics, and pragmatics can indisputably play a large role in the dynamics of turn-taking, we have argued elsewhere that natural turn-taking is partially driven by a content-free planning system[2] (Thórisson, 2002b). People rely on signals and contextual cues that from the vantage point of humans are fairly primitive, e.g. prosody, speech loudness, gaze direction, facial expressions, etc. (Goodwin, 1981). In humans, recognition of prosodic patterns, based on the timing of speech loudness, silences, and intonation, is a more light-weight process than either word recognition, syntactic, or semantic processing (Card et al., 1986). Processing load between semantic processing and contextual/turn-signal processing is even more pronounced for artificial perception (the former being more computationally intensive than the latter), and therefore such cues represent prime candidates for inclusion in the process of recognizing turn signals in artificial dialogue systems. While in the future we intend to address the full scope of human turn management contextual cues, at present even these obvious ones present challenges to architectural and system design for real-time performance that must be overcome, and are therefore continuously addressed in our work.

In natural interactions, mid-sentence pauses are a frequent occurrence. Humans have little difficulty recognizing these from proper end-of-utterance silences,[3] and reliably determine the time at which it is appropriate to take turn—even on the phone, when no visual information is available. Temporal analysis of conversational behaviors in human discourse shows that turn-transitions in natural conversations most commonly take between 0 and 250 msecs (Stivers, 2009; Wilson and Wilson, 2005; Ford and Thompson, 1996; Goodwin, 1981) in face-to-face conversation. Silences in telephone conversations—when visual cues are absent—are at least 100 msecs longer on average (Bosch et al., 2005). In a study by Wilson and Wilson (2005), response time is measured in a face-to-face scenario where both parties always

---

[2] We use the term "planning" in the most general sense, referring to any system that makes a priori decisions about what should happen before they are put in action. By "content-free"' we mean, in short, virtually without consideration for the particular dialogue topic of a conversation.

[3] Silences are often not needed to signal end-of-turn in free-form human dialogue because the interlocutor derives it from other cues, such as prosody and content, often resulting in zero silence between turns (Goodwin, 1981).

had something to say. They found that 30% of between-speaker silences (turn-transitions) were shorter than 200 msecs and 70% shorter than 500 msecs. Within-turn silences, that is, silences where the same person speaks before and after the silence, are on average around 200 msecs but can be as long as 1 second, which has been reported to be the average "silence tolerance" for American-English speakers (Jefferson, 1989); longer silences are thus likely to be interpreted by a listener as a "turn-giving signal".[4] Tolerance for silences in dialogue varies greatly between individuals, ethnic groups, and situations; participants in a political debate exhibit a considerably shorter silence tolerance than people in casual conversation—this can further be impacted by social norms (e.g. relationship of the conversants), information inferable from the interaction (type of conversation, semantics, etc.), and internal information (e.g. mood, sense of urgency, etc.). To be on par with humans in turn-taking efficiency, a system thus needs to be able to predict, given an observed silence, what the interlocutor intends to do.

The motivation for the present work is to develop a complete conversational agent that can learn to interact and adapt its interaction behavior to its conversational partners, in a short amount of time. The agent may not know a lot about any particular topic of discussion, but it would be an "expert dialoguer", whose topic knowledge could be expanded as needed for various applications and as permitted by the artificial intelligence techniques under the hood. The Ymir Turn-Taking Model (YTTM) dialogue model (Thórisson, 2002b) proposes a framework for separating envelope control from topic control, making such an approach tractable. As a first step in this endeavour we are targeting a cooperative agent that can take turns, ideally with no speech overlap, yet achieves the shortest possible silence duration between speaker turns. Our approach is intended to achieve four key goals. *First*, we want to use on-line open-mic and natural speech when communicating with the system, integrating continuous acoustic perceptions as basis for decision making. We do not want to assume that the human must change their speech style or approach the system any differently than they do another human they might talk to. *Second*, we want to model turn-taking at a higher level of detail than previous attempts have done by including incremental perception and generation in a unified way. *Third*, because of the high individual variability in interaction style and pace, we want to incorporate *learning* from the outset, allowing the system to adapt to every person it interacts with

---

[4] "Turn-giving signals" are in quotes because they are not true "signals" in the engineering sense of the term, but rather socially conditioned "contexts"—combinations of features which together constitute "polite", "improper", "rude", or otherwise connotated contexts for the behavior of the interlocutors' behaviors.

on the fly. Fourth, we have argued elsewhere (Thórisson, 2008) that conversational skills—and by extension cognitive skills—allow for a high interconnectivity between its many functions; that they are a large, heterogeneous, densely coupled system (HeLD). The design of such HeLDs requires new architectural principles—standard software development methods will simply not suffice as they result in rigid systems and require more manpower for longer extended periods than any typical university or research lab is capable of securing. As a result, both the underlying software and conceptual architecture[5] must be highly modular, expandable and malleable. This approach puts a greater emphasis on methodology than is typical, but we believe it to be one of the few ways of actually achieving the integration of the many mechanisms necessary for creating a system approaching the flexibility and generality of real-world real-time human dialogue. It may also be considered of a "practical" nature, as it makes continuous expansion of the architecture more tractable for a small team. We have found architectural structure and makeup to greatly influence not only what kinds of operations it supports but also the speed of development and manageability. We see architectural design as a *necessary* part of any effort to develop dialogue systems intended to (incrementally) approach human dialogue skills.

The architecture described below thus rests on three main theoretical pillars. The first is a distributed-systems perspective,[6] the second relates to architectural software methodology, and the third is an underlying theory of turn-taking in multimodal real-time dialogue, outlined in Thórisson (2002b), emphasizing real-time negotiation as a key principle in turn-taking. In our approach, turn-taking negotiation is managed by time-dependent "cognitive contexts" (also called "fluid states" and "schema") that, for each participant, hold which perceptions and decisions are relevant or appropriate at each particular point in time, and represent the disposition of the system at any point in the dialogue, e.g. whether we might expect the other to produce a certain turn-taking cue, whether it is relevant to generate a particular behavior (e.g. volume increase in the voice upon interruption by the other, etc.).

---

[5] By "architecture" we mean the structure and operation of the system as a whole, containing many identifiable interacting parts whose organization essentially dictates how the system acts as a whole. The difference between software architecture and conceptual architecture is often subtle, but essentially is a separation between the operation of the particular software on the particular hardware and the behavior of the dialogue system it implements.

[6] By "distributed" we mean a system with multiple semi-independent processes that can be run on multiple CPUs, computers, and/or clusters.

Our current version of the system learns to become better at taking cooperative turns in real-time dialogue while it is up and running, improving its own ability to take turns *correctly* and *quickly*, with minimal speech overlap. The results are in line with prior versions of the system, where the system interacted with itself over hundreds of trials (Jonsdottir et al., 2008). Evaluation including human subjects so far includes a within-subjects study of 5 minutes of continuous interaction with each user (a total of 50 minutes), in three different conditions: (1) A closed, noise-free, setup with a very consistent interlocutor—another instance of itself ("Artificial" condition). (2) An open-mic setup, using Skype, where the system repeatedly interviews a fairly consistent interlocutor—the same human ("Single person" condition). (3) An open-mic setup, using Skype, with individual inconsistencies where the agent interviews 10 different human participants consecutively ("10 people" condition). The system adapts quickly and effectively (linearly) within 2 minutes of interaction, a result which, in light of most other machine-learning work on the subject—many of which require thousands of hand-picked training examples—is exceptionally efficient.

The rest of this chapter is organized as follows: First, we review related work, then we detail the architecture and learning mechanisms. A description of the evaluation setup comes next, followed by the results, summary, and future work.

2. Related Work

Models of dialogue produced by a standard divide-and-conquer approach can only address a subset of a system's behaviors, and are even quite possibly doomed at the outset. This view has been presented in our prior work (Thórisson, 2008) and is echoed in other work on dialogue architectures (cf. Moore, 2007). Requiring a holistic approach to a complex system such as human real-time dialogue may seem to be impossibly difficult. In our experience, and perhaps somewhat counterintuitively, when taking a breath-first approach to the creation of an architecture that models any complex system—where most of the significant high-level features of the system to be addressed are taken into account—the set of likely contributing underlying mechanisms will be greatly reduced (Schwabacher and Gelsey, 1996), quite possibly to a small, manageable set, thus greatly simplifying the task. It is the use of levels of abstraction that is especially important for cognitive phenomena: Use of hierarchical approaches is common in other scientific fields such as physics; for example, behind models of optics lie more detailed models of electromagnetic waves (Schaffner, 2006). A way to address the problem of building more complete models of dialogue is thus to take an interdisciplinary approach, bringing results from a number of sources to the table at various levels of abstraction and detail. This is essentially our approach.

When dealing with the modeling of complex phenomena, building architectures for systems that integrate multimodal data and exhibit heterogeneous real-time behaviors, it seems sensible to try to constrain the possible design space from the outset. One powerful way to do this is to build multilevel representations (cf. Schwabacher and Gelsey, 1996; Gaud et al., 2007; Dayan, 2000; Arbib, 1987); this may, in fact, be the only way to get our models right when trying to understand complex systems such as natural human dialogue. The thrust of this argument is not that multiple levels are "valid" or even "important", as that is a commonly accepted view in science and philosophy, but, rather, that to map correctly to the many ways sub-systems interact in such systems they are a *critical necessity*—that, unless our simulations are built at fairly high levels of fidelity, we cannot expect manipulations (expansions, modifications) by its designers to the architecture at various levels of detail to produce valid results. Modularity in the architecture is thus highly desirable as it brings transparency and openness to the architecture, making the modelling of a highly complex system tractable. However, gross modularity does not allow the kind of fine-grain representation that we argue is important for such systems. One drawback of fine-grain modularity is that decoupling components results in essence in a more distributed architecture, which calls for non-centralized control schemes. The kind of modularity and methodology one adopts is critical to the success of such decoupling.

Many of the existing methodologies that have been offered in the area of distributed agent-based system construction (cf. Wood and Deloach, 2000; Wooldridge et al., 2000) suffer from lack of actual use-case experiences, especially for artificial intelligence projects that involve construction of single-mind systems. We have built our present model using the Constructionist Design Methodology (CDM) (Thórisson et al., 2004) which helps us create complex multi-component systems at a fairly high level of fidelity, without losing control of the development process. CDM proposes nine iterative principles to help with the creation of such systems and has already been applied in the construction of several systems, both for robots and virtual agents (cf. Thórisson et al., 2004; Ng-Thow-Hing et al., 2007; Thórisson and Jonsdottir, 2008). CDM assumes a relatively manual construction process whereby a large number of pieces are integrated, for example speech recognition, animation, planning, etc., some of which may be off-the-shelf while others are custom-built. As such systems have to be deconstructed and reconstructed often, CDM proposes blackboards as the backbone for such integration. This makes it relatively easy to change information flow, add or remove computational functionality, etc., even at runtime, as we have regularly done.

As far as dialogue management and turn-taking are concerned, modular or distributed approaches are scarce. Among the few is the YTTM (Thórisson, 2002b), a model of multimodal real-time turn-taking. YTTM proposes that processing related to turn-taking can be separated, in a particular manner, from the processing of content (i.e. topic). Echoing the CDM, its architectural approach is distributed and modular and supports full-duplex multi-layered input analysis and output generation with natural response times (real-time). One of the background assumptions behind the approach, which has been reinforced over time by systems built using the approach (Thórisson et al., 2008; Jonsdottir, 2008; Ng-Thow-Hing et al., 2007), is that real-time performance calls for the incremental processing of interpretation and output generation.

The J.Jr. system (Thórisson, 1993) was a real-time communicative agent that could take turns in real-time casual conversation with a human. It was controlled by a finite state-machine architecture, similar to the Subsumption Architecture (Brooks, 1986). The system did not process the *content* of a user's speech, but instead relied on an analysis of prosodic information to make decisions about when to ask questions (i.e. take turn) and when to interject back-channel feedback. While modular, this architecture turned out to be difficult to expand into a larger, more intelligent architecture (Thórisson, 1996), especially when confronted with features at different time scales and levels of abstraction and detail (prosodic, semantic, pragmatic). Subsequent work on Gandalf (Thórisson, 1996) incorporated mechanisms from J.Jr. into the Ymir architecture, but presented a much more expandable, modular system of perception modules, deciders, and action modules in a holistic architecture that addressed content (interpretation and generation of meaning) as well as envelope phenomena (process control). A descendant of this architecture and methodology was recently used in building an advanced dialogue and planning system for the Honda ASIMO robot (Ng-Thow-Hing et al., 2007).

Raux and Eskenazi (2008) presented data from a corpus analysis of an online bus scheduling/information system, showing that a number of dialogue features, including speech act type, can be used to improve the identification of speech endpoint, given a silence. The authors tested their findings in a real-time system: Using information about dialogue structure—speech act classes, a measure of semantic completeness, and probability distribution of how long utterances go (but not prosody)—the system improved turn-taking latency by as much as 50% in some cases, but significantly less in others. This work reported no benefits from prosody for this purpose, which is surprising given that many studies have shown the opposite to be true (cf. Gratch et al., 2006; Schlangen, 2006; Thórisson, 1996; Traum and Heeman, 1996; Pierrehumbert and Hirschberg, 1990; Goodwin, 1981). We suspect one reason could be that the pitch and intensity extraction methods they used did not work very well on the data selected for analysis. Prosodic information has successfully been used to determine back-channel feedback in real-time. The Rapport Agent (Gratch et al., 2006) uses gaze, posture, and prosodic perception, among other things, to detect backchannel opportunities. The Ymir/Gandalf system (Thórisson, 1996) also used prosody, adding analysis of semantic, syntactic (and to a small extent even pragmatic) completeness to determine turntaking behaviors. Unfortunately evaluations of its benefit, for the purpose of turn-taking per se, are not available. The major lesson that can be learned from Raux and Eskenazi, echoing the work on Gandalf, is that turn-taking can be improved through an integrated, coordinated use of various features *in context.*

The problem of utterance segmenting for the purpose of proper turn-taking has been addressed to some extent in prior work. Of all the data sources informing dialogue participants about the state of the dialogue, prosody is the most prominent among the non-semantic ones. From the prior work reviewed, this seems like the most obvious place to start when attempting to design turn-taking mechanisms. Sato et al. (2002) use a decision tree to classify when silence signals that a turn should be taken. They annotated various features in a large corpus of human-human conversation to train and test the tree. The results show that semantic and syntactic categories, as well as understanding, are the most important features. These experiments have so far been limited to annotated data of a single, task-oriented domain. Applying their methods to a casual real-time conversation using today's speech recognition methods would inevitably increase the recognition time beyond any practical use because of an increased vocabulary—the content interpretation results could simply not be produced fast and reliably enough for making turn-taking decisions at sub-second speeds (Jonsdottir et al., 2007).

The introduction of learning into a dialogue system gives its designers yet another complex dimension which can affect everything and anything in the architecture's organization. Schlangen (2006) has successfully used machine learning to categorize prosodic features from corpus, showing that acoustic features can be learnt. Traum and Heeman (1996) have addressed the problem of utterance segmenting, showing that prosodic features such as boundary tones do play a role in turn-taking. As far as we know, none of this work has been applied to real-time situations. Bonaiuto and Thórisson (2008) demonstrate a system of two simulated interacting dialogue participants that learn to exploit each other's multimodal behaviors (that is, modality-independent multi-dimensional behaviors) to achieve a cooperative interaction where minimizing speech overlaps and speech pauses are the shared goals (as is the standard situation in amicable interactions between acquaintances, friends, and family—shared with the present work). Using a neuro-cognitive model of learning, the work shows that emergent properties of dialogue, pauses, hesitations, interruptions— i.e. negotiations of turn—can be learned via the general framework provided by YTTM, and its fluid states, coupled with Bonaiuto and Arbib's ACQ model of learning (Bonaiuto and Arbib, 2010). While Bonaiuto and Thórisson's system was based on the YTTM, the implementation of the learning mechanisms was neither meant to run on-line nor in real-time.

In summary, no prior system has implemented a comprehensive dialogue system capable of on-line learning of turn-taking skills, and allowed it to adapt to its interlocutors in real-time. The turn-taking model presented here is an extended version of the YTTM (Thórisson, 2002b) with the simplification that the communicative channel is limited to the speech modality. Turn-taking is modeled as an agent-oriented negotiation process with eight turn-taking, semi-global "cognitive contexts" or fluid states that define the perceptual and behavioral disposition of the system at any point in the dialogue, as already mentioned. These contexts support, in effect, a distributed planning and control system for both perception and action; the distributed learning scheme we present below implements a negotiation-driven tuning of real-time turn-taking behaviors within this framework.

3. System Architecture

Our multi-module dialogue system is capable of real-time dialogue with human users speaking naturally, with no artificial constraints on the process of interaction. As mentioned above, the architecture follows the principles of modularity outlined above, as specified by the CDM methodology (Thórisson et al., 2004; Thórisson, 2008), and enables us to introduce learning into the architecture in a modular

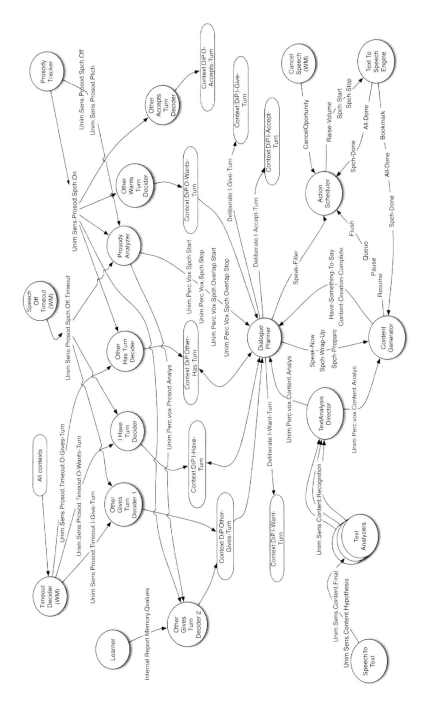

Figure 1. Flat layout of message passing between modules.

way.[7] As an indication of the architecture's present scope, Figure 1 presents a full (flat) view of the system's gross architecture. (Note that while a flat view is informative of the architecture's scope, it belies its naturally hierarchical nature—an important feature of our system that allows us to build a complex architecture with a very small team of developers.) We will discuss the architecture's various components below. A complete introduction to the architecture is beyond the scope of this chapter; the main focus of this chapter will be on the parts of turn-taking needed to support learning of efficient turn-taking.

Following the Ymir architecture (Thórisson, 1996), our system's modules are categorized based on their functionality; perception-, decider-, and action modules, at the coarsest granularity (see Figure 2). We will now describe the modules of these types that relate to the turn-taking system.

3.1 Perception

As already mentioned, although the architecture is inherently a multimodal system (as shown in prior work (c.f. Bonaiuto and Thórisson, 2008; Ng-Thow-Hing et al., 2007; Thórisson, 2002b, 1996)), the current system's input is limited to audio input. There are two main perception modules that deal with prosodic features in the system, the Prosody Tracker and the Prosody Analyzer. The Prosody Tracker is a low-level perception module whose input is a raw audio signal (Nivel and Thorisson, 2008). It computes speech signal levels and determines information about speech activity, producing time-stamped Speech-On and Speech-Off messages. It also analyzes the speech pitch incrementally (in steps of 16 msecs) and produces pitch values, in the form of a continuous stream of pitch message updates.

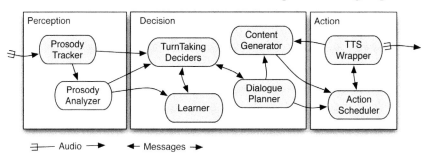

Figure 2. System components, each component consists of one or more modules.

---

[7] This is important because it not only allows the architecture to be more easily expanded in the future, but also the learning mechanisms, which we will show in future papers.

Similar to Thórisson (2002a), pitch is further analyzed by a Prosody Analyzer perception module to compute a more compact representation of the pitch pattern in a discrete state space, in our case to support the learning: The most recent tail of speech right before a silence, the last 300 msecs, is analyzed to detect minimum and maximum values of the fundamental pitch to produce a tail-slope pattern of the pitch. Slope is split into semantic categories; in the present implementation we have used three categories for slope: *Up, Straight* and *Down* according to Formula 1 and three for the relative value of pitch right before silence: *Above, At* and *Below*, as compared to the average pitch according to Formula 2.

$$m = \frac{\Delta pitch}{\Delta msecs} \begin{cases} if\ m > 0.05 \rightarrow slope = Up \\ if\ (-0.05 \le m \le 0.05) \rightarrow slope = Straight \\ if\ m < 0.05 \rightarrow slope = Down \end{cases} \qquad 1$$

$$d = pitch_{end} - pitch_{avg} \begin{cases} if\ d > Pt \rightarrow end = Above \\ if\ (-Pt \le d \le Pt) \rightarrow end = At \\ if\ d < Pt \rightarrow end = Below \end{cases} \qquad 2$$

where *Pt* is the average ± 10, i.e. pitch average with a bit of tolerance for deviation.

The primary output of the Prosody Analyzer is a symbolic representation of the particular prosody pattern identified in this tail period (see Figure 3). More features could be added into the symbolic representation, with the obvious side effect of increasing the state space.

The Speech-To-Text module and Text Analyzers deal with speech recognition. Speech recognition is done incrementally with the best

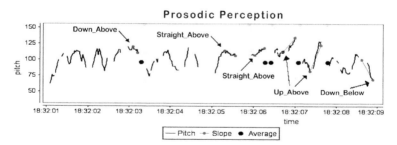

Figure 3. A window of 9 seconds of spontaneous speech, which includes speech periods and silences, categorized into descriptive groups for slope and end position relative to the average pitch. Only slope of the fundamental pitch during the immediate 300 msecs preceding a silence (indicated by the gray area) is categorized (into Up, Straight, and Down). (Abscissa: Voice F0 in Hz, as produced in near real-time by Prosodica; mantissa: Time-Hours/minutes/seconds.)

score hypothesis being available to the rest of the system during interlocutors' speech, but final utterance is not calculated until at least one second of silence has been detected.

3.2 Deciders

Our detailed turn-taking model consists of eight dialogue states (see Figure 4). This represents the states taken when the turn switches hands. The dialogue states are modeled with a distributed semi-global context system, implementing what can (approximately) be described as a distributed finite state machine that selectively applies to the activation and de-activation of most modules in the system. Context transition control ("state transitions") in this system is managed by a set of deciders (Thórisson, 2008). There is no theoretical limit to how many deciders can be active for a single given system-wide context. Likewise, there is no limit to how many deciders can manage identical or non-identical transitions. Reactive deciders (IGTD, OWTD, ...) are the simplest, with one decider per transition. Each contains at least one rule about when to transition, based on both temporal and other information. Transitions are made in a pull manner: the Other-Accepts-Turn-Decider, e.g. transits to context Other-Accepts-Turn (see Figure 4).

The Dialogue Planner (DP) and Learning modules (see further description below) can influence the dialogue state directly by sending context transition messages I-Want-Turn, I-Accept-Turn, and I-Give-Turn; however, all these decisions are under the supervisory control of the DP: If the Content Generator (CG) has some content ready to be communicated, the agent might want to signal that it wants a turn and it may want to signal I-Give-Turn when the content queue is empty (i.e. have nothing to say). Decisions made by these modules override decisions made by other turn- taking modules. The DP also manages the content delivery; that is, when to start speaking, withdraw, or raise one's voice. The CG is responsible for creating utterances incrementally, in "thought chunks", typically of durations shorter than 1 second. We are developing a dynamic content generation system at present; based on these principles the CG currently simulates its activity by selecting thought units to speak from a pre-defined list. It signals when content is available to be communicated and when content has been delivered.

In the present system, the module Other-Gives-Turn-Decider-2 (OGTD-2) uses the data produced by the Learner module to change the behavior of the system. At the point when the speaker stops speaking, the challenge for the listening agent is to decide how long to wait before starting to speak (OGTD-1 has a static behavior of transitioning

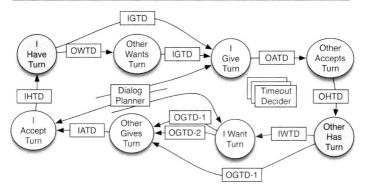

Figure 4. The heart of turn-taking control in the system consists of a set of eight semi-global context-states and 11 deciders. In context-state I-Have-Turn (IHT), both I-Give-Turn-Decider (IGTD) and Other-Wants-Turn-Decider (OWTD) are active. Unlike other modules, the Dialog Planner (DP) can transition independently from the system's current context-state and override the decisions from the reactive deciders. A Timeout-Decider handles transitions if one of the negotiating contexts is being held unacceptably long (but its transitions are not included in this diagram; also not shown are which modules are active during which contexts).

to Other-Gives-Turn after a two-second silence). If the agent waits too long, and the speaker does not continue, there will be an unwanted silence; if he starts too soon and the speaker continues speaking, overlapping speech will result. We solve this by having OGTD-2 use information about past prosody, which occurs right before the latest silence, to select an optimal silence tolerance window (STW), as will now be described in detail.

4. The Learner

The learning mechanism is implemented as an independent component (Learner module) in the modular architecture described above. It is based on the Actor-Critic distribution of functionality (Sutton and Barto, 1998), where one or more actors make decisions about which actions to perform and a critic evaluates the effect of these actions on the environment; the separation between decision and action is important because in our system a decision can be made to act in the future. In the highly general and distributed learning mechanism we have implemented, any module in the system can take the role of an actor by sending out decisions and receiving, in return, an updated decision policy from an associated Learner module. A decision consists of a state-action pair: the action being selected and the evidence used in making that action represents the state. Each actor follows its own action-selection policy, which controls how it explores its actions; various methods such as å-greedy exploration, guided exploration, or confidence value thresholds can be used (Sutton and Barto, 1998).

In our system, the Learner module takes the role of a critic. It consists of the learning method, reward functions, and the decision policy being learnt. A Learner monitors decisions being made in the system and calculates rewards based on a reward function, a list of decision/event pairs, and signals from the environment—in our case overlapping speech and long silences—and publishes an updated decision policy (the environment consists of the relevant modules in the system), which any actor module can subsequently use to base its decision on.

We use a delayed one-step Q-Learning method according to the formula:

$$\mathbf{Q(s, a) = Q(s, a) + \alpha[reward - Q(s, a)]} \qquad 3$$

where $Q(s,a)$ is the learnt estimated return for picking action a in state s, and $?$ is the learning rate. The reward functions—what events following what actions lead to what reward—is pre-determined in the Learner's configuration in the form of rules: A *reward* of x if *event y* succeeds at *action z*. Each decision has a lifetime in which system events can determine a reward, but the reward can also be calculated in absence of an event, after its given lifetime has passed (e.g. no overlapping speech). Each time an action gets a reward, the return value is recalculated according to Formula 3 and the Learner broadcasts the new value.

In the current setup, Other-Gives-Turn-Decider-2 (OGTD-2) is an actor in Sutton's sense (Sutton and Barto, 1998); it decides essentially what its name implies. This decider is only active in the state I-Want-Turn. It learns an "optimal" STW, which prevents it from speaking on top of the other, while minimizing the lag in starting to speak, given a silence. Each time a Speech-Off signal is detected, OGTD-2 receives analysis of the pitch in the last part of the utterance preceding the silence from the Prosody Analyzer. The prosody information is then used to represent the state for the decision; a predicted safe STW is selected as the *action* and the Decision is posted. The end of the STW determines when, in the future, the participant who currently doesn't have the turn will start speaking (take the turn). In the case where the interlocutor starts speaking again before this STW closes, the decider doesn't signal Other-Giving-Turn, essentially canceling the plan to start speaking (see Figure 5). This leads to a better reward, since no overlapping speech occurs. If he starts talking just after the STW closes, after the decider signals Other-Gives-Turn, overlapping speech will likely occur (keep in mind that, due to processing time, once a decision has been made it can take time before it is actually executed), leading to negative reinforcement for this size of STW given to the particular prosodic information observed.

This learning strategy is based on the assumption that both agents want to take turns cooperatively ("politely") and efficiently. We have already begun expanding the system to be able to interrupt dynamically and deliberately—i.e. be "rude"—and the ability to switch back to being polite at any time, without destroying the learned data. This research will be discussed at a later date.

5. Quantitative Evaluation of Learning

We will look at system performance across three dependent measures:

- The system's ability to select an appropriate STW. Given a silence in the user's speech, the selection of an STW is based on the type of prosody pattern perceived right before the silence. If turn-giving indicators are perceivable to the system, we should find clear variations in STW lengths based on the pattern perceived. If no evidence of turn-giving is detected by the system, we should find an even distribution of STW size between patterns.

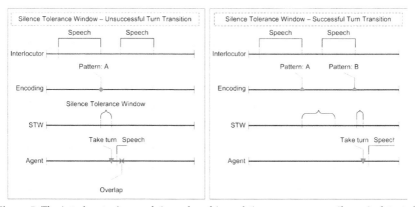

Figure 5. The interlocutor's speech is analyzed in real-time; as soon as a silence is detected the prosody preceding the silence is decoded. The system makes a prediction by selecting an STW, based on the prosody pattern perceived in the interlocutor. This window is a prediction of the shortest safe duration to wait before taking turn: A window that is too short will probably result in overlapping speech while a window that is too large may cause unnecessary or unwanted silences.

- How quickly the agent takes its turn. We evaluate this by measuring the length of the silence before each successful turn-transition (from other to the agent) and compare the results to human data.

- Frequency of overlapping speech. Because the agent should be learning to be polite—i.e. not speak on top of the other—the number of overlaps should decrease over time. (Note: In our Speaking-with-Self condition we use a closed sound loop (no open mic), but an open mic setup when the system speaks with humans.)

5.1 Hypotheses and statistics

To evaluate the learning mechanism, we used linear regression on the single-case data sessions (*Artificial*—talking to itself (a copy of itself in the interviewee role) for 10 consecutive sessions with 30 questions each; *Single person*—talking to one person for 10 consecutive sessions with 30 questions each). For the *10-person* condition (asking 10 different people 30 questions each), we used within-subject *t*-tests between the first five sessions and the second five sessions. In all cases the dependent variables are: (a) Taking Turn in less than 500 msecs, (b) Taking Turn in less than 300 msecs, and (c) Number of Overlaps.

The hypotheses are:

- H1: Frequency of taking turn within less than 500 msecs should increase as a function of number of turns.

- H2: Frequency of taking turn within less than 300 msecs should increase as a function of number of turns.

- H3: Frequency of overlapping speech should be higher in the first half of the interviews than in the second half of the interviews.

5.2 Interview setup

The agent is configured to ask 30 pre-defined questions, using, among other things, STW to control its turn-taking behavior during the interlocutor's turn (see Figure 5). Each interaction takes approximately five minutes. We have run three different evaluation conditions with the system.

1. **The system interviewing itself ("Artificial").** Having a single artificial interlocutor interacting with a non-learning instance of itself gives us a very consistent behavior in a setup with no background noise, providing a baseline for the real-world evaluations.

2. **The system interviewing a single person ("Single person").** A single person should be fairly consistent in behavior, but some external noise is inevitable since the communication is through Skype. Significant results with a single person would show that the system can adapt with a very small set of learning data—a highly desirable feature for such systems.

3. **The system interviewing 10 people ("10 people").** This is the most complex condition, as there is both individual variation between participants as well as background noise. Individual variations could be a confounding factor; getting significant results in this condition would mean that the system shows robustness to individual variation. Improvement over time indicates that the system can learn from, and in spite of, individual differences.

In all conditions, the system is learning to take turn in a "polite" cooperative manner while striving for the shortest possible silence between turns. Each evaluation consists of 10 consecutive interviews. Our learning system, named Askur for convenience, begins the first interview with no knowledge, and gradually adapts to its interlocutors throughout the 10 interview sessions.

The goal of the learning system is to learn to take turns with no speech overlap, yet achieve the shortest possible duration of silence between speaker turns. To eliminate variations in STW due to lack of something to say, we have chosen an interview scenario where the learning agent is the interviewer, in which case it always has something to say (until it runs out of questions and the interview is over).

We are aiming at an agent that can adapt its turn-taking behavior to dialogue in a short amount of time, using incremental perception. In the evaluations we focus exclusively on detecting turn-giving indicators in deliberately generated prosody, leaving out the topic of turn-opportunity detection (i.e. turn transition without prior indication from the speaker that she's giving the turn), which would, for example, be necessary for producing human-like interruptions.

A sample of 11 Icelandic volunteers took part in the experiment, none of whom had interacted with the system before. All subjects spoke English to the agent, with varying amounts of Icelandic prosody patterns, which differ from native English-speaking subjects. The study was done in a partially controlled setup; all subjects interacted with the system through Skype using the same hardware (computer, microphone, etc.) but the location was only semi-private and some background noise was present in all cases.

5.3 Parameter settings

The main goal of the learning task is to differentiate silences in real-time based on partial information of an interlocutor's behavior (prosody only) and predict the best reciprocal behavior. For best performance, the system needs to find the right tradeoff between shorter silences and the risk of overlapping speech. To formulate this as a Reinforcement Learning problem, we need to define states and actions for our scenario.

Using single-step Q-Learning, the feature combination in the prosody preceding the current silence becomes the *state* and the length of the STW becomes the *action* to be learned. For efficiency, we have split the continuous action space into discrete logarithmic values (see Table 1), starting with 10 msecs and doubling the value up to 1.28 seconds (the maximum STW where the system takes the turn by default). The action selection policy for OGTD-2 is ε-greedy with 10% exploration, always selecting the shorter STW if two or more actions share the top spot.

The reward given for decisions that do not lead to overlapping speech (i.e. successful transitions) is the milliseconds in the selected STW; a 100 msec STW will receive a reward of −100 if successful and STW of 10 msecs will receive −10 points. If, however, overlapping speech results from the decision (i.e. the action is unsuccessful), a fixed reward of −2000 (i.e. more than waiting the maximum amount of time) is given. This is to simulate that when two STWs are without overlap, the smaller is better. Every reward in the learning system is negative, resulting in unexplored actions being the best option at each time, since return starts at 0.0 for unexplored actions, and once a reward has been given the return can only decrease. In the beginning, the agent

Table 1.   Discrete actions representing STW size in msecs.

Action (STW)	Reward: Successful transition	Reward: Unsuccessful transition
10	−10	−2000
20	−20	−2000
40	−40	−2000
80	−80	−2000
160	−160	−2000
320	−320	−2000
640	−640	−2000
1280	−1280	−2000

is only aware of actions 1280 and 640 and only explores shorter STWs for patterns where the lowest available STW is considered the best.

6. Results

To reiterate, there are three conditions: Artificial, Single person, and 10 people. First we will answer the question of whether the system is learning; then we will look at the above dependent measures in more detail.

6.1 Is the system learning?

The system showed significant learning effects for the Artificial condition, both for reaction time (simple regression $F = 12.83$; $p < 0.0005$) and overlaps (simple regression $F = 10.41$; $p < 0.0047$). The system also showed significant learning effects for the 10-person condition, for reaction time (see Table 2), and overlaps (see Table 3). Although an 89 msec gain in STW may seem small, it makes a big qualitative difference for most average dialogue participants, essentially changing an automatic dialogue system from being obviously inadequate and sometimes annoyingly slow to not being so. The system starts each interview with previous learning and thus optimal STW based on another person's prosody patterns instead of beginning with a "safe" 1-2 second STW. To shorten this previous optimal STW, at the same time as overlaps drop from 24% to 10%, shows that the agent is learning new skills on the fly, becoming increasingly more "polite" (efficient and cooperative) by improving its reaction time and speech overlap performance between- as well as within-interviews.

Table 2. Paired one-tail *t*-test: Interviewing 10 consecutive people.

Turn	Observation (*N*)	Mean	St.Dev
Turn 1–15	10	655 msecs	137.25
Turn 16–30	10	566 msecs	73.83
T-value = 2.46, *P*-value = 0.018, DF = 9			

Table 3. Paired one-tail *t*-test: Overlaps when interviewing 10 consecutive people.

Turn	Observation (*N*)	Mean	St.Dev
Turn 1–15	10	0.24	0.11
Turn 16–30	10	0.10	0.09
T-value = 4.16, *P*-value = 0.0012, DF = 10			

In the single-person condition, overlaps get continually fewer (simple regression $F = 3.39$; $p < 0.08$), but improvement in reaction time is not statistically significant although still indicative of the same trends as observed in the other conditions. The observed improvements are nevertheless in the expected direction, indicating that the system is, in fact, improving during its interactions with this particular individual, as it did with statistical significance for the more consistent artificial interlocutor. In future we will seek to improve the performance for individuals, since a noticeable adaptivity at the individual level is a worthy, quite impressive goal to reach.

6.2 Silence tolerance window (STW) by pattern

We look for turn-giving intonation patterns in the last 300 msecs of speech before each silence. Tail pattern of the pitch is currently categorized into nine semantic categories based on slope (Up, At, Down) and final pitch compared to average (Above, At, Below).

To begin with, we will analyze the distribution of these patterns before the silences that mark end of turn, and before the silences that are within turn. In both the artificial interviewee evaluation and single-person evaluation, the pattern *Down_Below* (representing a final fall in pitch) is most widely used at end of a turn (see Table 4). This harmonizes well with previous research (Pierrehumbert and Hirschberg, 1990), which has associated a final fall in pitch with a turn-giving signal. Furthermore, the person and the artificial interviewee have a very similar distribution of patterns at the end of a turn. The same cannot be said about prosody patterns perceived before silences that do not lead to turn transition. Prosody before silences within

Table 4.   Distribution between prosody patterns.

Pattern	Artificial Interlocutor		Human Interlocutor	
	At end	Within	At end	Within
Down_Below	58.6%	0.4%	42.0%	12.6%
Straight_Below	10.3%	0.1%	14.1%	17.2%
Up_Below	8.6%	2.4%	7.3%	3.6%
Down_At	8.4%	20.6%	10.4%	14.6%
Up_Above	5.6%	38.1%	5.0%	14.4%
Straight_At	2.7%	10.6%	6.5%	15.7%
Straight_Above	2.2%	13.7%	7.3%	9.7%
Down_Above	2.1%	10.2%	3.4%	4.6%
Up_At	1.5%	4.1%	3.9%	7.6%

turn are much more evenly distributed between categories in the person's speech than in the artificial interviewee's speech. The artificial interviewee is as stated before very consistent, he decides what to say beforehand and sticks to that. After listening to the recordings of the person speaking there is a lot more variation occurring; decisions are being made and changed at the spur of the moment leading to more inconsistencies in prosody. An example of that is a person giving a short answer with prosody that can be perceived as giving turn and then adding to the answer and again ending with a give-turn prosody (e.g. "My favorite actor is Will Smith. and Ben Affleck.").

When the agent interviews 10 consecutive people, we analyzed which patterns were most widely used at the end of a turn. We found that four patterns out of nine are seen in up to 80% of turn-transitions (see Figure 6). None of these patterns have an end pitch above session average. This might be due to the fact that people are not asking the agent any questions—and questions tend to end on a higher-than-average pitch.

We further analyzed the use of the final fall pattern *Down_Below*, both as turn-transition and within turn. The use of final fall, both at end of turn and within turn, varies considerably between participants. The person that uses final fall the most at end of turn uses it in 41% of the ends of turns, while the person that uses it the least only uses it in 2.7% of cases (see Table 5). This is surprising as the pool of participants are all from the same cultural background and we would thus speculate more similarities in behavior.

Table 5.   Usage of Down Below in the 10-person study.

Participant	At end	Within
1	7.7%	14.9%
2	14.8%	7.3%
3	34.8%	6.7%
4	6.3%	9.1%
5	2.7%	7.1%
6	27.3%	15.4%
7	41.0%	8.7%
8	18.8%	5.0%
9	11.1%	2.5%
10	25.0%	19.2%

6.3 Silence length

A study on human behavior by Wilson and Wilson (2005) measured silences in face-to-face conversations where participants always had something to say. They reported response time to be shorter than 500 msecs in 70% of turn-transitions and shorter than 200 msecs in 30% of turn-transitions.

Our study was conducted over a relatively low-quality voice connection (Skype) and not face-to-face, and thus allows only for voice cues to communicate envelope feedback regarding turns. The studies are compatible in the sense that our agent always has something to say, while people might have to think a bit before they answer. Silences in telephone conversations tend to average about 100 msecs longer than in face-to-face conversations (Bosch et al., 2005), so we have measured silences shorter than 300 msecs and shorter than 500 msecs.

Table 6.   Average silences for each condition.

Condition	Shorter than 500 msecs	Shorter than 300 msecs
Artificial	53.1%	32.2%
Single person	44.0%	16.3%
10 people	43.7%	8.4%

Our agent takes its turn in less than 500 mescs in 53.1% to 43.7% of turns for our three conditions (see Table 6). This is the average for the last nine interviews, eliminating the first interview due to the preset STW of 1280 and 640 msecs, which would interfere with obtaining results based on real-time interactions.

When looking at how silence length evolves during the series of interviews, it is obvious that Askur (the interviewer) adapts relatively quickly in the beginning in all cases. In the first session, when the agent interviews a copy of itself in the interviewee role, it is obviously interviewing the most consistent interviewee; the agent gets constantly better with only minor lapses until it reaches about 70% of silences shorter than 500 msecs and around 40% of silences shorter than 300 msecs (see Figure 7). When interviewing a single person for 10 consecutive interviews, the system cannot learn as well as when interviewing a copy of itself since there is more variation in behavior.

When interviewing 10 people, Askur has reached about 50% of before-turn silences shorter than 500 msecs (see Figure 7), compared to 70% in the human-human comparison data. There are two distinct dips in performance in interviews four and eight. These can be attributed to differences in the prosody patterns used by participants (see Figure

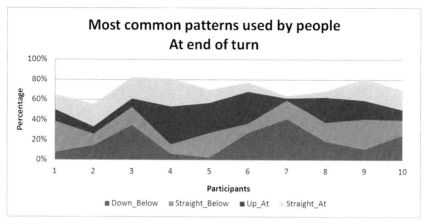

Figure 6. Four prosody categories out of nine are seen in up to 80% of turn-transitions before the agent takes turn.

6). In the case of interviewee four, the agent needs to learn that *Up_At* is a turn giving signal (used in 37.5% of 4's turns), but in the case of participant eight it is not as obvious. While examining overlaps, it can be seen that a lot of overlaps occurr in interview eight and at the beginning of interview nine, indicating that the agent is making mistakes (see Figure 8).

6.4 Turn overlaps

The final evaluation of success is to view the overlapped turns in each condition. In the first condition when interviewing self (Artificial), the overlaps are mostly in the first half of the evaluation. After that, overlaps drop considerably and stay low throughout the remainder of the sessions. This is due to the consistency of the interlocutor, the system learns how to interact with the interlocutor, and makes very few mistakes towards the end of the evaluation. In the second condition (Single person) when interviewing a single person for 10 sessions, overlaps are around 10% or below for all interviews except at the beginning of third and fifth interviews. In the last scenario where the system interviews 10 different people, overlaps occur more randomly due to differences in participants.

It is not surprising that most overlaps are perceived in the last condition, when the system interviews 10 different people (17%). It is, however, surprising that fewer overlaps are perceived when interviewing a single person over an open microphone than when interviewing an artificial interlocutor in a closed (sound card to sound card) setting (see Table 7). The artificial interlocutor always selects one to three sentence fragments and inserts artificial "think pauses"

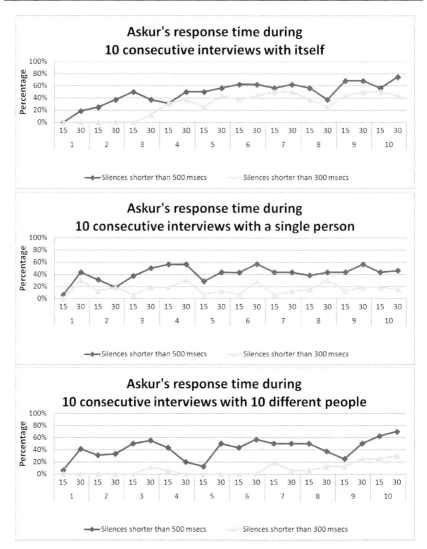

Figure 7. Proportion of silences with human speed characteristics. The graphs show 10 consecutive interviews in three different conditions. Each session is 10 consecutive interviews, each interview is 30 turns.

Table 7. Average silences for each condition.

Condition	Overlapped turns
Artificial	15.3%
Single person	10.3%
10 people	17.0%

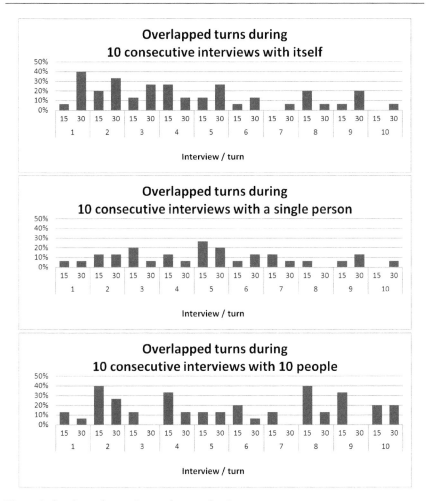

Figure 8. Overlapped turns in our three evaluations.

with a length 0 to 1000 msecs between them; people tend to answer in shorter sentences, not allowing for as many opportunities for mistakes.

7. Conclusions and Future Work

Our system learns to optimize STW and minimize speech overlaps and awkward silences, using prosody analysis to predict interlocutor behavior. It learns this on the fly, in a full-duplex "open-mic" (dynamic interaction) setup, and can take turns very efficiently in dialogues with copies of itself and with people, in relatively human-like ways.

The system finds prosodic features that can serve as predictors of human turn-giving behavior, and employs incremental (real-time) perception to work in as close to human natural dialogue speeds as possible. As the system learns on-line, it is able to adjust quite quickly to the particulars of individual speaking styles. At present, the system strongly targets the temporal characteristics of human-human dialogue, something that is mostly considered irrelevant by prior and related work on dialogue systems, as the above discussion shows. While the results are encouraging, there is room for significantly more work to be done in this direction.

At present, the system is limited in two main ways: it assumes a small set of turn-taking circumstances where content does not play a role and a single shared goal of cooperative "polite" conversation is assumed, where both parties want to minimize speech overlaps. Silences caused by outside interruptions—e.g. barge-in techniques and deliberate interruption techniques—are therefore a topic for future study. The system is highly expandable, however, as it was built as part of a much larger system architecture that addresses multiple topic- and task-oriented dialogue, as well as multiple communication modes such as gesture and facial expression. In the near future, we expect to expand the system to more advanced interaction types and situations. The learning mechanism described here will be expanded to learn not just the shortest durations but also the most efficient turn-taking techniques in multimodal interactions under many different conditions.

Because of the distributed nature of the architecture, the turn-taking system is constructed in such a way as to allow a mixed-control relationship with outside processes. This means that we can expand it to handle situations where the goals of the dialogue may be very different from being "friendly", even adversarial, as, for example, in on-air open-mic political debates. How easy this is remains to be seen; the main question revolves around the learning systems—how to manage learning in multiple circumstances without negatively affecting prior training.

Acknowledgement

This work was supported in part by research grants from RANNIS, Iceland, and by a Marie Curie European Reintegration Grant within the 6th European Community Framework Programme. The authors wish to thank Yngvi Björnsson for his contributions to the development of the reinforcement mechanisms.

REFERENCES

Allen, J.F., G. Ferguson and A. Stent. 2001. An architecture for more realistic conversational systems. Proceedings of the 6th international conference on Intelligent user interfaces, pp. 1–8.

Arbib, M. 1987. Levels of modeling of visually guided behavior. *Behavioral and Brain Sciences*, **10(3)**:407–465.

Bonaiuto, J. and K.R. Thórisson. 2008. Towards a neurocognitive model of realtime turn taking in face-to-face dialogue. In Wachsmuth, I., Lenzen, M. and Knoblich, G. (eds.), Embodied Communication in Humans and Machines. UK: Oxford University Press, pp. 451–485.

Bonaiuto, J. and M. Arbib. 2010. Extending the mirror neuron system model, II: what did I just do? A new role for mirror neurons. *Biological Cybernetics*, **102(4)**:341–359.

Bosch, L.T., N. Oostdijk and L. Boves. 2005. On temporal aspects of turn taking in conversational dialogues. *Speech Communication*, **47(11-2)**:80–86.

Brooks, R.A. 1986. A robust layered control system for a mobile robot. *IEEE Journal of Robotics and Automation,* **2(1)**:14–23.

Card, S.K., T.P. Moran and A. Newell. 1986. The model human processor: An engineering model of human performance. In K.R. Boff, L. Kaufman and J. P. Thomas (eds.), Handbook of Perception and Human Performance. Vol. 2: Cognitive Processes and Performance, 1986, pp. 1–35. Handbook of Human Perception, volume II. New York: John Wiley and Sons.

Dayan, P. 2000. Levels of analysis in neural modeling. Encyclopedia of Cognitive Science. London, England: MacMillan Press.

Ford, C. and S.A. Thompson. 1996. Interactional units in conversation: Syntactic, intonational, and pragmatic resources for the management of turns. In Ochs, E., Schegloff, E. and Thompson, S.A. (eds.), Interaction and Grammar, pp. 134–184. Cambridge: Cambridge University Press.

Gaud, N., F. Gechter, S. Galland and A. Koukam. 2007. Holonic multiagent multilevel simulation: Application to real-time pedestrian simulation in urban environment. *Procedings of International Joint Conference on Artificial Intelligence*, pp. 1275–1280.

Goodwin, C. 1981. Conversational Organization: Interaction between Speakers and Hearers. New York: Academic Press.

Gratch, J., A. Okhmatovskaia, F. Lamothe, S. Marsella, M. Morales, R.J. van der Werf, Louis-Philippe Morency. 2006. Virtual rapport. *Proceedings of International Virtual Agents*, pp. 14–27.

Jefferson, G. 1989. Preliminary notes on a possible metric which provides for a standard maximum silence of approximately one second in conversation. In Derek Roger and Peter Bull (eds.), Conversation: An Interdisciplinary Perspective, pp. 166–196.

Jonsdottir, G.R. 2008. A Distributed Dialogue Architecture with Learning. Master's thesis. Reykjavik University.

Jonsdottir, G.R., J. Gratch, E. Fast and K.R. Thórisson. 2007. Fluid semantic back-channel feedback in dialogue: Challenges and progress. In *IVA '07*, pp. 154–160. Springer.

Jonsdottir, G.R., K.R. Thórisson and N. Eric. 2008. Learning smooth, human-like turntaking in realtime dialogue. In IVA '08: Proceedings of the 8th international conference on Intelligent Virtual Agents, pp. 162–175. Berlin, Heidelberg: Springer-Verlag.

Moore, R. 2007. Presence: A human-inspired architecture for speech-based human-machine interaction. *IEEE Trans. Comput.*, **56(9):**1176–1188.

Ng-Thow-Hing, V., T. List, K.R. Thórisson, J. Lim and J. Wormer. 2007. Design and evaluation of communication middleware in a distributed humanoid robot architecture. In Prassler, E., Nilsson, K., Shakhimardanov, A. eds.), IEEE/RSJ Int. Conf. on Intelligent Robots and Systems (IROS'07) Workshop on Measures and Procedures for the Evaluation of Robot Architectures and Middleware.

Nivel, E. and K.R. Thórisson. 2008. Prosodica Realtime Prosody Tracker. Technical Report. Reykjavik University Department of Computer Science. Technical Report RUTR-CS08001.

Pierrehumbert, J. and J. Hirschberg. 1990. The meaning of intonational contours in the interpretation of discourse. In Cohen, P.R., Morgan, J. and Pollack, M. (eds.), Intentions in Communication, pp. 271–311. Cambridge, MA: MIT Press.

Raux, A. and M. Eskenazi. 2007. A multi-layer architecture for semi-synchronous event-driven dialogue management. In *ASRU*, pp. 514–519, Kyoto, Japan.

Raux, A. and M. Eskenazi. 2008. Optimizing endpointing thresholds using dialogue features in a spoken dialogue system. In Proceedings of the 9th SIGdial Workshop on Discourse and Dialogue, pp. 1–10. Columbus, Ohio: Association for Computational Linguistics.

Sato, R., R. Higashinaka, M. Tamoto, M. Nakano and K. Aikawa. 2002. Learning decision trees to determine turn-taking by spoken dialogue systems. In ICSLP '02, pp. 861–864.

Schaffner, K.F. 2006. Reduction: The Cheshire cat problem and a return to roots. In *Synthese*, pp. 377–402. The Netherlands: Springer. Schlangen, D. 2006. From reaction to prediction: Experiments with computational models of turn-taking. In Proceedings of Interspeech 2006, Panel on Prosody of Dialogue Acts and Turn-Taking, Pittsburgh, USA.

Schwabacher, M. and A. Gelsey. 1996. Multi-level simulation and numerical optimization of complex engineering designs. In 6th AIAA/NASA/USAF Multidisciplinary Analysis & Optimization Symposium, AIAA-96-4021.

Stivers, T., N.J. Enfield, P. Brown, C. Englert, M. Hayashi, T. Heinemann, G. Hoymann, F. Rossano, J.P. de Ruiter, K.-E. Yoon and S.C. Levinson. 2009. Universals and cultural variation in turn taking in conversation. Proceedings of the National Academy of Sciences, **106(26):**10587–10592.

Sutton, R.S. and A.G. Barto. 1998. Reinforcement Learning: An Introduction. Cambridge, MA: The MIT Press.

Thórisson, K.R. 1993. Dialogue control in social interface agents. In Inter- CHI Adjunct Proceedings, pp. 139–140. ACM Press, New York.

Thórisson, K.R. 1996. Communicative Humanoids: A Computational Model of Psycho-Social Dialogue Skills. Ph.D. thesis. Massachusetts Institute of Technology.

Thórisson, K.R. 2002a. Machine perception of multimodal natural dialogue. In McKevitt, P., Nulláin, S.Ó. and Mulvihill, C. (eds.), Language, Vision & Music, pp. 97–115. John Benjamins Publishing Co., Amsterdam, The Netherlands.

Thórisson, K.R. 2002b. Natural turn-taking needs no manual: Computational theory and model, from perception to action. In Granström, Björn, D. House, I. Karlsson (eds.), Multimodality in Language and Speech Systems, pp. 173–207. Dordrecht, The Netherlands: Kluwer Academic Publishers.

Thórisson, K.R. 2008. Modeling multimodal communication as a complex system. In Wachsmuth, I. and Knoblich, G. (eds.), Modeling Communication with Robots and Virtual Humans, pp. 143–168. Springer, Berlin, Germany.

Thórisson, K.R., H. Benko, A. Arnold, D. Abramov, S. Maskey and A. Vaseekaran. 2004. Constructionist design methodology for interactive intelligences. *A.I. Magazine*, **25(4):**77–90.

Thórisson, K.R. and G.R. Jonsdottir. 2008. A granular architecture for dynamic realtime dialogue. In Proceedings of the 8th International Conference on Intelligent Virtual Agents, pp. 131–138. Springer, Berlin, Germany.

Thórisson, K.R., G.R. Jonsdottir and E. Nivel. 2008. Methods for complex single-mind architecture designs. In Proceedings of the 7th International Joint Conference on Autonomous Agents and Multiagent Systems (AAMAS), pp. 1273–1276, Richland, S.C. International Foundation for Autonomous Agents and Multiagent Systems.

Thórisson, K.R., T. List, J. DiPirro and C. Pennock. 2004. OpenAIR: A Publish-Subscribe Message and Routing Specification 1.0. Technical Report.

Traum, D.R. and P.A. Heeman. 1996. Utterance units and grounding in spoken dialogue. In Proc. ICSLP '96, pp. 1884–1887, Philadelphia, PA.

Wilson, M. and T.P. Wilson. 2005. An oscillator model of the timing of turntaking. *Psychonomic Bulletin Review,* **38(12):**957–968.

Wood, M.F. and S.A. Deloach. 2000. An overview of the multiagent systems engineering methodology. In The First International Workshop on Agent-Oriented Software Engineering (AOSE-2000), pp. 207–221.

Wooldridge, M., N.R. Jennings and D. Kinny. 2000. The gaia methodology for agent-oriented analysis and design. *Autonomous Agents and Multi-Agent Systems,* **3(3):**285–312.

CHAPTER 13

TTS-driven Synthetic Behavior Generation Model for Embodied Conversational Agents

Izidor Mlakar, Zdravko Kačič and Matej Rojc

1. Introduction

Communication is a concept of information exchange. It incorporates speech perception, visual perception, and understanding. These concepts are correlated by sets of complex and interspersed conscious and subconscious processes. The complementary nature of multimodal information suggests that verbal information is accompanied by spontaneous and planned body language, especially by gesturing using lips, hands, and arms. In fact, the perception of a speech sound is affected by a non-matching lip/mouth movement (Sargin et al., 2006). Most researchers agree that speech-synchronized coverbal behavior is essential in communication. Subsequently, artificial interfaces should also integrate natural speech-gesture production to, at least, some degree. The communication between different entities, whether it be human or synthetic, commonly reflects personality and emotions (Wallbott, 1998), idiosyncratic features (Gallaher, 1992), and communicative functions (Allwood et al., 2007a). These non-verbal features are regarded as coverbal behavior. Coverbal behavior benefits the understanding process of the person being addressed

(Bavelas and Chovil, 2000; Kendon, 2004), and the cognitive process of the speaker (Pine et al., 2007; Kita and Davies, 2009). Information expressed by co-aligned verbal and non-verbal behavior is better understood and faster in achieving its purpose (e.g. inducing a social response). Embodied Conversational Agents (ECAs), like in (Kopp and Wachsmuth, 2002; Mlakar and Rojc, 2011; Poggi et al., 2005; Thiebaux et al., 2008), present a paradigm of artificial bodies that can control and move different body-parts, and are capable of communicating by using their voices, faces, hands, and arms (or full body). ECAs may represent the coverbal behavior in the form of communicative function and/or by directly representing the semiotic nature of the spoken dialog (e.g. as iconic/metamorphic representation, stressing the importance of spoken segments, or simply as to regulate the flow of information exchange). The believability of synthetic coverbal behavior closely relates to the term expressivity, namely, to the ability of performing continuous and smooth, context adaptable communicative acts, emulating natural movement tendencies and dynamics, and synchrony with the situational context and/or the verbal flow. The interaction incorporating expressive ECAs is proven to provide visual meaning, and benefit an understanding of the spoken words and actions performed in multimodal interfaces. Although ECAs and synthetic 'communicative' behavior have been well researched, the co-alignment of speech and non-verbal expressions still represents an important and challenging task. Most of the behavior overlaid by such agents is, therefore, often limited to lip-sync (Tang et al., 2008; Zoric and Pandžić, 2008), and facial expressions (Lankes and Bernhaupt, 2011), or is based on behavior generation/realization engines that incorporate scenarios and/or semantically tagged text (Čereković and Pandžić, 2011; Krenn et al., 2011; Nowina-Krowicki et al., 2011; van Oijen et al., 2012). In general, the correlation between verbal and non-verbal signals originates from the semantic, pragmatic and temporal features of the multimodal-content (Jokinen, 2009; Kendon, 2000; McNeill, 1992). Some coverbal gestures like iconic expressions (Hadar and Krauss, 1999; Straube et al., 2011), symbolic expressions (Barbieri et al., 2009), and mimicry (Holler and Wilkin, 2011) are also tightly interlinked with speech. These gestures may be identified by the linguistic (semantic) properties of the input text, like word-type, word-type-order, word-affiliation, etc. Other coverbal gestures, especially those representing communicative functions (e.g. indexical and adaptive expressions, Allwood, 2001), have less (if at all) evident semantic or linguistic alignment with the text. However, they may still be identified by linguistic fillers (Grenfell, 2011), turn-taking, and directional signals. Although speech and coverbal gestures are manifested by the same underlying process (e.g. different sides of the same coin), the information is conveyed by each modality in a different way. For instance, two gestures produced together do not necessarily represent a gesture expressing complex meaning. Gestures are also not completely of a linguistic nature. Several gestures may represent a similar meaning, whereas similar gestures may represent a totally different meaning. There are also no grammatical rules on the movement structure by which a gesture is propagated. The language, on the other hand, has grammar and order.

2. Related Work

Research into human-like ECAs' communicative behavior represents a very hot topic. In general, the theories and concepts rely on the understanding of communicative behavior (Allwood et al., 2007a; McNeill, 2005). These concepts are then transformed into specifications of ECAs' embodied (virtual) movement represented as coverbal gestures. The growth point theory (McNeill, 1992) suggests the representation of 'verbal' thinking in the form of idea units. The interaction between two active models of the thinking, linguistic and imagistic, triggers the language forms and manifestation of the coverbal gesture. These gestures are generated from spatio-motoric processes that interact on-line with the speech production process. In contrast to the imagistic knowledge, the featural model (Krauss et al., 2000; de Melo and Paiva, 2007) is based on propositional and non-propositional features. The behavior models deduced from behavior theories are mostly based on a motor planner, and lexical retrieval. Additional language-dependent models further suggest that the produced gestures are influenced by the speaker's language (Kita and Özyürek, 2003). The SAIBA framework (Kopp et al., 2006; Vilhjalmsson et al., 2007) is a framework for multimodal behavior generation and re-creation by using ECAs. The framework suggests the usage of knowledge structures that describe the form of communicative behavior, and the life-span of synthetic behavior at different levels of abstraction. The three levels of abstraction represent interfaces between those processes used for: (a) behavioral planning (e.g. planning of a communicative event), (b) multimodal realization planning (e.g. describing the multimodal channels used for the realization of communicative events), and (c) realization of planned behaviors on an ECA. The concept of the three-layered SAIBA behavioral model represents a structure that can be adopted by any multimodal behavior generation system. The three processing stages are modular and relatively independent. Each stage may introduce a wide-specter of different algorithms that produce the desired results. However, the transitions between stages are bi-directional in their natures. One stage delivers the input to the next stage and also accepts the feedback data running back to it. The processing within each stage and its internal structure is largely treated as a 'black box', left to be further researched. The interface between stages (a) and (b), the intent planning and behavior planning stages, describes the communicative and expressive intent of the communicative event with reference to the physical behavior. The intent planning stage basically describes the function of verbal and coverbal behavior within a communicative event. It, therefore, provides a semantic description that accounts for those aspects that are relevant and may influence the planning of verbal and non-verbal behavior. The FML (Function Mark-up Language) (Heylen et al., 2008) is used to specify the semantic data. The FML description defines the basic semantic units associated with a communicative event and allows the annotation of these units with properties that further describe the communicative function of multimodal behavior (e.g. expressive, affective, discursive, epistemic, or pragmatic). The interface between stages (b) and (c), the behavior planning and behavior realization stages, describes the physical features of multimodal behaviors as to be realized by the final stage of the SAIBA behavioral model. The BML (Behavior Mark-up Language) (Vilhjalmsson et al., 2007) suggests language should be used for such meditation. Most of the existing behavior realization (animation) engines are capable of realizing every aspect of behavior (verbal, gestural, phonological, etc.) the behavior planner may specify.

Current research into coverbal expression-synthesis, incorporating ECAs, mostly agrees with the SAIBA architecture, providing independent systems for: content-planning system (what to display), behavior-planning system (how to display), and behavior-realization system (physical/virtual generation of the artificial behavior). Speech and gesture production are, in general, addressed as two independent processes, and synchronized prior to execution. For instance, the Articulated Communicator Engine (ACE) (Kopp and Wachsmuth, 2004) is a behavior-realization engine that allows for the modeling of virtual animated agents using the MURML gesture description language (Kransted et al., 2002). ACE is independent of a graphics platform and can synthesize multimodal utterances containing prosodic speech synchronized with body and hand gestures, or facial expressions. Its smart scheduling techniques, blending/co-articulation of gesture and speech, are enabled by a connection between behavioral realization, planning, and behavioral execution. ACE also allows a user to define synchronization points between channels, but automatically handles the animation (based on the movements' trajectories, and other motion functions). ECAs generated by the concepts presented in this work rely on key-poses, and an automatic key-frame-based generation of interpolated motion between successive motion-segments. Each group of such independent motion-segments is regarded as an independent animation stream, which can be reconfigured (paused, stopped, remodeled, etc.) at any given frame of the animation.

The Behavior Expression Animation Toolkit (BEAT) (Cassell et al., 2001) is a rule-based system that can, based on the input text, generate non-verbal behavior. It is a text-driven behavior model that enables visual speech synthesis and coverbal behavior generation. For example, it allows for the animators to input typed text that they wish to be spoken by an animated human figure. The output is obtained as appropriate and synchronized non-verbal behaviors and synthesized speech, both provided in a form that can be sent to a number of different animation systems (realization engines). In terms of the SAIBA behavior model, the BEAT toolkit performs the intent planning and behavior planning stages. The system uses linguistic and contextual information contained within the text, to identify the communicative intent and to plan its physical realization, e.g. to control the movements of the hands, arms, and face, and the intonation of the voice. The mapping between text to facial, intonational and body gestures is contained in sets of rules derived from the state of the art in non-verbal conversational behavioral research. The intent planning phase is implemented by using *Language tagging* and *Behavior suggestion* modules. The *Language tagging* module annotates the input text with the linguistic and contextual information. The automatic language-tagging process proposed by the toolkit is quite similar to the TTS-based tagging processes that allow TTS systems to produce appropriate word phrases along with the prosodic features (e.g. intonation, accents, etc.).

Similarly, Poggi et al. (2005) and Hartmann et al. (2005) suggest a concept for the expressive gesture synthesis of an ECA named GRETA. Their gesture production model is based on the concept of gestural dictionary that stores different shapes obtained during the annotation process. These gestural affiliates are then accessible online by using APML behavior specification language (De Carolis et al., 2004), and semantically described input text sequences.

Most of the related research, therefore, separates speech and gesture production, either externally or internally. The co-alignment of speech and coverbal movement is regarded as an implicit process of semantically matching words (word patterns) and shapes based on rules or different context-processing techniques performed by an external behavior planning engine. Further, behavior planning engines rarely rely on what is directly being said as primary source of information (text-data), and rather incorporate different concepts and techniques for specifying the alignment at a higher abstract level (e.g. semantic tagging, manual planning, planning through annotation, etc.), where different additional contexts (e.g. emotions, personality, and even communicative intent) are integrated. However, Bergmann and Kopp (2008) describe a computational model for the joint production of speech and gesture. The model is based on an interface between imagistic and propositional knowledge at the level of content representation. In Romportl et al. (2010), a system is presented that is capable of speech and gesture alignment through FML behavioral description language. Text-driven (or TTS-driven) approaches are often limited to lip-sync processes, and facial animation (Malcangi, 2010; Zoric and Pandžić, 2008; Wang et al., 2011). The major disadvantage of systems that generate behavior based on raw text is the almost absence of natural synchronization due to the lack of contextual factors. However, in Ng-Thow-Hing et al. (2010), a text-driven approach is suggested that is already based on text processing and semiotic indicators, as content and planning indicators. In order to specify and generate the coverbal expressions and speech, their gesture-model processes text. By POS-tagging and by processing additional semantic tags within text, their gesture model identifies the duration and the form of the overlaid gestures. The aforementioned system transforms an ECA-based gesture production model on to a robotic unit. Similarly, the authors of an expressive gesture model (Le et al., 2011) generate an ECA-based behavior that is further constrained and adapted in order to be represented on a robotic platform NAO.

3. Architecture of the TTS-driven Conversational Behavioral Generation System

The TTS-driven conversational behavioral generation system proposed in this chapter describes an understanding of speech and coverbal gestures based on McNeill (2005), and also includes several prosodic features generated by the PLATTOS TTS engine (Rojc and Kačič, 2007). The proposed algorithms search for meaningful text sequences based on semiotic grammar. In this way the meaningful text sequences may be represented by one (or even none) meaningful conversational expression. Each meaningful text sequence is further identified by words (or word phrases) that carry meaning. These meaningful words (word phrases) are represented by a gesture, if a word (or phrase) is emphasized, or primary-accented. The non-verbal meaning is then manifested at its full span at the end of the emphasized or primary-accented syllable. And the prosodic features of speech define the dynamics of coverbal gesture in the form of movement-phases (Kendon, 2004). The PLATTOS, therefore, synchronizes the meaning expressed by gesture to be compatible with the one expressed by the connected speech fragment (e.g. the shape of the dominant expression manifested whilst synthesizing verbal indicators).

In the system, features driving the coverbal behavior are deduced from raw text, and the processing steps for the planning and generating of non-verbal behavior incorporate semiotic grammar, gesture dictionary, and lexical affiliation. Semiotic grammar, on the one hand, incorporates rules by the use of which the most meaningful parts of a text are selected, and by using which the dynamical alignment (in the forms of movement-phases (Kendon, 2004)) regarding verbal and non-verbal behavior can be performed. The gesture dictionary and lexical affiliation, on the other hand, incorporate the semantic relation between meaning (meaningful word) and meaning representation (shape manifestations). Additionally, the PLATTOS system synchronizes the coverbal expressions in such a way that the meaningful part of a gesture (the so-called stroke) co-occurs with the most prominent segment of the accompanying generated speech, as proposed in McNeill (1992). Furthermore, the PLATTOS system also transforms the generated coverbal expressions into a form understandable to an ECA-based behavior realization engine supporting mark-up languages, such as BML (Vilhjalmsson et al., 2007) and EVA-SCRIPT (Mlakar and Rojc, 2011).

The TTS-driven coverbal behavior synthesis engine is based on the following baseline (Allwood, 2001; Esposito and Esposito, 2011; Kendon, 2004; McNeill, 2005). The *first part* of the human gesture production process relies on the lexicon and lexical affiliates, and can, therefore, be linked directly to words, word phrases (or even to smaller utterances, such as syllables, etc.). Pierce has defined that such bodily movements can be linguistically grouped into semiotic types (Peirce, 1958). The properties of semiotic types may be identified by key-utterances (mostly words, word phrases, word types, and word-type-orders) (Deacon, 2003). The *second part* of the human gesture production process relies on the paralinguistic and extra-linguistic information conveyed by speech and gestures (Wallbott, 1998). Meaning is, in addition to linguistic relation, also conveyed through non-lexical expressions carrying specific communicative values, such as: turn-taking, feedback, speech-repair, and sequencing. These expressions are generated based on a complex mechanism of communicative functions (Kipp, 2001; McNeill, 1992; Peirce, 1958).

The link between such verbal and coverbal information is established over linguistic and prosodic features, such as pauses, vocalization, and nasalization. In many cases these signals formulate speech-flow and may also be identified by different sounds (e.g. fillers such as 'hmm…' 'ehhh…'). Further, the coverbal content selection and non-verbal behavior planning (coverbal alignment), as described in this chapter, are fused into a common engine in an efficient and flexible way. The core represents the text-to-speech synthesis engine (TTS) (Rojc and Kačič, 2007) that is enriched with additional modules devoted to non-verbal expressions' synthesis. The realization engine, which then reproduces the generated verbal and non-verbal behavior (including co-articulation), is a proprietary modular, hierarchically oriented ECA-based engine (Mlakar and Rojc, 2011), capable of animating procedural, key-frame (key-shape)-based animations.

The architecture of the proposed system for generating coverbal conversational behavior is presented in Figure 1. The language-dependent resources are separated from the language-independent engine, by using FSM formalism. All modules allow easy integration of new algorithms into the common queuing mechanism. This well-established mechanism also allows for flexible, efficient and easy integration of additional modules. This is especially useful in our case, since in addition to those modules used for speech synthesis (tokenizer, part-of-speech tagger, grapheme-to-phoneme conversion, symbolic and acoustic prosody, unit selection, concatenation, and acoustic processing), several modules have to be integrated into

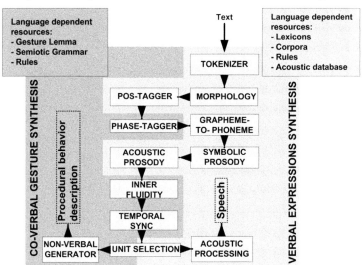

Figure 1. The PLATTOS TTS-driven behavior generation system's engine.

the structure (for synthesis of non-verbal expressions, symbolically and temporarily aligned with speech). The following modules are proposed for adding to the core TTS engine: phase-tagger, inner-fluidity, temporal-sync, and non-verbal generator. The *phase-tagger* is needed for the symbolical synchronization of verbal and non-verbal behavior (meaning identification and selection). The *inner-fluidity* module is needed for specifying the inner-fluidity and dynamics of the meaningful coverbal gesture sequence. Then the *temporal-sync* module is needed in order to temporally co-align the propagation of coverbal gestures with the generated verbal content's pronunciation and prosodic features (Kröger et al., 2010). Finally, the *non-verbal generator* module is used for transforming the generated 'behavior plan' into abstract behavior specification that can be animated by an ECA within an animation engine. In this way, the system is completely TTS-driven. It, therefore, benefits from the core TTS system and its underlying extracted and the predicted linguistic and prosodic features, generally used for the generation of speech signal from text (e.g. stress, prominence, phrase breaks, segments' durations, pauses, etc.). Further, in order for the system's outputs to be fully synchronized and used by different virtual/physical interfaces at interactive speeds, it is most efficient when the TTS-driven approach synthesizes coverbal gestures and speech signals, simultaneously. In the PLATTOS system, this is achieved by fusing the coverbal gesture synthesis stream with the engine's verbal expressions synthesis stream. The verbal expressions synthesis stream is responsible for determining linguistic and several prosodic features on general input text, and to synthesize the speech signal. The coverbal gesture synthesis stream is then responsible for identifying the meaning, selecting visual content and type of the presentation, co-aligning it with the generated speech signal, and for transforming the coverbal behavior (sequence of gestures) into a form that is understandable by synthetic ECAs.

All modules within the system are based on three flexible and efficient data structures: deques that are used for the flexible linking of several of the engine's processing steps, heterogeneous relation graphs that are used for storing extracted and predicted linguistic and prosodic data on general input text, and finite-state machines that are used for separating the language-dependent resources from the system (Rojc and Kačič, 2007).

4. Grammar

The lexical affiliation and gesture dictionaries (Gesticons) have been described by Schegloff (1985) as: "word of words deemed to correspond most closely to a gesture in meaning". The affiliates are used in order to establish explicit relationships between linguistic units (word, word phrases, word types, word-type-order patterns), and conversational shapes (gestures). The most natural way to establish these relations is by observing natural conversational behavior (e.g. gesture analysis by annotation of multimodal corpus). The annotation scheme used for building the Gesticon used within the PLATTOS system had already been proposed in Mlakar and Rojc (2012). The gesture topology and the formal model used are presented in Figure 2. Each conversational expression is encoded as a series of consecutive (representative) poses, encoded with additional linguistic information. For example, the formal model of the annotation captures morphological correlation separately for each body-part, and in the form of movement phases, and movement phrases. Each pose is also described with a semiotic tag, as by meaning the coverbal expression (speech + gesture) carries. Although there are some significant differences between speaker-styles (e.g. intensity, frequency, particular shapes used, etc.), some general trends can be identified (Ng-Thow-Hing et al., 2010). For instance, specific gesture classes have a tendency to co-occur with certain words, phrases, or parts-of-speech (semantically and morphologically).

Movement phrase encapsulates sequential segments of coverbal behavior (sequence of gestures) into a meaningful sequence of synchronized speech fragments and gestures, e.g. conversational acts. Each *movement phrase* is further segmented into *movement phases*. A *movement phase* describes the five stages of movement propagation of a single gesture. The most important *movement phase* is the stroke phase carrying the majority of meaning. The stroke phase is a mandatory part of any gesture (similarly to noun, verb or pronoun being mandatory for grammatically correct sentence formulation in most languages). The stroke phase may be followed by either the post-stroke hold movement phase, or the retraction phase. The hold phases are defined as a short stop of movement prior/after the stroke phase. They are used to synchronize the beginnings of verbal and coverbal information, and to extend the meaning conveyed through the stroke movement phase. The hold movement phases are, in speech, indicated by short pauses (*sil* elements), and by the residue (part not following stroke and not assigned to stroke) of the 'meaningful word (phrase)'. The holds are most evident in iconic expressions, especially those

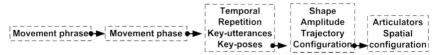

Figure 2. The formal-model of expressively oriented annotation scheme.

emphasizing concept and adaptive expressions (e.g. longer pauses following linguistic fillers). Finally, the retraction phase drives the observed body-part to its initial/another rest position.

In the context of the PLATTOS system, the retraction phase is generated after each movement phrase. As shown in Figure 2, a *movement phase* is defined by its temporal features (duration), key-utterance sequences, repetition, and by key-poses. The temporal features define the duration of the phase. These features highly depend on to the pronunciation rate and the accentuation of the utterances spoken. The key-utterances (e.g. key-words, key-word sequence) represent the verbal information that is simultaneous to the movement propagation. The repetition describes how (if at all) the stroke phase is repeated within the observed conversational expression. The repetition feature is represented by the repetition of a similar stroke, and post-stroke hold combinations. Key-poses are also part of the movement phase description. Key-poses describe spatial configurations (and manifestations) that can be observed at the starting and ending points of *movement phases*. Whilst describing the manifestations at these points, each key-pose is defined by shape, amplitude, trajectory, and configuration. The temporal component of the lexical affiliate ensures that a gesture retains its dynamical features even after it has been temporarily adjusted to the speech fragments. As shown in the topology diagram (Figure 2), each key-pose can be further adjusted in terms of distinct spatial adjustment of the articulators (Mlakar and Rojc, 2012).

4.1 Measuring the similarity of shapes

Similarity of movement is defined (based on the Hamming distance (Hogrefe et al., 2011)) as a quantitative measure of resemblance between two manifestations. The similarity of gestures/shapes is used in order to identify distinct gesture units and, based on similarity, grouping them into distinct gesture classes. The gesture classes can then be further transformed into different lexical affiliates (e.g. pragmatics, semiotic relation, word-type, semantics, etc.). The Hamming distance generally measures the difference between two hand gestures in the forms of the following features: hand-shape, palm-orientation, finger direction, movement, and repetition. Two gestures are similar, when the Hamming distance value is low. In order to be able to form a body-parts' based expressive gesture dictionary, the Hamming similarity measure is extended for each body-part, and applied against each body-part individually. The hand-shapes within the PLATTOS system are encoded and classified based on Ham-No-Sys symbolic shape (referent hand shape). Next, the similarity between two arm gestures is defined, as proposed by Kipp (2001), based on the palm-orientation, radial-orientation of the elbow (e.g. far-out, side, front, inward, etc.), height of the palm position (e.g. head, shoulders, chest, abdomen, belt, etc.), distance of palm from torso (e.g. fully extended, normal, close, touch, etc.), and the elbow inclination (e.g. orthogonal, normal, etc.). The similarity of head movements is defined based on gaze and radial orientation of the head, whereas the similarity of facial expressions is based on FAPS, and the MMI facial expressions database (Lankes and Bernhaupt, 2011). If spatial configurations are described separately for different body-parts and connected through similarity measure, additional conversational expressions may be formed (or attuned) by substituting similar manifestations. As a result the synthetic agent's gestural inventory becomes more flexible and diverse. Each synthetic non-verbal behavior manifestation can, therefore, be better aligned within the context of situations and verbal information.

4.2 Defining the meanings of words—semiotic grammar

The meaningful words and meaningful shapes represented by the coverbal movement are, in the PLATTOS system, identified based on semiotic grammar. The semiotic grammar's rules are based on the properties of semiotic gestures (Allwood, 2001; Breslow et al., 2010; Deacon, 2003), and annotations of video corpora (Allwood et al., 2007b; Luke et al., 2012; Mlakar and Rojc, 2012). The semiotic grammar identifies the following gesture types, accompanied by the corresponding rules. *Symbolic expressions* are gestures that represent some sort of a conventional expression and/or conventional sign. These gestures are indicated and triggered by the conventionalized key-word phrases. These gestures are, in general, co-aligned with the beginnings and ends of conventionalized phrases. For instance, a 'waving' hand gesture usually accompanies the word phrase 'Bye, Bye'. *Indexical expressions* are gestures that represent determination, reference to the addressee, a direction, and turn-taking communicative functions within the dialog (Allwood, 2001). 'Determination' gestures are triggered by typical determiners, key-words, such as your, his, mine, etc. Next, the references to objects/persons (addressee) are triggered with isolated nouns, pronouns, and verbs, and keywords, such as you, me, this, that, etc. These gestures are usually directly aligned with the trigger word. And the turn-taking gestures are triggered by turn-taking signals (Luke et al., 2012). They also expect some-sort of co-speakers response (e.g. longer pauses, after phrases intonated as questions, e.g. alright? / ok?, etc.). These gestures are co-aligned over the whole trigger phrase and, in general, signal the existence of a pause at the end of the conversational expression. *Adaptors (manipulators)* are gestures that address/outline speech-flow and interruptions in the flow. The more common interruptions are: speakers searching for words in the lexical dictionary, instances of thinking or re-thinking, etc. These gestures are usually triggered by fillers (phrases), such as hmm..., oh..., well..., etc. They are co-aligned with the triggering filler. Additionally, these gestures also maintain bodily manifestation during the silent period (if it exists), following the trigger until the next conversational expression is produced. The word phrases, such as 'ohh...', 'right...', etc., are usually accompanied by a gestural movement. Mappings between words (word-types) and adaptors are stored in the form of FSM-based gestural dictionary relations. *Iconic expressions* are gestures illustrating objects, processes, or qualitative meaning. Gestures that indicate (outline) concrete objects are usually triggered by preposition-noun (PN), adjective-noun (AN), adverb-adjective-noun (RAN), and other combinations involving a noun as primary indicator. Gestures illustrating processes are triggered by similar compositions as object gestures; however, the noun is replaced by a verb as the primary indicator. Mappings between words, word-types and word-type-orders (semiotic patterns), and iconic expressions are also stored in the form of a gestural dictionary. Most languages require paring of at least a noun (subject, object, topic, etc.) and verb (predicate, operation, process, comment, etc.) in order to constitute a well-formed sentence. In languages, e.g. English and Slovenian, pronouns can also replace nouns. Further, the communicative dialog also contains words/word phrases that address/outline speech flow, interruptions in the flow (e.g. speakers searching for words in the lexical dictionary, instance of thinking or re-thinking, etc.), and outlining directionality. The most probable word indicators of such adaptive movement are linguistic fillers (phrases), such as hmm..., oh..., well..., etc., and determiners (my, this, your, etc.). Therefore, for the semiotic grammar used in the PLATTOS system, nouns, verbs, pronouns, and linguistic fillers are those word-types that indicate a high probability of movement. In addition to semiotic grammar, also rules indicating implicit (semantic) relations between words (word phrases) are established. For instance, a Slovenian word phrase 'in tako naprej' (*and so forth*) is represented by a circular gesture reaching its maximum over the stressed fragment of the word 'tako' (*so*) and propagating over the whole phrase (not only over the word 'tako'). Similarly, a concept of enumeration is better explained by implicit rules than directly by semiotics. For instance, when a speaker is enumerating activities, e.g. phrase 'Mislim, gledam in pišem.' (*I think, watch and write*), the coverbal sequence represents an enumerative gesture composition rather than direct representation of words in the form of a sequence of indexical (or iconic) expressions, as it would be suggested directly by the semiotic rules.

5. Algorithms for Behavior Planning and Coverbal Synchronization

This section describes in detail the TTS-driven architecture (as proposed in Figure 1), based on the PLATTOS TTS system. The TTS system is a corpus-based speech synthesis system, using a concatenative approach and TD-PSOLA speech synthesis algorithm (Rojc and Kačič, 2007). All deques are tied together into a common engine, using the heterogeneous relation graph structure (HRG) for representation of all linguistic information (Figure 3). All modules contribute to the linguistic information used for generating both the speech signal and the coverbal gestures. One common HRG structure is used, and is made accessible by all modules in the system. In this way, processes in given modules are able to access, change, store or enrich linguistic information when appropriate. The HRG structure gathers all the linguistic information extracted from the input sentence. In the final system's modules it is used for generation of the speech signal (acoustic processing deque), and for generating EVA-SCRIPT-based behavior descriptions (non-verbal generator deque).

The HRG structure demonstrates the use of both types of relation structures, in the form of linear lists and trees. Those relation

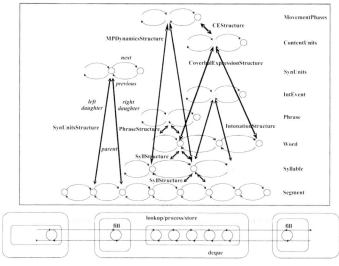

Figure 3. A complete queuing mechanism with an HRG structure.

structures in the form of linear lists are: Segment, Syllable, Word, Phrase, IntEvent, SynUnits, ContentUnits, and MovementPhases. And those relation structures in the forms of trees are: SyllableStructure, PhraseStructure, IntonationStructure, SynUnitsStructure, CEStructure, CoverbalExpressionStructure, and the MPDynamicsStructure. All these relation structures contain, in general, different numbers of specific linguistic objects, depending on the nature of the process that creates and uses them. In this way, several complex features can easily be specified and constructed, and several additional features beneficial for processing steps within the system can be added. As can be seen, the CEStructure relates to units in the MovementPhases relation layer, and units in the ContentUnits relation layer. The SyllableStructure relates units in the Segment relation layer, and units in the Syllable relation layer. It also relates to units in the Syllable relation layer, and units in the Word relation layer. The MPDynamicsStructure relates to units in the MovementPhases relation layer, and units in the Syllable relation layer. The CoverbalExpressionStructure relates to units in the ContentUnits relation layer, and units in the Word relation layer. And the PhraseStructure relates to units in the Phrase relation layer, and units in the Word relation layer. Naturally, all these relation structures can easily be changed and adapted to different structures, following the processing needs of the modules and algorithms in the engine.

5.1 Phase tagging deque

The *phase tagging deque* is responsible for the symbolical synchronization (Figure 4). Namely, it identifies the semiotic phrases, the words carrying most meaning (the meaningful words), and the shapes that could illustrate the meaning of the indicated words. In order to identify meaningful words/word phrases, the semiotic grammar is used, closely interlinked with a gesture lexicon (gesture affiliate). The phase tagging deque's input is already a POS-tagged text. Further, the *phase tagging deque* creates a new ContentUnit relation layer in the HRG structure. This layer serves for storing several 'Content units'. These units are then used to establish the relation between the shapes manifested within the movement-phases. Although the stroke movement-phase represents that part of the gesture where the content/intention is identified with the greatest clarity, the visual content re-presented within a stroke phase may not always refer to the words (speech fragments) over which the stroke is being propagated. The *phase tagging deque*, therefore, performs symbolic synchronization (verbal-trigger indication, and content selection) by implementing the following procedures: semiotic tagging, semiotic processing, and matching.

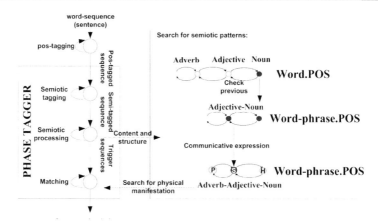

Figure 4. The phase tagging deque, based on FSMs and HRG structures.

These processes are implemented by using FSMs and common HRG structures (right-hand side in Figure 4). Each word within the HRG structure is stored in a Word relation layer. The deque processes POS-tagged word sequences, and matched them against semiotic grammar's rules and relations stored in the off-line constructed gestural dictionary. Several FSMs are used in order to identify the meaning, and to select the physical representation of the meaning. Furthermore, FSMs also predict how the propagation of meaning could be performed over the specific text sequence (particularly during the preparation movement-phase, and within the concepts of repetitive, circular, enumerative gestures). In this way, the deque performs synchronization of the form. The process of synchronization of the form proceeds as follows. Firstly, the semiotic tagging process tags those words that could trigger coverbal movements (mandatory word-type-units). At the morphology level (based on semiotic grammar's rules), the algorithm searches for nouns, verbs, pronouns, fillers and determiners.

Next, the mandatory word-type-units and their neighbors are grouped into word phrases, and compared against semiotic word-patterns, in order to define the content and suggest the structure of the coverbal expressions. As seen in Figure 4 (right-hand-side), the semiotic processing is traversing through the HRG's Word relation layer, and by accessing preceding words (e.g. p.pos, pp.pos, ppp. pos), and succeeding words (e.g. n.pos, nn.pos, nnn.pos) detect those patterns that match one (or more) of the word-type-orders defined within semiotic rules. In this way, e.g. the noun represents the mandatory word-type-unit that indicates existence of coverbal movement according to the semiotic grammar. By combining the noun POS tag with its predecessors' POS tags (in this case the adjective (p.pos), and the adverb (pp.pos)), and by matching them against semiotic patterns, a semiotic pattern of an iconic gesture is indicated. The general patterns, such as adverb-adjective-noun, represent an emphasis on the physical dimension of the object (a noun). For instance, the Slovenian verbal phrase 'zelo velika prednost' (*very great advantage*) emphasizes how huge the advantage was. At this level, the words and word phrases are also matched against implicit rules. These rules can override/replace the decisions based on semiotics. A semiotic pattern verb-verb-verb will result in three isolated verbs, indicating three 'pointing' gestures (e.g. thinking, watching, and writing in the text sequence 'I was thinking, watching, and writing'). However, an implicit rule for enumeration will also indicate that such a sequence represents an enumerative expression. Instead of 'pointing' gestures, the system will indicate the context of enumeration. The final processing step of the deque represents the semantic matching process. In this step, lexical affiliates are chosen for those words/ word phrases identified as carriers of meaning. This process performs the look-up in the gesture dictionary, and selects an appropriate physical representation of the word(s). The FSM-based engine hosting the relations firstly filters the available expressions for the semiotic type. Secondly, the engine filters the remaining expressions based on the meaningful word and morphological patterns. If more than one conversational expression is available at the end, the matching process selects such a conversational expression that matches the implicit rule, or semiotic patterns of the phrase, or represents the closest matching pattern (e.g. at least adjective-noun pattern). Further, the initial pose, within the preparation phase of the gesture, is chosen based on the most probable initial state of the meaningful manifestation. The initial pose is selected based on the following rules: (a) the pose that has been observed to precede the meaningful manifestation, and (b) the pose that has been observed to precede the meaningful manifestations co-align-induced by similar semiotic patterns. The properties of the coverbal movement are then stored as units within the Word relation layer. As presented in Figure 5, the deque is able to identify the left- and the right-word-boundaries of the coverbal expression (e.g. words 'vedno' (*always*) and 'iste' (*the same*)). The new information within the Content units is stored in the form of attribute-name/ values specification. Each attribute has a predefined set of values that subsequent processes and deques can select from. For instance, the set of the predefined values for the attribute *semi* is based on the semiotic types the system is able to classify (e.g. iconic, adaptor, symbolic, etc.). The attribute *mwords* holds those key-words that may, in the observed sequence, represent meaning (e.g. word 'iste' (*the same*)).

And the Content unit's attribute *pwords* indicate those words that preparation movement phases may be triggered by (e.g. word 'vedno' (*always*)). Further, Word units are enriched with additional attributes, describing the symbolical coverbal alignment, and movement-structure (propagation) of shapes of coverbal gestures. These attributes are: movement phase, right-hand-shape (rhshape), right-arm-shape (rashape), left-hand-shape (lhshape), and left-arm-shape (lashape). These shape attributes hold physical descriptions of the pose stored as an affiliate in the gesture dictionary. Figure 5 represents in detail the selection and co-alignment of coverbal expressions based on semiotic properties of the input text, and the relations between words, word phrases, and physical manifestations established within the gesture dictionary. As can be seen in the Word relation layer, the processed word sequence *'vedno iste in samo iste obraze'* (*always the same and only the same faces*) is tagged as: R(adverb), P(pronoun), C(conjunction), Q(particle), P(pronoun), and N(noun). The first process, *semiotic tagging,* in this case recognizes three mandatory word-types: the two pronouns and the noun. The second process *semiotic processing* indicates two semiotic patterns: adverb-pronoun and particle-pronoun-noun that may trigger a coverbal expression (two iconic *Content* units). In both semiotic patterns, the pronoun carries the meaning that may be reflected by the physical manifestation.

The pronoun also indicates the position of the stroke movement-phase. The adverb in the first pattern and the particle in the second pattern indicate the preparation movement-phase, and the noun within the second pattern indicates the hold movement-phase. And after the *matching process* (by considering semiotic relations, implicit rules, and lexical affiliation) is the coverbal movement for each semiotic pattern identified as 1-handed (the unit in Figure 5) chosen as K7 (for the hand shape), and Fr | Ab | Cl | O (front, abdomen, close, orthogonal for the position of the right arm). The deque identifies what non-verbal content can be represented, and the starting and ending points in the Word relation layer that drive the manifestation of the physical shapes. It, therefore, identifies the word(s) that carries the meaning (or

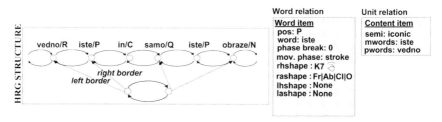

Figure 5. The relation between words and coverbal movement in the HRG structure.

functional role), and the physical manifestation the meaning may be represented by. The symbolic synchronization of meaning, therefore, only indicates a sequence of shapes that adequately illustrate the meaning of the inputted text. For instance, in the Slovenian phrase *'zelo velika prednost'* (*a very big advantage*), the word *'velika'* (*big*) indicates the meaning; however, in the context of the movement-phases, the stroke phase would be co-aligned with the word *'zelo'* (*very*), and not the word that is indicated as meaningful *'velika'*. The inner fluidity of coverbal expressions in form of movement-phases is within the PLATTOS system driven by the symbolic and acoustic prosodic deques (on the syllable level). The next section describes in detail how movement structure is then co-aligned with words.

5.2 Inner fluidity deque

The inner fluidity deque performs the temporal alignment of movement-phases regarding the coverbal gesture, and the input-text on the level of syllables. It extracts and processes the information in these HRG layers: Content unit relation, Phrase breaks, Word relation, and Syllable relation. Additionally, it creates a new layer, named Movement Phases, for describing the relation between the propagation of coverbal movement, and input text on the syllable level. The functioning of this deque and corresponding HRG structure are presented in Figure 6. Firstly, the starting and ending points of Content units are compared against prosodic word phrases (logical content segments), as predicted by the TTS engine's symbolic prosody deque. In the symbolic prosody deque, the CART-based phrase-break prediction model uses a B3 label for labeling major phrase breaks, and a B2 label for minor phrase breaks. A meaningful phrase inducing coverbal gesture is indicated by prosodic word phrase. Further, sentences with more than one prosodic word phrase indicate either the existence of multiple sequential gestures, or additional explanation, emphasis, or even negation of the preceding meaning.

The first process *'Search for word phrase breaks'* adjusts (extends, or even removes) the starting and ending points of the *Content units* based on phrase-break patterns. It aligns *Content units* with their predicted prosodic counterparts. A rule applied in this process is that each prosodic word phrase can contain only one (or none) coverbal gesture. The movement-propagation of each coverbal expression must also be maintained within the indicated prosodic word phrase. As seen in Figure 6, the *inner fluidity deque* processes a Slovenian text sequence *'vedno iste in samo iste obraze'* (*always the same, and only the same faces*). The starting and ending points of both Content units, CU-1 and CU-

2, are determined by the *phase tagging deque* as: *'vedno* iste' (*always the same*) (CU-1), and *'samo iste obraze'* (*only the same faces*) (CU-2). The prosodic word phrases, as determined by the predicted phrase breaks, demand some adaptation of these content units. For example, the predicted word phrases are *'vedno iste'* (WP-1), and *'in samo iste obraze'* (WP-2). The first *Content unit* CU-1 obviously matches the WP-1, whereas the second *Content unit* CU-2 has to be extended to also include the word *'in'*. When, however, the CUs include more words than the predicted WPs, those words are automatically disregarded. The second process *'Searching of emphasized words'* identifies the emphasized words. As the emphasized word is identified, those words contain a syllable assigned with a PA label (primary accent). The primary accent is assigned by the *symbolic prosody deque* (using CART models) to the most accentuated syllable inside the intonation prosodic phrase. In the context of movement phases, the stroke-movement phase is performed, co-aligned with the word containing the PA syllable. When, however, the CU has no PA syllable (and no emphasized word), the CU is disregarded. Further, the words preceding the emphasized word represent preparation, or pre-stroke-hold movement-phases. And words following the emphasized word represent the post-stroke-hold, or retraction movement-phases. On other syllables within the intonation prosodic phrase, the secondary accents (NA) are predicted. The NA prominence labels in general identify the preparation movement-phase (when the word that contains the NA syllable precedes the prosodic gesture trigger), or the retraction movement-phase (when this is the last coverbal gesture in the sequence, or a longer pause is indicated). Finally, the silences (*sil*), when predicted by the *acoustic prosody deque*, indicate the existence of the hold-movement-phases. The third process *'Searching of stressed syllables'* aligns the stroke movement phases with speech fragments, according to the PA syllable. The starting and ending points of the stroke phase are determined by the beginning of the emphasized word, and by the end of the PA syllable. In Figure 6, there are two emphasized words, each containing a PA syllable. The first PA syllable *'is'* (emphasized word *'iste'* (*the same*)) identifies the position of the first meaningful shape. The second PA syllable *'mo'* (emphasized word *'samo'*) identifies the position regarding the manifestation of the second meaningful shape. The fourth process *'Align stroke'* then aligns the stroke movement-phases to the speech fragments, as indicated by the stressed syllables. The fifth process *'Align preparation'* aligns the preparation movement phase. The starting and ending points of the preparation phase are identified by the NA syllable, and the end of the 'preparation' word. In Figure 6, the words *'vedno'* and *'in'* are indicated as 'preparation' words for the first and second prosodic word

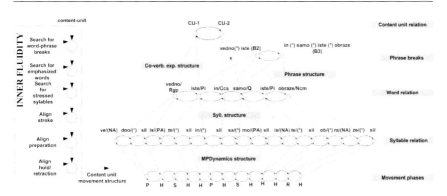

Figure 6. The functioning of the inner fluidity deque, using HRG structure.

phrases. In the case of CU-1, the NA syllable is *'ve'*, and the starting and ending points of the movement phase are determined by the syllables *'ve'* and *'dno'*. In the case of CU-2, there is no word preceding the prosodic gesture trigger that would also contain an NA syllable. As by the aforementioned rule, this indicates that no preparation movement phase is required. However, since both CU-1 and CU-2 represent similar meaning (in terms of semiotics and gesture affiliation), the preparation phase over the word *'in'* is artificially inserted into the movement propagation scheme. Finally, the sixth process *'Align hold/ retraction'* aligns the hold (both pre- and post-stroke), and retraction movement phases. The retraction movement phase is identified by the last meaningful word phrase by the word that contains the NA syllable and precedes the B3 phrase break or precedes a longer pause. In Figure 6, the starting and ending points are NA syllable *'ra'* and syllable *'ze'* of the word *'obraze'*. The hold movement phase is then determined by the pauses (*sil* units predicted and stored within the Syllable relation layer), and by the residual of the meaningful words not yet aligned with the stroke movement-phase. In addition to the *sil* units, the residual of word phrase WP-1 is the unused syllable *'te'*, and the residuals of the word phrase WP-2 are the syllables *'is'*, *'te'*, *'ob'*. As shown in the Movement Phases relation layer within the HRG structure, the residual content and the *sil* units are assigned as hold movement-phases. The detailed movement structure, as generated by the *inner fluidity deque*, is presented in Figure 7. The movement phases, preparation (P), stroke (S), and retraction (R), indicate a physical manifestation of a (new) shape, whereas the hold (H) movement phase indicates the maintaining of the closest previous physical manifestation.

As can be seen in Figure 7, the conversational expression CE-1 (gesture 1) is aligned with the prosodic boundaries (phrase breaks) of the word phrase WP1 (*'vedno iste'* (*always the same*)). During

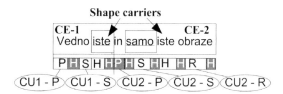

Figure 7. The generated movement structure, and sequence of shapes for conversational expressions.

propagating, two shapes will be manifested. The first shape CU1-P acts as an initial pose and will manifest itself over the syllables *'ve'* and *'dno'*. The preparation movement phase is followed by a hold, indicated by the *sil* unit. The stroke shape CU1-S will manifest itself over the syllable *'is'*. The residual syllable *'te'* indicates the post-stroke hold phase (H). An additional post-stroke hold phase (H) is inserted due to the presence of the predicted *sil* unit after the word *'iste'*. Similarly, we can structurally describe the conversational expression CE-2 aligned with the prosodic boundaries (phrase breaks) of the word phrase WP2. The resulting movement-phases layer stores the movement structure of those gestures performed within the observed conversational act. However, this module does not specify any temporal information for the coverbal gestures.

5.3 *Temporal-sync deque*

Finally, the *temporal-sync deque* temporally aligns speech fragments and coverbal gestures. As presented in Figure 8, the temporal-sync deque benefits from these HRG's layers: Content unit relation layer, Segments relation layer, and Movement phases relation layer. The deque creates new units, named Phase units stored within the Movement Phases relation layer of the HRG structure, and enriches the Content units, stored within Content unit relation layer, with additional attributes. The first process in the *temporal-sync deque* is the 'filter process' that searches for those *sil* units that have predicted durations 0 ms (can be *sil* units following words, and/or phrase-breaks). In order to filter unnecessary holds, the vertical HRG's relations between the *sil* unit (Segments relation layer) and the hold movement-phase are suspended, and the corresponding Phase unit removed from the Movement Phases relation layer. In the case shown in Figure 8, the first *sil* unit positioned between syllables 'dno' and 'is' has a predicted duration of 0 ms. Therefore, the connected hold movement phase unit has to be removed from the Movement-Phase sequence, and a new relation between neighboring preparation and stroke movement-phases has

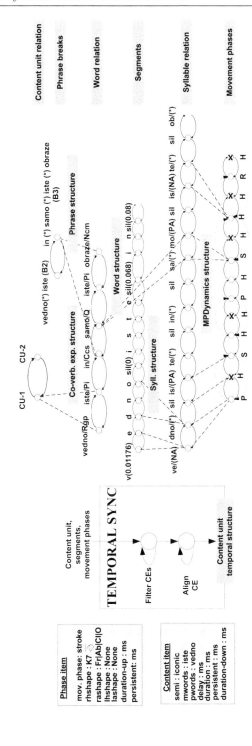

Figure 8. The functionality of the temporal-sync deque, by using HRG structure.

to be established (dotted links). Further, this process also merges the repeated hold-movement phases into a single hold.

The second process *'Align CE'* then temporally aligns each conversational expression (CE) with the input text. Each *Content* unit (CU) is temporally defined by the following attributes: *delay, duration, persistence,* and *duration-down*. And the Phase unit (PI), stored in the *Movement Phases relation* layer, is temporally defined by the following attributes: *duration-up* and *persistence*. The temporal values of the Content units and Phase units are calculated from the temporal values of their children (phoneme and *sil* units), segment units stored within the *Segment* layer (predicted by the *acoustic prosody deque*). The following equations are used:

$$CE_i.delay = \left(\sum_{j=1}^{i} CE_j.delay + CE_j.duration \right); \qquad 1$$

$$CE_i.persistence = t_n(MP_H); \qquad 2$$

$$CE_i.duration = \left(\sum_{i=1}^{n} t(MP_i) \right) - CE_i.persistence; \qquad 3$$

$$CE_i.duration_{down} = t(MP_R) = \left(\sum_{i=k}^{n} t(s_i) \right); \qquad 4$$

$$CE_i.duration_{up} = t(MP_{SP}) = \left(\sum_{i=k}^{n} t(s_i) \right); \qquad 5$$

$$PI_i.persistence = t(MP_H) = \left(\sum_{i=k}^{n} t(s_i) \right); \qquad 6$$

The *delay* attribute (equation 1) represents the absolute duration, measured from the beginning of the input text sequence, for which the non-verbal behavior is withheld prior to execution (i represents observed content item). If the first coverbal gesture does not start with the beginning of the text sequence, a default delay is inserted, with the duration equal to the sum of the speech fragments preceding the first indication of a coverbal gesture. The duration of the hold-movement phase for each CE represents the value of the *persistence* attribute for the CE. The value is calculated by equation 2. This attribute represents the duration for which the shape is maintained prior to the execution of the next CE (if last movement, then persistence = "hold", else persistence = 0). The *duration* attribute (equation 3) represents the overall duration of the coverbal gesture (*n* represents the number of movement phases). The *duration-down* attribute (equation 4) represents the duration of the retraction phase (*n* represents the number of segments, *k* represents the first segment). The Phase item (PI), stored within the Movement Phases relation layer, in addition to shape manifestation, also stores the temporal values of the preparation/ stroke movement-phase and the temporal values of the holds that follow the observed preparation/stroke movement-phase. Attribute *duration-up* (equation 5) represents the duration of the preparation/ stroke movement-phase; that is how long the transformation between shapes has to last (*n* represents the number of segments, *k* represents the first segment). As can be seen, the attribute's value is calculated as a sum of the predicted temporal values on those segment units (phonemes/visemes) contained within the specific preparation/stroke movement-phase. The attribute *persistence* (equation 6) is determined by the hold phase that follows the observed preparation/stroke movement-phase. This attribute's value is calculated as the sum of the temporal values for those segments contained within the hold phase. Figure 9 represents the creation of conversational expressions, and the temporal alignment of movement on the level of phonemes/visemes.

Figure 9 presents the complex relation that can be efficiently and flexibly established between Segment relation layer, and the Movement phases relation layer. The Left-hand side in Figure 9 presents the relations between coverbal gestures, words, movement-phases, and phase-items. As can be observed, the conversational expression CE-1 consists of two phase-items, marked as PI1 and PI2. The execution of CE-1 is finished by a hold (H) movement-phase. The CE-1 is temporally described by equations 1–3. Since the CE-1 in this case starts with the first segment of the input text sequence, the CE-1.delay attribute is set to 0. The CE.*persistence* attribute's value is determined by the duration of its last hold movement-phase (equation 2). In this case the value for CE-1.*persistence* is 0.068 s. The CE-1.duration attribute's value (equation 3) is determined by the sum of the PI-1 and PI-2 durations. The PI duration is determined by the predicted duration of segments it encapsulates. In the case of CE-1, the PI-1.duration attribute is determined as follows: $t('v') + t('e') + t('d') + t('n') +$

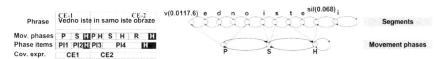

Figure 9. Temporal alignment of movement on phoneme/viseme level.

t('o'). And the PI-2.duration attribute's value is determined as: t('i') + t('s'). Since none of the PI phases encapsulates a hold movement-phase in this case, the PI.persistence attributes' values are, in both cases, set at 0. The results of the temporal sync are two types of units that store all the information necessary for the re-creation of the generated coverbal gestures and their transformation into behavior description. The Content unit stores the global temporal structure for the corresponding conversational expressions, whereas the Phase unit stores local information regarding overlaid shapes. In the following section, we discuss the last deque that has to transform Content units and corresponding Phase units into procedural descriptions of synthetic behavior (behavior-plans), as required for virtual re-creation of coverbal behavior (including hand/arm gestures, and lip-sync) on synthetic ECAs.

5.4 Non-verbal generator deque

Most of the ECA-based animation engines re-create non-verbal behavior based on procedural animation description mark-up languages. Within the PLATTOS system, the *non-verbal generator deque* is used to transform the coverbal gestures into a form understandable to the animation engine. Therefore, the deque transforms the HRG structure into EVA-SCRIPT-based behavioral descriptions supporting both lip-sync and coverbal gesture animation processes (Mlakar and Rojc, 2011). Nevertheless, since the HRG structure stores very detailed information on coverbal gestures, the transformation into other mark-up languages is also simple and quite straightforward. Figure 10 demonstrates the functionality of this deque. The deque needs Content units and Phase units as input. The output of the deque is a behavioral script, written in EVA-SCRIPT animation descriptive mark-up language. Further, in order to re-create (animate) the conversational expressions as stored in the HRG structure, those shape models determined by the Phase units' attributes (*rhshape, rashape, lhshape,* and *lashape*) are selected in the gestural dictionary. These models already contain the movement controllers' configurations (shape description), and must only be temporally aligned, according to the temporal specification, as specified by the Content (CU) and Phase (PI) units. Each CU unit is represented by a '*bgesture*' tag (Figure 10). The obligatory attributes of the '*bgesture*' tag are: *name, type,* and *delay*. The '*name*' attribute is formed generically, in this case, as: "*gestureCU-*" + *CU id*, as used in the HRG structure. The '*name*' attribute is, in the context of animation, used only for easier access to the selected procedure on the animation graph. Further, the '*type*' attribute is generically set to '*arm_animation*', since hand and arm

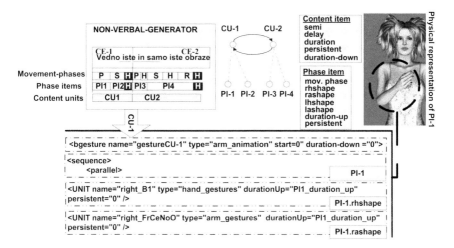

Figure 10. Transforming HRG structure into procedural animation specification.

gestures are to be animated, as specified by the phase-items. The '*start*' attribute is also important in the context of the TTS-driven behavior generation. For example, this attribute is an absolute temporal value that represents how long the animation of the described behavior has to be withheld prior to execution. It is equivalent to the '*delay*' attribute, as stored in the Content unit. And, the '*durationDown*' attribute describes the duration of the transition from the excited to a neutral (or previous) state. It is equivalent to the '*duration-down*' attribute also stored in the Content units.

Further, each PI unit in the HRG structure represents a sequence of body-parts moving in a parallel manner. Such a sequence has to be represented in the form of 'sequence-parallel' XML formation encapsulating behavioral description. The obtained behavioral description encapsulated within the 'sequence-parallel' formation then represents the specification of temporally defined shapes represented by a 'UNIT' tag. When the PI units' attributes (rhshape, rashape, lhshape, and lashape) are set (not 'None'), then each of them represents one 'UNIT' tag, starting with the '*name*' attribute. This attribute describes a gesture affiliate. By using it, the animation engine can easily recall the movement controllers defined as manifesting the synthetic shape. The temporal specification of the 'UNIT' is provided by specifying the following two obligatory EVA-SCRIPT-based attributes. The '*durationUp*' attribute represents absolute temporal value when describing how long it takes for the specified shape to be fully overlaid on the synthetic agent (e.g. the duration of the transition between shapes). It is equivalent to the PI's *duration-up* attribute. And the *persistence* attribute determines for how long, after the transition between sequential shapes is completed, the overlaid shape (excited state) has to be maintained. It is equivalent to the PI's persistence attribute. The *non-verbal generator deque* also forms the needed description of the facial expressions required for the lip-sync process. In this case, the visual sequence of visemes/phonemes is described in the form of 'viseme' EVA-SCRIPT tags, encapsulated within the 'speech' tag. The needed information is stored in Segment units of the Segment relation layer. For example, each segment (including *sil*) represents one 'viseme' tag within lip-sync specification and, at the same time, a gestural affiliate for a facial expression that within the mouth region overlays the pronunciation of the segment. The duration of each segment is also available. It is equivalent to the *duration* attribute of the EVA-SCRIPT's 'viseme' tag. The proper transition between segments is handled internally by the EVA animation engine, as proposed in Rojc and Mlakar (2011).

6. Evaluation

In order to test the algorithms of the proposed TTS-driven behavioral generation system, and to evaluate the quality and naturalness of the generated output, we annotated over 35 minutes of the proprietary video corpus, and created the gestural dictionary containing 300 distinct conversational configurations of arms and hands, described in the form of EVA-SCRIPT shape models. The shape models varied in structure (base represented shape), and intensity of the shape. These shape configurations are (based on the annotation and literature) manually linked to the verbal information (words and phrases) and can be automatically selected and temporally adjusted by the behavioral generation process. For the reliability of the resources, we rely on the findings presented in literature (e.g. Kita et al., 1998; Loehr, 2004); however, we have also ensured that meaningful words (manually identified meaningful words, serving only for the purpose of semiotic grammar evaluation) in the experimental sequences had at least one representative affiliate stored within the gesture dictionary that can be accessed based on either semiotic or implicit rules. Further, we re-created and animated several text-sequences. For the following text sequence: 'Vedno iste, in samo iste obraze' (*Always the same and only the same faces*), the TTS-driven system first runs the *phase-tagging-deque*. The HRG structure at this level already contains morphological information about the input text sequence, stored within the Word relation layer: 'Vedno/R iste/P- ,/XCOMMA in/C samo/Q iste/P- obraze/N ./ XPERIOD'. Based on word-type order (semiotic patterns), the system is then able to identify the following two meaningful phrases:

adverb-pronoun (CU-1) and particle-pronoun-noun (CU-2). For each meaningful phrase, the system creates a corresponding Content unit within the ContentUnits relation layer (CU-1 and CU-2 in Figure 11). In this case, the meaningful word-type in both meaningful phrases is identified as a pronoun. The meaningful word also provides the same meaning in both cases. As shown in Figure 11 (conversational shapes), the meaningful word is described by the meaningful hand shape right_K7, and arm configuration named Fr I Ab I Cl I O (front, abdomen, close, orthogonal). The most probable initial state of the meaningful shape is right_B1 for handshape, and Fr I Ce I No I O (front, chest, normal, orthogonal) for the arm configuration. Then follows the *inner fluidity deque*. At this level the data stored in the Phrase and Syllable relation layers can be used in order to adjust (if necessary) the extent of each CU, and in order to create a movement plan (structure) for the propagation of shapes within each CU. The Phrase layer in this case contains the following predicted phrase breaks: Vedno(*) iste(B2), (*) in(*) samo(*) iste(*) obraze(B3). As a result, the CU-2 item is extended (in order to also include the word 'in'). On the Syllable relation layer, the *inner fluidity deque* searches for emphasized syllables—those including syllables with PA (primary accent) and NA (secondary accent) labels. The syllables labeled as PA represent utterances over which the stroke is performed. In the system, the stroke phase will be co-aligned with the beginning of the PA-word, and the ending of its PA-syllable. In Figure 11, the CU-1's stroke phase starts with the beginning of the word 'iste' (*the same*), and ends on the phoneme 's'. During this interval, the ECA will perform transition from shape, identified within the Phase item PI-1 (right_B1, Fr I Ce I No I O), to the shape identified by PI-2 (right_K7, Fr I Ab I Cl I O). Similarly, the shape transition from PI-3 to PI-4 (first image of the PI-4 sequence) is

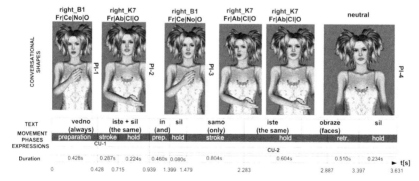

Figure 11. Transforming HRG structure into EVA-SCRIPT-based procedural animation specification.

(Color image of this figure appears in the color plate section at the end of the book.)

performed over the whole word 'samo' (*only*), since the ending of the PA-syllable also identifies the ending of the word. The NA-syllables are used to identify the duration of the preparation/retraction movement phases. The interval of a retraction/preparation phase is determined by the beginning of the retraction/preparation word's (word phrase) NA-syllable, and the word's (word phrases) ending. The retraction/ preparation phases of the sequence are the words 'vedno' (*always*) and 'in' (*and*), and the utterance 'raze'. At the end of each phase, the ECA will overlay the designated articulated configuration (e.g. pre-stroke shape, or neutral shape). The *temporal-sync deque* then temporarily aligns the movement phases in regard to the pronunciation rate of the phonemes/visemes, and silences. This is implemented by using the proposed equations (1–6), and predicted temporal information, as stored in the Segments layer. Figure 11 represents the absolute durations, and corresponding relative temporal positioning (in regard to the beginning of the text sequence). Therefore, the duration of the preparation phase of CU-1 is determined to be 0.428 s. It is followed by a stroke that is determined to last 0.287 s. Finally, the stroke shape of CU-1 has to be maintained for 0.224 s. The execution of CU-2 has to be withheld for the duration of CU-1, and will start after 0.939 s. The CU-2 will then begin with a preparation (0.460 s), and a pre-stroke-hold (0.080 s). This sequence represents the PI-3 unit, and the shape overlaid is described as right_B1, Fr|Ce|No|O. In a similar way, the temporal information is added to all other PI and CU units.

In order to re-create the non-verbal behavior as generated by the system and presented in Figure 11 (conversational shapes), the system at the end generates EVA-SCRIPT-based behavior description, containing information about lip-sync and gesture. This final step is performed by the *non-verbal generator deque*. The output is then synthesized by the conversational agent EVA, as symbolically and temporally co-aligned communicative behavior. The synthetic coverbal gestures were also evaluated by staff members and students. They evaluated lip-sync, the symbolic representations of meaningful words, and the alignment of coverbal gestures with the synthesized speech. All of the evaluators agreed that the speech and visual pronunciation was in temporal sync; however, 35% of them suggested improvement of the correlation between visual and audio stressing, 55% of the observed sequences adequately represented the verbal content, and 30% of the sequences were observed as a meaningful word mismatch. Based on verbal information, the evaluators expected another word to be represented. However, when the meaningful word was suggested to them, most of them agreed that the representation was adequate, and appeared more natural. Finally, 15% of the sequences were evaluated as out-of-sync in either preparation, and/or stroke movement phase.

The evaluation showed that most of the generated behavior of the proposed PLATTOS system can be assessed as viable, and close to human-like.

7. Conclusion

This chapter presented a novel TTS-driven non-verbal behavior system for coverbal gesture synthesis. The system's architecture and grammar, used to synchronize the non-verbal expressions with verbal information in symbolical and temporal domain, were presented in detail. Further, we have presented how meaningful parts of verbal content are identified and selected based on word-type-order, and the semiotic patterns within the proposed system. We have also described how a visual representation of meaning can be selected, how the structure of its propagation can be generated as sequence movement-phases (based on lexical affiliation and semiotic rules), and how movement-phases and durations of movements can be aligned with the verbal content. Finally, we have explained how a procedural script is formed that drives the synchronized verbal and coverbal behavior. The generated synthetic behavior already reflects a very high-degree of lip-sync and iconic, symbolic, and indexical expressions, as well as adaptors. As proven by the evaluation, most of the generated behavior appears 'natural', and may adequately represent the verbal content. As part of our future works we will investigate the possibility of dynamically adjusting the degree of visual pronunciation in regard to accenting the phonemes (visemes). Since EVA-SCRIPT and ECA-EVA already support this feature, most of the investigation will be oriented towards expressive TTS models. In order to improve the rules stored within semiotic grammar, we will further annotate our video corpora in order to further fine-tune existing rules, especially regarding the movement dynamics, and additional shapes representing the meanings of words and word phrases. Some effort will also be directed towards further richness of the presented gestural dictionary. By annotating additional segments of video corpora, we will create additional gesture-instances. This will further contribute to the diversity (that is typical for naturalness), and expressive capabilities of ECAs. Additionally, we will try to incorporate several other part-of-speech attributes for each word-type. This should not only enrich the non-verbal behavior but also improve those processes used for selecting the meaningful word, the semiotic type of movement, and the position of a meaningful shape (stroke phase).

Acknowledgement

Operation part financed by the European Union, European Social Fund.

REFERENCES

Allwood, J. 2001. Dialog Coding—Function and Grammar: Göteborg Coding Schemas. *Gothenburg Papers in Theoretical Linguistics*, **85**:1–67.

Allwood, J., E. Ahlsén, J. Lund and J. Sundqvist. 2007a. Multimodality in own communication management. *Current Trends in Research on Spoken Language in the Nordic Countries*, **2**:10–19.

Allwood, J., L. Cerrato, K. Jokinen, C. Navarretta and P. Paggio. 2007b. The MUMIN coding scheme for the annotation of feedback, turn management and sequencing phenomena. *J. of Language Resources and Evaluation*, **41(3)**:273–287.

Barbieri, F., A. Buonocore, R.D. Volta and M. Gentilucci. 2009. How symbolic gestures and words interact with each other. *J. of Brain and Language*, **110(1)**:1–11.

Bavelas, J.B. and N. Chovil. 2000. Visible acts of meaning: An integrated message model of language in face-to-face dialogue. *J. of Language and Social Psychology*, **19(2)**:163–194.

Bergmann, K. and S. Kopp. 2008. Multimodal Content Representation for Speech and Gesture Production. Symposium at the AISB Annual Convention: Multimodal Output Generation, pp. 61–68.

Breslow, L.A., A.M. Harrison and J.G. Trafton. 2010. Linguistic Spatial Gestures. In Proc. of 10th International Conference on Cognitive Modeling, Philadelphia, PA: Drexel University, pp. 13–18.

Cassell, J., H. Vilhjálmsson and T. Bickmore. 2001. BEAT: The Behavior Expression Animation Toolkit. In Proc. of SIGGRAPH 2001, pp. 477–486.

Čereković, A. and I. Pandžić, 2011. Multimodal behavior realization for embodied conversational agents. *J. of Multimedia Tools and Applications*, **54(1)**:143–164.

Deacon, T.W. 2003. Universal grammar and semiotic constraints. Language Evolution: The States of the Art. Oxford University Press, Oxford, 7, pp. 111–139.

De Carolis, B., C. Pelachaud, I. Poggi and M. Steedman. 2004. APML, a mark-up language for believable behavior generation. In Life-like Characters: Tools, Affective Functions, and Applications, pp. 65–85.

Esposito, A. and A.M. Esposito. 2011. On Speech and Gestures Synchrony. Analysis of Verbal and Nonverbal Communication and Enactment. The Processing Issues, LNCS 6800, pp. 252–272.

Gallaher, P.E. 1992. Individual differences in nonverbal behavior: Dimensions of style. *J. of Personality and Social Psychology*, **63(1)**:133–145.

Grenfell, M.J. 2011. Bourdieu, Language and Linguistics. Continuum International Publishing Group, January 2011, ISBN: 9781847065698.

Hadar, U. and R.K. Krauss. 1999. Iconic gestures: The grammatical categories of lexical affiliates. *J. of Neurolinguistics,* **12(1)**:1–12.

Hartmann, B., M. Mancini and C. Pelachaud. 2005. Implementing expressive gesture synthesis for embodied conversational agents. In Proc. of the 6th International Conference on Gesture in Human-Computer Interaction and Simulation (GW'05), pp. 188–199.

Heylen, D., S. Kopp, S. Marsella, C. Pelachaud and H. Vilhjálmsson. 2008. The Next Step towards a Function Markup Language. Intelligent Virtual Agents, LNCS 5208, pp. 270–280.

Hogrefe, K., W. Ziegler and G. Goldenberg, 2011. Measuring the formal diversity of hand gestures by their hamming distance. Integrating Gestures: The Interdisciplinary Nature of Gesture, **8**:75–88.

Holler, J. and K. Wilkin, 2011. Co-speech gesture mimicry in the process of collaborative referring during face-to-face dialogue. *J. of Nonverbal Behavior,* **35**:133–153.

Jokinen, K. 2009. Gaze and Gesture Activity in Communication. Universal Access in Human-Computer Interaction, Intelligent and Ubiquitous Interaction Environments, LNCS 5615, pp. 537–546.

Kendon, A. 2000. Language and gesture: Unity or duality. Language and Gesture, Cambridge University Press, **2**:47–63.

Kendon, A. 2004. Gesture: Visible Action as Utterance. Cambridge University Press, September 2004, ISBN: 9780521542937.

Kipp, M. 2001. From human gesture to synthetic action. In Proc. of the Workshop on Multimodal Communication and Context in Embodied Agents, Fifth International Conference on Autonomous Agents, pp. 9–14.

Kita, S. and T.S. Davies. 2009. Competing conceptual representations trigger co-speech representational gestures. *Language and Cognitive Processes,* **24(5)**:761–775.

Kita, S. and A. Özyürek, 2003. What does cross-linguistic variation in semantic coordination of speech and gesture reveal? Evidence for an interface representation of spatial thinking and speaking. *J. of Memory and Language,* **48(1)**:16–32.

Kita, S., I. van Gijn and H. van der Hulst. 1998. Movement phases in signs and co-speech gestures, and their transcription by human coders. Gesture and Sign Language in Human-Computer Interaction. **1371**:23–35.

Kopp, S., B. Krenn, S. Marsella, A.N. Marshall, C. Pelachaud, H. Pirker, K. Thorisson and H. Vilhjalmsson. 2006. Towards a common framework for multimodal generation: The behavior markup language. In Proc. of the 6th international conference on Intelligent Virtual Agents (IVA '06), pp. 205–217.

Kopp, S. and I. Wachsmuth. 2002. Model-based animation of co-verbal gesture. *In Proc. of the Computer Animation,* pp. 252–257.

Kopp, S. and I. Wachsmuth. 2004. Synthesizing multimodal utterances for conversational agents, *J. of Computer Animation and virtual Worlds,* **15(1)**:39–52.

Kransted, A., S. Kopp and I. Wachsmuth. 2002. MURML: A multimodal utterance representation markup language for conversational agents. In Proc. of the AAMAS Workshop on Embodied Conversational Agents—Let's Specify and Evaluate Them.

Krauss, M., Y. Chen and R. Gottesman. 2000. 13 Lexical gestures and lexical access: A process model. *Language and Gesture*, **2**:261–283.

Krenn, B., C. Pelachaud, H. Pirker and C. Peters. 2011. Embodied conversational characters: Representation formats for multimodal communicative behaviours. Emotion-Oriented Systems, pp. 389–415.

Kröger, B.J., S. Kopp and A. Lowit. 2010. A model for production, perception, and acquisition of actions in face-to-face communication. *Cogn. Process*, **11(3)**:187–205.

Lankes, M. and R. Bernhaupt. 2011. Using embodied conversational agents in video games to investigate emotional facial expressions. *J. of Entertainment Computing*, **2(1)**:29–37.

Le, Q.A., S. Hanoune and C. Pelachaud. 2011. Design and implementation of an expressive gesture model for a humanoid robot. 11th IEEE-RAS International Conference on Humanoid Robots (Humanoids), pp. 134–140.

Loehr, D. 2004. Gesture and intonation. Doctoral Dissertation, Georgetown University.

Luke, K.K., S.A. Thompson and T. Ono. 2012. Turns and increments: A comparative perspective. *Discourse Processes*, **49(3–4)**:155–162.

Malcangi, M. 2010. Text-driven avatars based on artificial neural networks and fuzzy logic. *J. of Computers*, 4(2), pp. 61–69.

McNeill, D. 1992. Hand and Mind—What Gestures Reveal about Thought. The University of Chicago Press, August 1992, ISBN: 9780226561325.

McNeill, D. 2005. Gesture and Thought. The University of Chicago Press, November 2005, ISBN: 9780226514628.

de Melo, C.M. and A. Paiva. 2007. Modeling gesticulation expression in virtual humans. *New Advances in Virtual Humans*, **140**:133–151.

Mlakar, I. and M. Rojc. 2011. Towards ECA's Animation of Expressive Complex Behaviour. Analysis of Verbal and Nonverbal Communication and Enactment. The Processing Issues, LNCS 6800, pp. 185–198.

Mlakar, I. and M. Rojc. 2012. Capturing form of non-verbal conversational behavior for recreation on synthetic conversational agent EVA. *WSEAS Transactions on Computers*, **7(11)**:216–226.

Ng-Thow-Hing, V., L. Pengcheng and S. Okita, 2010. Synchronized gesture and speech production for humanoid robots. In Proc. of International Conference on Intelligent Robots and Systems (IROS '10), pp. 4617–4624.

Nowina-Krowicki, M., A. Zschorn, M. Pilling and S. Wark. 2011. ENGAGE: Automated Gestures for Animated Characters. In Proc. of Australasian Language Technology Association Workshop '11, pp. 166–174.

van Oijen, J., L. Vanhée and F. Dignum. 2012. CIGA: A Middleware for Intelligent Agents in Virtual Environments. Agents for Educational Games and Simulations, **7471**:22–37.

Peirce, C.S. 1958. Collected Papers of Charles Sanders Peirce, 1931–1958, 8 vols., MA: Harvard University Press.

Pine, K., H. Bird and E. Kirk. 2007. The effects of prohibiting gestures on children's lexical retrieval ability. *Developmental Science,* **10(6)**:747–754.

Poggi, I., C. Pelachaud, F. de Rosis, V. Carofigilo and B. de Carolis. 2005. Greta. A believable embodied conversational agent. Multimodal Intelligent Information Presentation, **27**:3–25.

Rojc, M. and Z. Kačič. 2007. Time and Space-Efficient Architecture for a Corpus-based Text-to-Speech Synthesis System. *Speech Communication,* **49(3)**:230–249.

Rojc, M. and I. Mlakar. 2011. Multilingual and multimodal corpus-based text-to-speech system—PLATTOS. *Speech and Language Technologies/Book,* **2**:129–154.

Romportl, J., E. Zovato, R. Santos, P. Ircing, J.R. Gil and M. Danieli. 2010. Application of Expressive TTS Synthesis in an Advanced ECA System. In Proc. of ISCA Tutorial and Research Workshop on Speech Synthesis, pp. 120–125.

Sargin, M.E., O. Aran, A. Karpov, F. Ofli, Y. Yasinnik, S. Wilson, E. Erzin, Y. Yemez and A.M. Tekalp, 2006. Combined Gesture-Speech Analysis and Speech Driven Gesture Synthesis. IEEE International Conference on Multimedia and Expo, pp. 893–896.

Schegloff, E.A. 1985. On some gestures' relation to talk. Structures of Social Action: Studies in Conversation Analysis, Cambridge University Press, pp. 266–298.

Straube, B., A. Green, B. Bromberger and T. Kircher. 2011. The differentiation of iconic and metaphoric gestures: Common and unique integration processes. *J. of Human Brain Mapping,* **32(4)**:520–533.

Tang, H., Y. Fu, J. Tu, M. Hasegawa-Johnson and T.S. Huang, 2008. Humanoid audio-visual avatar with emotive text-to-speech synthesis. *IEEE Transactions on Multimedia,* **10(6)**:969–981.

Thiebaux, M., S. Marsella, A.N. Marshall and M. Kallmann, 2008. SmartBody: behavior realization for embodied conversational agents. In Proc. of the 7th International Joint Conference on Autonomous Agents and Multiagent Systems-Volume 1 (AAMAS '08), pp. 151–158.

Vilhjalmsson, H., N. Cantelmo, J. Cassell, N.E. Chafai, M. Kipp, S. Kopp and R. van der Werf. 2007. The behavior markup language: Recent developments and challenges. Intelligent Virtual Agents, **4722**:99–111.

Wallbott, H.G. 1998. Bodily expression of emotion. *European Journal of Social Psychology,* **28(6)**:879–896.

Wang, L., W. Han, F.K. Soong and Q. Huo. 2011. Text driven 3D photo-realistic talking head. In Proc. of INTERSPEECH 2011, pp. 3307–3308.

Zoric, G. and I. Pandžić, 2008. Towards real-time speech-based facial animation applications built on HUGE architecture. In Proc. of International Conference on Auditory-Visual Speech Processing.

CHAPTER 14

Modeling Human Communication Dynamics for Virtual Human

Louis-Philippe Morency, Ari Shapiro and Stacy Marsella

1. Introduction

Face-to-face communication is a highly interactive process where participants mutually exchange and interpret linguistic and gestural signals. Communication dynamics represent the temporal relationship between these communicative signals. Even when only one person speaks at a time, other participants exchange information continuously amongst themselves and with the speaker through gesture, gaze, posture and facial expressions. The transactional view of human communication shows an important dynamic between communicative behaviors where each person serves simultaneously as speaker and listener (Watzlawick et al., 1967). At the same time you send a message, you also receive messages from your own communications (individual dynamics) as well as from the reactions of the other person(s) (interpersonal dynamics) (DeVito, 2008).

Individual and interpersonal dynamics play a key role when a teacher automatically adjusts his/her explanations based on the student nonverbal behaviors, when a doctor diagnoses a social disorder such as autism, or when a negotiator detects deception in the opposite team. An important challenge for artificial intelligence researchers in the 21st century is in creating socially intelligent robots and computers, able to recognize, predict and analyze verbal and nonverbal dynamics during face-to-face communication. This will not only open up new avenues for human-computer interactions but create new computational tools for social and behavior researchers—software able to automatically analyze human social and nonverbal behaviors, and extract important interaction patterns.

Human face-to-face communication is a little like a dance, in that participants continuously adjust their behaviors based on verbal and nonverbal behaviors from other participants. We identify four important types of dynamics during social interactions:

- **Behavioral dynamic:** A first relevant dynamic in human communication is the dynamic of each specific behavior. For example, a smile has its own dynamic in the sense that the speed of the onset and offset can change its meaning (e.g., fake smile versus real smile). This is also true about words when pronounced to emphasize their importance. The behavioral dynamic needs to be correctly represented when modeling social interactions.

- **Multimodal dynamic:** Even when observing participants individually, the interpretation of their behaviors is a multimodal problem in that both verbal and nonverbal messages are necessary to a complete understanding of human behaviors. Multimodal dynamics represent this influence and relationship between the different channels of information such as language, prosody and gestures. Modeling the multimodal dynamics is challenging since gestures may not always be synchronized with speech and the communicative signals may have different granularity (e.g., linguistic signals are interpreted at the word level while prosodic information varies much faster).

- **Interpersonal dynamic:** The verbal and nonverbal messages from one participant are better interpreted when put into context with the concurrent and previous messages from other participants. For example, a smile may be interpreted as an acknowledgement if the speaker just looked back at the listener and paused while it could be interpreted as a signal of empathy if the speaker just confessed something personal. Interpersonal dynamics represent this influence and relationship between multiple sources (e.g., participants). This dynamic is referred as micro-dynamic by sociologists (Hawley, 1950).

- **Societal dynamic:** We categorize the organizational (often referred as meso-level) and societal (often referred as macro-level) dynamics in this general category which emphasize the cultural change in a large group or society. While this proposal does not focus on societal dynamics, it is important to point out the bottom-up and top-down influences. The bottom-up approach emphasizes the influence of micro-dynamics (behavioral, individual and interpersonal) on large-scale societal behaviors (e.g., organizational behavior analysis based on audio micro-dynamics (Pentland, 2004)). As important is the top-down influence of society and culture on individual and interpersonal dynamics.

In this book chapter, we first discuss techniques to model the behavior dynamics of virtual human as well as human participants. We then address the challenge of multimodal dynamic and, more specifically, the synchrony between speech and gestures for virtual human. Finally, we present approaches to model the interpersonal dynamic between speaker and listeners using state-of-the-art machine learning. We finally conclude by discussing the future challenges related to societal and cultural dynamics.

2. Behavior Dynamic: Virtual Human Animation

Virtual humans can be used to express portray a wide range of behaviors, including synchronized speech, gestures and facial expressions. In order to generate such expressions on a virtual human, an animation must be synthesized and replayed on a digital character. Such an animation can be generated from a variety of sources, such as motion capture, hand-designed by a digital artist, or procedurally generated from a motion synthesis algorithm. The synthesized motion can be coordinated with audio and lip motion on the digital character if the gesture is associated with an utterance. Digital 3D characters that are human-like in appearance and responses are termed virtual humans.

The use of digital characters is common in video games, simulations and live action and animated feature films. The most widely used method of generating a 3D character in motion involves combining a 3D geometric mesh, its surface colors or images, called textures, with a set of joints combined in a hierarchy, called a skeleton. The skeleton is then used to modify the 3D geometry via a deformation, or skinning process, where each joint of the skeleton is related to one of more of the faces of the geometry and modifies the geometry as the position and orientation of the joint changes. Thus, 3D characters that consist of a 3D mesh can move, stretch and twist such a mesh in a way that appears somewhat natural by simply rotating and translating the joints in the underlying skeleton. Thus, it is often sufficient to generate the motion for a 3D character in order to appropriately express a gesture or movement when using these common 3D character animation techniques. Thus, the challenge of synthesizing gestures is generally associated with acquiring and synthesizing skeletal motion.

Other methods of generating digital characters and motions exist, such as the use of image-based techniques, where an entire mesh, texture and movement are captured at the same time. However, at the time of this writing, such methods are not mature and are not widely supported.

There are several different kinds of gesture and expression architectures that can be used, each offering a different level of quality, different data requirements, different flexibility and varying complexity of use. In general, the gesture architectures that offer the highest level of quality are those that use motion capture data explicitly and replay it without modification. Those with the lowest level of quality segment gestures into smaller phases then synthesize motions procedurally from various algorithms. Gesture architectures with the greatest level of control and precision generally favor the reverse; procedurally generated, phased gesture motion provides better control than replayed motion capture clips. The following sections describe variations in a gesture architecture based on these ideas.

2.1 Motion capture session

In general, a motion capture session requires a motion capture system based on cameras or inertial sensors. The session typically requires transferring the data onto a virtual character during the capture session. The motion capture process typically synthesizes data onto an existing skeleton, which does not match exactly the proportions and sizes of the real human actor. Thus, the captured data needs to be retargeted or transferred onto the skeleton which models the virtual character. Motion capture data is typically segmented into clips, each clip representing semantically related content. Similarly, longer motion capture clips can be segmented into smaller ones during a post-processing phase, where data is refined and edited. The specifics of capturing motion via motion capture can be found in other sources. Below we describe several capture and motion synthesis strategies.

Figure 1. An actor using an LED-based motion capture suit [PhaseSpace]. In this session, the body motion is being captured, but not the facial performance or audio/utterance.

(Color image of this figure appears in the color plate section at the end of the book.)

Figure 2. A hub-and-spoke gesture architecture. An underlying idle pose is created (center character), from which a set of gestures can be played. This allows individual gestures to be used for different utterances. Note that a different set of gestures must be generated for each underlying pose, which can vary from standing, to sitting, to standing with various hand poses and body lean.

(Color image of this figure appears in the color plate section at the end of the book.)

2.2 Full performance architectures

The highest level of quality can be obtained by simultaneously capturing both the utterance and the gesture. Thus, a performer will act out an utterance in combination with its associated body movements, including gestures. This performance will then be replayed in its entirety on a virtual human. The advantage to this approach is that the virtual human will be able to faithfully replay the human performance, notwithstanding the retargeting necessary to fit the captured performance onto the virtual human's skeleton. However, one of the difficulties of using this method is that the audio of the utterance in combination with the associated facial movements must be synchronized with the body animation. In other words, the body movement, facial movement and audio track must be synchronized together exactly as they were captured. This means capturing the facial animation performance at the same time as the body movement (Stone et al., 2004), which typically requires three different capture systems: a motion capture system for the body, a separate one for the face, and a third audio capture for the utterance. In order to simplify this process, the facial performance can be captured during a separate performance, but doing so risks a lack of synchronization with the original body performance. The best level of quality is achieved through capture and playback of the performance through these three capture systems. However, each system requires different processes in order to obtain data that can be reused on a virtual character. Thus, it takes a large amount of effort to integrate synchronized data from three separate systems together. In addition, the high level of quality comes at the expense of specificity; the performance is only meaningful in contexts similar to those during the recorded session. For example, consider an actor whose performance in a dialogue is captured and synthesized onto a virtual character. By only capturing one of the two actors and synthesizing that performance onto a virtual human, you risk misapplying the subtleties of the actor during the performance that are in response to the presence or movements of the other actor. A recording interaction might be subtlety different when used in a different conversation with a different person. Conversational energy, timing, backchannelling and even gaze can be different with different partners. Thus, the high level of quality achieved by replaying an actor's performance can be limited in its use outside of the original context.

2.3 Hub-and-spoke architectures

As an alternative to reusing a motion captured performance directly, a hub-and-spoke architecture can be used to achieve greater reuse of speech, gestures and facial performance. This method uses an underlying base, or idle pose, and blends gesture motions that start and end in a similar position as the idle pose (Shapiro, 2011). For example, an actor will perform a number of gestures starting from the base pose, performing the gesture, and then returning to the base pose. Thus, each gesture can be replayed on a virtual human in different order with other gestures, each starting and ending from the same base pose, which is usually implemented as a continuous idle posture. This method allows you to synthesize an arbitrary sequence of gestures, while maintaining a high level of motion quality, since each individual gesture maintains the nuances of the original performance. A drawback of this method is the large number of gesture performances needed for each base pose or posture. For example, a different performance would be needed for gestures when standing up with your hands relaxed at your side, versus standing up with your hands on your hips, versus standing up with only one hand resting on your hip, and so forth. Posture changes, such as sitting down, crossing your legs and so forth, would also require a new set of gestures for each posture. Note that speech and facial movements can be simultaneously recorded with the gestures when using the hub-and-spoke method, or synthesized at a later time. Also note that using such a method limits the types of performances that can be captured; the actor must return to the same position that he started from, perhaps causing a lack of continuity between utterances. However, this approach's strength is in its ability to sequence gestures in any order or time as needed.

2.4 Blending gestures to achieve motion variation

While the best quality reproduction will be achieved by replaying a performance exactly as it were captured, it is often impractical or unfeasible to capture all the movements that would be replayed on a virtual human. For example, a pointing gesture can take many forms: accusatory, informational, subtle and so forth. Thus, variations in gesture can be generated by multiple performances of an actor of those gesture variations. This will result in a high-quality reproduction of the gesture. This also means that additional motion-captured resources will be needed to capture the additional gesture variation, and itself will only be reusable for one additional synthesized performance that matches the new variation. To remedy this problem, similar gesture motions can be blended together in order to give a range of variations between various example motions. For example, an energetic beat gesture can be combined with a slow, deliberate beat gesture to generate a gesture motion that appears halfway between an energetic and a slow gesture. Likewise, directional gestures, such as pointing gestures, can be generated with directional variations, then recombined in order to form a pointing gesture in a direction not explicitly captured, but rather synthesized as a combination of two or more other pointing gestures. There are several limitations to this approach; gestures that are to be blended together need to have compatible characteristics in order for the blending to work properly. In most cases, this means that the motions must have the same number and type of phases. One motion cannot, for example,

Figure 3. A simplified hierarchy for synthesizing motion. The rectangles indicate the parts of the body controlled during each state, starting from the larger rectangles and going inward towards the smaller rectangles. The entire body motion is synthesized for, in this case, a sitting virtual human. Next, the upper body gesturing is synthesized by layering gesture movements on top of the lower body. Next, spine movements controlling posture and gaze are layered on top of the gesture. Then, head movements are added for backchannelling and movement during speaking. Next, facial movements to express emotion and coordinate lip movement, then eye movement including saccades and eyelid positioning.

(Color image of this figure appears in the color plate section at the end of the book.)

have three small shakes of the hand, while the other only has two. In addition, large movements across the body or broad variations in poses across the gestures blend together poorly. Blended poses can vary in completion time; it is unlikely that any two motion-captured gestures will take the same amount of time. Gestures are blended together by first timewarping, that is stretching or compressing, the motions to the desired time, and then combining the various motions together. A large difference in time between any two motions will lead to poor quality blends, since one or both motions will need to be lengthened or shortened to match the other, typically changing the dynamics of motion that are embedded within the original captured motion. Thus, a gesture that is synthesized from one or more blends maintains the highest level of fidelity when the individual blended gestures have matching phases, and similar timings. There are many different ways to blend motions together, offering various trade-offs between execution time and memory (Kovar and Gleischer, 2004; Huang and Kallmann, 2010) as well as tradeoffs between precision and smoothness (Pettre and Laumond, 2006; Rose et al., 1998). An overview of common blending techniques can be found in Feng et al. (2012).

2.5 Hierarchical gesture models

One model for achieving variations in gestures is to use a model of hierarchical of control over the virtual human movement (Kallmann and Marsella, 2005). In this model, virtual human movement is divided into a generalization/specialization hierarchy. Thus, movement is first performed for the entire body, usually a sitting or standing pose, then a gesture movement using the arms and torso is performed, then a separate head and neck movement, then facial and eye movement. By layering such animations together, it becomes possible to achieve a large variation in gesture performance. For example, the same gesture can be combined with several different head or face movements, producing differing performances. In addition, the hierarchical nature allows the integration of procedural elements such as gaze control (Thiebaux et al., 2008) to override specific parts of the body in order to modify the underlying motion for the specific context in which the motion in used. The drawback to using such architecture is the loss of fidelity of the resulting motion; since the synthesized motion was never originally captured on a human, the dynamics and subtleties of the synthesized motion will differ from that of a human actor performing the same motion. Thus, using a hierarchy to generate gesture performance yields a large variation in performance, at the expense of motion quality.

3. Multimodal Dynamic: Speech and Gestures Generation

The generation of multimodal behavior for a virtual human faces a range of challenges. Most fundamentally is the question of what behaviors to exhibit. Nonverbal behaviors serve a wide variety of communicative functions in face-to-face interaction. They can regulate the interaction: a speaker may avert gaze to hold onto the dialog turn and may hand off the turn by gazing at a listener. The speaker can use nonverbal behaviors to convey propositional content: a nod can convey agreement; raising eyebrows can emphasize a word. The propositional content of the nonverbal behavior can stand in different relations to the verbal content, providing information that embellishes, substitutes for, and even contradicts the information provided verbally. In other words, the nonverbal behavior is not simply an illustrator of the verbal information. Nonverbal behaviors also convey a wide range of mental states and traits: gaze aversions can signal increased cognitive load, blushing can suggest shyness and facial expressions can reveal emotional states.

Another challenge here is that this mapping between communicative function and behaviors is many-to-many. One can emphasize aspects of the dialog using a hand gesture, a nod or eyebrow raise. On the other hand, a nod can be used for affirmation, emphasis or to hand over the dialog turn. The context in which the behavior occurs can transform the interpretation, as can subtle changes in the dynamics of the behavior: head nods signaling affirmation versus emphasis typically have different dynamics. Further, behaviors can be composed with each other, further transforming their interpretation.

Additionally, the behaviors are often tightly synchronized and changes in this synchronization can lead to significant changes in what is conveyed to a listener. For instance, the stroke of a hand gesture, a nod or an eyebrow raise individually or together are often used to emphasize the significance of a word or phrase in the speech. To achieve that emphasis, the behavior must be closely synchronized with the utterance of the associated words being emphasized. Alteration of the timing will change what words are being emphasized and consequently change what is conveyed to a listener.

Achieving such synchronization in a virtual human can be difficult, especially in the case of behaviors such as hand gestures that involve relatively large-scale motion and multiple phases. Consider a beat gesture, a staccato, often downward stroke of the hand that can be used to provide emphasis. To perform a downward motion, the hand must be raised in preparation for the stroke. After the stroke, the hand can be held in a pose to provide further emphasis, followed by a relaxation to a rest position. This sequence of behaviors occur in alignment with the speech, so there must be sufficient time to prepare for the stroke; the stroke and any post-stroke hold must be tightly coordinated with the parts of the dialog that is being emphasized. Further the relaxation may need to take into account co-articulation that there will be subsequent gestures to be performed.

In addition to synchronization of speaker's behaviors, there is synchronization between speaker and listeners as the speaker's utterance unfolds. Listeners exhibit a variety of behaviors. These behaviors include *generic* responses that signals the listener is attending or broadly comprehending the speaker as well as *specific* responses tightly coupled to a deeper understanding of, and reaction to, the details of what the speaker is saying at that moment (Bavelas et al., 2000). The speaker in turn will dynamically adapt what they are saying in response to this feedback from the listener.

Such dynamic adjustments raise challenges in generating both the verbal and nonverbal behaviors dynamically and incrementally for a virtual human. Ideally a virtual human listener should respond to a human speaker, providing generic feedback signaling attention as well as more specific feedback that signals comprehension and reaction to the speaker's unfolding utterance. This requires a natural language system that can incrementally understand the human speaker's utterance. Conversely, the virtual human as it is speaking should be aware of a human listener's behavior, responding to nonverbal signals such as confusion. Together such dynamic interaction suggests capabilities such as interruption of behavior as well as incremental understanding and generation of verbal and nonverbal behavior.

3.1 An approach to nonverbal behavior generation

A range of systems have tackled various aspects of these challenges (refs). Here we discuss one of the approaches: The Nonverbal Behavior Generator (NVBG) (Lee and Marsella, 2006; Wang et al., 2011). NVBG automates the generation of physical behaviors for virtual humans, including nonverbal behaviors accompanying the virtual humans dialog, responses to perceptual events as well as listening behaviors. It takes input from the virtual human's knowledge of its task, dialog, emotional reactions to events and perceptual processes. Modular processing pipelines transform the input into behavior schedules, written in the Behavior Markup Language (BML, Kopp et al., 2006) and then passed to a character animation system (SmartBody, Thiebaux et al., 2008).

Figure 1 depicts three pipelines used to generate behavior for the virtual human including coverbal nonverbal behavior, reaction to events and listening behavior. The sections below discuss these three

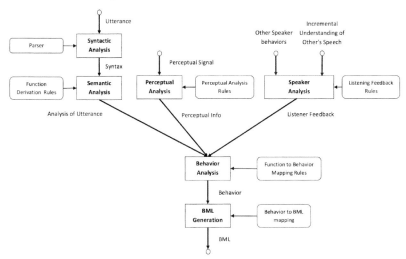

Figure 4. Overview architecture for verbal and nonverbal behavior generation in a virtual human.

pipelines. Note the initial processing differs but eventually merges at the behavior analysis.

3.2 *Processing the virtual human's utterances*

The utterance pipeline (left side in Figure 4) analyzes the surface text of the virtual human's utterance to infer appropriate nonverbal behavior. The processing in NVBG does not make any strong assumption about the input's markup of the agent's communicative intent or internal state (e.g., affective state, attitude). When such information is missing, the system attempts to infer it (essentially falling back on the more limited role of illustrating and embellishing the language channel). For instance, in the absence of detailed markup of the virtual human's communicative intent, such as points of emphasis or emotion, NVBG analyzes the surface text to support the generation of believable nonverbal behaviors.

To this end, the sentence is first parsed to derive the syntactic structure. Then a semantic analysis phase attempts to infer aspects of the utterance's communicative function using inference rules to build up a hierarchical structured lexical, semantic and pragmatic analysis. Examples of these communicative functions include affirmation, inclusivity, intensification, etc. (Lee and Marsella (2006) for details). NVBG then goes through a behavior analysis stage, in which a set of *nonverbal behavior rules* map from communicative functions to classes of nonverbal behaviors. A BML generation phase then maps those behavior classes to specific behaviors, described in BML. This mapping can use character-specific mappings designed to support differences including personality, culture, gender and body types. Conflict resolution occurs at several phases in the overall process. For example, if there are two or more rules overlapping with each other causing conflict, NVBG resolves the conflict by filtering out the rule with lower priority. The priority value of rules has been set through a study of human behaviors using video corpora. The final result is a schedule of behaviors that is passed to the character animation system.

Research on NVBG has explored several approaches to encoding the knowledge used in the function derivation and behavior mapping. Initial work on NVBG was based on an extensive literature review of the research on nonverbal behavior. This seeded the development of rules encoding the function derivation and behavior mapping rules. Then videos of real human face-to-face interactions were annotated and analyzed to verify the rule knowledge, embellish knowledge with dynamic information about behaviors and develop a conflict resolution system that is used to resolve conflicts between behavior suggestions. This annotation and analysis was critical because existing literature said little about dynamics of behaviors and further conflict resolution was to resolve potential conflicts both between the behaviors suggested by the rules as well as differences across literature sources.

More recently a variety of machine learning techniques have been explored, including Hidden Markov Models and Latent-Dynamic Conditional Random Fields to learn the mapping between features of an utterance and nonverbal behaviors using annotated data face-to-face interactions. In particular, Lee and Marsella (2010) contrasts several approaches to learning models of head and eyebrow movement as well as contrasting the results with the knowledge encoded in NVBG by the literature approach discussed above.

3.3 Perceptual processing

Perceptual messages are treated differently than generating nonverbal behavior for the virtual human's utterances. For the perceptual messages, NVBG is deciding on how to respond to signals about external events, including the physical behavior of objects, humans or other virtual humans. These responses, such as looking at a moving object, can in large measure be reflexive or automatic as opposed to having an explicit communicative intention like an utterance. Due to the differences between the perceptual and utterance use cases, NVBG's perceptual responses analyses use a different processing pipeline than the utterance handling.

Specifically, NVBG's response is determined by a Perceptual Analysis stage that leads into the Behavior Analysis and BML Generation stages discussed previously. The rules used during Perceptual Analysis take into account what is the perceived behavior and whether the perceived behavior is above some acceptance threshold (e.g., an object's speed, size and distance or an event's duration or magnitude).

3.4 Listener feedback

The listener feedback pipeline handles the virtual human's behavior while listening to a human or virtual human speaker. The approach makes a distinction between generic feedback and specific feedback, handling them using different rule sets. Generic feedback is driven by speaker behaviors including nods and pauses in speech. Specific feedback is driven by the virtual human's unfolding interpretation of, and reaction to, the speaker's utterance, which requires natural language technology that provides incremental interpretation of partial utterances such as the work of Devault et al. (2011) which provides a semantic interpretation, a measure of confidence in the current understanding and a measure of whether continued listening will lead to better understanding. The virtual human's reaction to the understanding is a valenced reaction to the evolving interpretation of the speaker's utterance. For example, if the virtual human interprets the speaker's partial utterance as deliberately proposing an action to harm the virtual human, then the reaction will be anger.

The model analyzes this information and triggers relevant listener feedback rules, which are mapped to appropriate nonverbal behaviors, such as nods for generic feedback and expressions of confusion, comprehension, happiness or anger for the specific feedback. These behaviors are also conditional on the listener's roles and goals. In particular, a listener that is the main addressee and has the goals of participating in and understanding the conversation will engage in mutual gaze with the speaker, nod to signal attention and signal comprehension and reaction to the content of the utterance. On the other hand, an eavesdropper that has the goal of avoiding participation in the conversation will avoid mutual gaze and signaling attention with nods.

4. Interpersonal Dynamic: Speaker and Listener Interaction

A great example of interpersonal dynamics is backchannel feedback, the nods and para-verbals such as "uh-huh" and "mm-hmm" that listeners produce as someone is speaking (Watzlawick et al., 1967). They can express a certain degree of connection between listener and speaker (e.g., rapport), a way to show acknowledgement (e.g., grounding) or they can also be used for signifying agreement. Backchannel feedback is an essential and predictable aspect of natural conversation and its absence can significantly disrupt participant's ability to communicate (Bavelas et al., 2000). Accurately recognizing the backchannel feedback from one individual is challenging since these conversational cues are subtle and vary between people. Learning how to predict backchannel feedback is a key research problem for building immersive virtual human and robots. Finally, there are still some unanswered questions in linguistic, psychology and sociology on what triggers backchannel feedback and how it differs from different cultures. In this chapter we show the importance of modeling both the multimodal and interpersonal dynamics of backchannel feedback for recognition, prediction and analysis.

The approach for modeling listener backchannel feedback based on speaker behaviors is depicted in Figure 5. We first describe the main ideas behind contextual prediction, then show some of the audiovisual features commonly used and finally describe recent prediction models.

4.1 Contextual prediction

Natural conversation is fluid and highly interactive. Participants seem tightly enmeshed in something like a dance, rapidly detecting and responding, not only to each other's words, but to speech prosody, gesture, gaze, posture, and facial expression movements. These "extra-linguistic" signals play a powerful role in determining the nature of social exchange. When these signals are positive, coordinated and reciprocated, they can lead to feelings of rapport and promote beneficial outcomes in such diverse areas as negotiations and conflict resolution (Drolet and Morris, 2000; Goldberg, 2005), psychotherapeutic effectiveness (Tsui and Schultz, 2005), improved test performance in classrooms (Fuchs, 1987), and improved quality of child care (Burns, 1984). Not surprisingly, supporting such fluid interactions has become an important topic of human-centered research.

In the contextual prediction framework, the prediction model automatically learns which subset of a speaker's verbal and nonverbal actions influences the listener's nonverbal behaviors, finds the optimal way to dynamically integrate the relevant speaker actions and outputs probabilistic measurements describing the likelihood of a listener

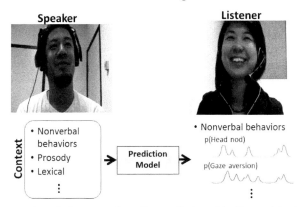

Figure 5. Contextual prediction example: online prediction of the listener's backchannel based on the speaker's contextual features. In the contextual prediction framework, the prediction model automatically (1) learns which subset of the speaker's verbal and nonverbal actions influences the listener's nonverbal behaviors, (2) finds the optimal way to dynamically integrate the relevant speaker actions and (3) outputs probabilistic measurements describing how likely listener nonverbal behaviors are.

(Color image of this figure appears in the color plate section at the end of the book.)

nonverbal behavior. Figure 5 presents an example of contextual prediction for the listener's backchannel.

The goal of a prediction model is to create online predictions of human nonverbal behaviors based on external contextual information. The prediction model learns automatically which contextual feature is important and how it affects the timing of nonverbal behaviors. This goal is achieved by using a machine learning approach wherein a sequential probabilistic model is trained using a database of human interactions. A sequential probabilistic model takes as input a sequence of observation features (e.g., the speaker's features) and returns a sequence of probabilities (e.g., of the listener's backchannel). Some of the most popular sequential models are the Hidden Markov Model (HMM) (Rabiner, 1989) and the Conditional Random Field (CRF) (Lafferty et al., 2001). A main difference between these two models is that the CRF is discriminative (i.e., tries to find the best way to differentiate cases where the human/agent produces a particular behavior or does not) while the HMM is generative (i.e., tries to find the best way to generalize the samples from the cases where the human/agent produces a behavior without considering the cases where no such behavior occurs). The contextual prediction framework is designed to work with any types of sequential probabilistic models.

At the core of the approach is the idea of context, the set of external factors which can potentially influence a person's nonverbal behavior.

4.2 Context (shallow features)

Conceptually, context includes all verbal and nonverbal behaviors performed by other participants (human, robot, computer or virtual human) as well as the description of the interaction environment. For a dyadic interaction between two humans (as shown in Figure 5), to predict the nonverbal behavior of the listener, the context will include the speaker's verbal and nonverbal behaviors, including head movements, eye gaze, pauses and prosodic features. What differentiates the computation framework from "conventional" multimodal approaches (e.g., audio-visual speech recognition) is that the influence of other participants (and the environment) on a person's nonverbal behavior is modeled directly instead of only modeling signals from the same source (e.g., the listener in Figure 5).

In previous work, four types of contextual features were highlighted: lexical features, prosody and punctuation features, timing information and gesture displays. Such features were used to recognize human nonverbal gestures, when the robot spoke to a human, or to generate a gesture for a virtual human given a human's verbal and nonverbal contributions in an interaction (Morency et al., 2008).

Shallow versions of each of these features were calculated either automatically or manually annotated from the dialogue manager of the robot (or virtual human) or directly from a human's action. For lexical features, individual words (unigrams) and word pairs (bigrams) provided information regarding the likelihood of gestural reaction. A range of techniques were used for prosodic features. Using Aizula system (Ward and Tsukahara, 2000), pitch, intensity and other prosodic features were automatically computed from the human's speech (Morency et al., 2008). With robots and virtual humans, the generated punctuation was used to approximate prosodic cues, such as pauses and interrogative utterances. Synthesized visual gestures are a key capability of robots and virtual humans, and they can also be leveraged as a context cue for gesture interpretation. The gestures expressed by the ECA influences the type of visual feedback from the human participant. For example, if the agent makes a deictic pointing gesture, the user is more likely to look at the location that the ECA is pointing to; in human-human dialogues, a critical gestural feature was where the speaker looked. This demonstrates that shallow, very simple features are reliably useful in predicting nonverbal gestures.

The shallow features used in previous work were easy to calculate or annotate and were used for both ECA-human and human-human interactions. However, the features' very simplicity allows them to capture only a small part of the information available in linguistic and gestural behavior.

4.3 Modeling latent dynamic

One of the key challenges with modeling the individual and interpersonal dynamics is to automatically learn the synchrony and complementarities in a person's verbal and nonverbal behaviors and between people. A new computational model called Latent-Dynamic CRF (see Figure 6) was developed to incorporate hidden state variables that model the sub-structure of a class sequence and learn dynamics between class labels (Morency et al., 2007). It is a significant change from previous approaches which only examined individual modalities, ignoring the synergy between speech and gestures.

The task of the Latent-Dynamic CRF model is to learn a mapping between a sequence of observations $x = \{x_1, x_2, ..., x_m\}$ and a sequence of labels $y = \{y_1, y_2, ..., y_m\}$. Each y_j is a class label for the jth frame of a video sequence and is a member of a set Y of possible class labels, for example, $Y = \{backchannel, other-gesture\}$. Each observation

x_j is represented by a feature vector x (x_j) in R^d, for example, the head velocities at each frame. For each sequence, a vector of "substructure" variables $h = \{h_1, h_2, ..., h_m\}$ is assumed. These variables are not observed in the training examples and will therefore form a set of hidden variables in the model.

Given the above definitions, the latent conditional model is defined as:

$$P(y|x,\theta) = \sum_h P(y|h, x,\theta)P(h|x,\theta) \qquad\qquad 1$$

where θ are the parameters of the Latent-Dynamic CRF model. These are learned automatically during training using a gradient ascent approach to search for the optimal parameter values. More details can be found in Morency et al. (2007).

The Latent-Dynamic CRF model was applied to the problem of learning individual dynamics of backchannel feedback. Figure 7 shows

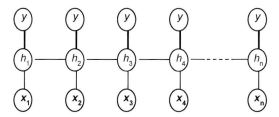

Figure 6. Graphical representation of the LDCRF model. x_j represents the jth observation (corresponding to the jth observation of the sequence), h_j is a hidden state assigned to x_j, and y_j the class label of x_j (i.e., positive or negative). Gray circles are observed variables.

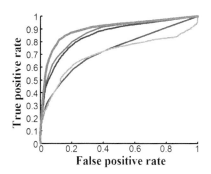

Figure 7. Recognition of backchannel feedback based on individual dynamics only. Comparison of the Latent-Dynamic CRF model with previous approaches for probabilistic sequential modeling.

(Color image of this figure appears in the color plate section at the end of the book.)

the LDCRF model compared to previous approaches for probabilistic sequence labeling (e.g., Hidden Markov Model and Support Vector Machine). By modeling the hidden dynamic, the Latent-Dynamic model outperforms previous approaches. The software was made available online on an open-source website (sourceforge.net/projects/hcrf).

4.4 Prediction model of interpersonal dynamics

In the contextual prediction framework, the prediction model automatically learns which subset of a speaker's verbal and nonverbal actions influences the listener's nonverbal behaviors, finds the optimal way to dynamically integrate the relevant speaker actions and outputs probabilistic measurements describing the likelihood of a listener nonverbal behavior. Figure 5 presents an example of contextual prediction for the listener's backchannel.

The goal of a prediction model is to create online predictions of human nonverbal behaviors based on external contextual information. The prediction model learns automatically which contextual feature is important and how it affects the timing of nonverbal behaviors. This goal is achieved by using a machine learning approach wherein a sequential probabilistic model is trained using a database of human interactions.

The contextual prediction framework can learn to predict and generate dyadic conversational behavior from multimodal conversational data, and applied it to listener backchannel feedback (Morency et al., 2008). Generating appropriate backchannels is a notoriously difficult problem because they happen rapidly, in the midst of speech, and seem elicited by a variety of speaker verbal, prosodic and nonverbal cues. Unlike prior approaches that use a single modality (e.g., speech), it incorporated multimodal features (e.g., speech and gesture) and devised a machine learning method that automatically selects appropriate features from multimodal data and produces sequential probabilistic models with greater predictive accuracy.

4.5 Signal punctuation and encoding dictionary

While human communication is a continuous process, people naturally segment these continuous streams in small pieces when describing a social interaction. This tendency to divide communication sequences of stimuli and responses is referred to as punctuation (Watzlawick et al., 1967). This punctuation process implies that human communication should not only be represented by signals but also with communicative acts that represents the intuitive segmentation of human communication. Communicative acts can range from a spoken word to a segmented gesture (e.g., start and end time of a pointing) or a prosodic act (e.g., region of low pitch).

To improve the expressiveness of these communicative acts, the idea of encoding dictionary was proposed. Since communicative acts are not always synchronous, they are allowed to be represented with various delay and length. In the experiments with backchannel feedback, 13 encoding templates were identified to represent a wide range of ways that speaker actions can influence the listener backchannel feedback. These encoding templates will help to represent long-range dependencies that are otherwise hard to learn using directly a sequential probabilistic model (e.g., when the influence of an input feature decays slowly over time, possibly with a delay). An example of a long-range dependency will be the effect of low-pitch regions on backchannel feedback with an average delay of 0.7 seconds (observed by Ward and Tsukahara (2000)). In the prediction framework, the prediction model will pick an encoding template with a 0.5 seconds delay and the exact alignment will be learned by the sequential probabilistic model (e.g., Latent-Dynamic CRF) which will also take into account the influence of other input features. The three main types of encoding templates are:

- **Binary encoding:** This encoding is designed for speaker features directly synchronized with listener backchannel.
- **Step function:** This encoding is a generalization of binary encoding by adding two parameters: width of the encoded feature and delay between the start of the feature and its encoded version. This encoding is useful if the feature influence on backchannel is constant but with a certain delay and duration.
- **Ramp function:** This encoding linearly decreases for a set period of time (i.e., width parameter). This encoding is useful if the feature influence on backchannel is changing over time.

It is important to note that a feature can have an *individual* influence on backchannel and/or a *joint* influence. An *individual* influence means the input feature directly influences listener backchannel. For example, a long pause can, by itself, trigger backchannel feedback from the listener. A *joint* influence means that more than one feature is involved in triggering the feedback. For example, saying the word "and"' followed by a look back at the listener can trigger listener feedback. This also means that a feature may need to be encoded more than one way since it may have an *individual* influence as well as one or more *joint* influences.

4.5 *Wisdom of crowds*

In many real-life scenarios, it is hard to collect the actual labels for training, because it is expensive or the labeling is subjective. To address this issue, a new direction of research appeared in the last decade, taking full advantage of the "wisdom of crowds" (Smith et al., 2005). In simple words, wisdom of crowds enables the fast acquisition of opinions from multiple annotators/experts.

Based on this intuition, wisdom of crowds was modeled using Parasocial Consensus Sampling paradigm (Huang et al., 2010) for data acquisition, which allows multiple crowd members to experience the same situation. Parasocial Consensus Sampling (PCS) paradigm is based on the theory that people behave similarly when interacting through a media (e.g., video conference).

The goals of the computational model are to automatically discover the prototypical patterns of backchannel feedback and learn the dynamic between these patterns. This will allow the computational model to accurately predict the responses of a new listener even if he/she changes her backchannel patterns in the middle of the interaction. It will also improve generalization by allowing mixtures of these prototypical patterns.

To achieve these goals, a variant of the Latent Mixture of Discriminative Experts (Ozkan et al., 2010) was proposed to take full advantage of the wisdom of crowds. The Wisdom-LMDE model is based on a two-step process: a Conditional Random Field (CRF) is learned first for each expert, and the outputs of these models are used as an input to a Latent Dynamic Conditional Random Field (LDCRF, Figure 6) model, which is capable of learning the hidden structure within the input. In the Wisdom-LMDE, each expert corresponds to a different listener from the wisdom of crowds. Figure 8 shows an overview of the approach.

Table 1 summarizes the experiments comparing the Wisdom-LMDE model with state-of-the-art approaches for behavior prediction. The Wisdom-LMDE model achieves the best f-1 score. The second best f-1 score is achieved by CRF Mixture of experts, which is the only model among other baseline models that combines the different listener labels in a late fusion manner. This result supports the claim that wisdom of clouds improves learning of prediction models.

5. Discussion

Modeling human communication dynamics enables the computational study of different aspects of human behaviors. While a backchannel

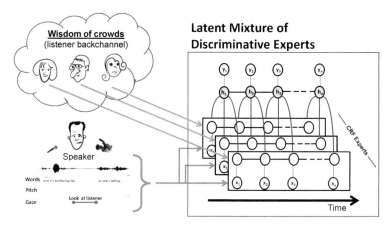

Figure 8. The approach for modeling wisdom of crowd: (1) multiple listeners experience the same series of stimuli (pre-recorded speakers) and (2) a Wisdom-LMDE model is learned using this wisdom of crowds, associating one expert for each listener.

Table 1. Comparison of the prediction model with previously published approaches. By integrating the knowledge from multiple listener, the Wisdom-LMDE is able to identify prototypical patterns in interpersonal dynamic.

Model	Wisdom of Crowds	Precision	Recall	F1-Score
Wisdom LMDE	Yes	0.2473	0.7349	**0.3701**
Consensus Classifier (Huang et al., 2010)	Yes	0.2217	0.3773	0.2793
CRF Mixture of Experts (Smith et al., 2005)	Yes	0.2696	0.4407	0.3345
AL Classifier (CRF)	No	0.2997	0.2819	0.2906
AL Classifier (LDCRF) (Morency et al., 2007)	No	0.1619	0.2996	0.2102
Multimodel LMDE (Ozkan et al., 2010)	No	0.2548	0.3752	0.3035
Random Classifier	No	0.1042	0.1250	0.1018
Rule Based Classifier (Ward et al., 2000)	No	0.1381	0.2195	0.1457

feedback such as head nod may, at first, look like a conversational signal ("I acknowledge what you said"), it can also be interpreted as an emotional signal where the person is trying to show empathy or a social signal where the person is trying to show dominance by expressing a strong head nod. The complete study of human face-to-face communication needs to take into account these different types of signals: linguistic, conversational, emotional and social. In all four cases, the individual and interpersonal dynamics are keys to a coherent interpretation.

As we have already shown in this book chapter, modeling human communication dynamics is important for both recognition and prediction. One other important advantage of these computational models is the automatic analysis of human behaviors. Studying interactions is grueling and time-consuming work. The rule of thumb in the field is that each recorded minute of interaction takes an hour or more to analyze. Moreover, many social cues are subtle, and not easily noticed by even the most attentive psychologists.

By being able to automatically and efficiently analyze a large quantity of human interactions, and detect relevant patterns, these new tools will enable psychologists and linguists to find hidden behavioral patterns which may be too subtle for the human eye to detect, or may be just too rare during human interactions. A concrete example is the recent work which studied engagement and rapport between speakers and listeners, specifically examining a person's backchannel feedback during conversation (Ward and Tsukahara, 2000). This research revealed new predictive cues related to gaze shifts and specific spoken words which were not identified by previous psycho-linguistic studies. These results not only give an inspiration for future behavioral studies but also make possible a new generation of robots and virtual humans able to convey gestures and expressions at the appropriate times.

REFERENCES

Bavelas, J.B., L. Coates and T. Johnson, 2000. Listeners as co-narrators. *Journal of Personality and Social Psychology*, **79(6)**:941–952.

Burns, M. 1984. Rapport and relationships: The basis of child care. *Journal of Child Care*, **4**:47–57.

DeVault, D., K. Sagae and D. Traum. 2011. Incremental interpretation and prediction of utterance meaning for interactive dialogue. *Dialogue & Discourse*, **2(1)**:143–170.

DeVito, J. 2008. The Interpersonal Communication Book. Pearson/Allyn and Bacon, 12th edition.

Drolet, A. and M. Morris. 2000. Rapport in conflict resolution: Accounting for how face-to-face contact fosters mutual cooperation in mixed-motive conflicts. *Experimental Social Psychology*, **36**:26–50.

Feng, A.W., Y. Huang, M. Kallmann and A. Shapiro. 2012. An analysis of motion blending techniques. The Fifth International Conference on Motion in Games, pp. 232–243.

Fuchs, D. 1987. Examiner familiarity effects on test performance: Implications for training and practice. Topics in Early Childhood Special Education, **7**:90–104.

Goldberg, S.B. 2005. The secrets of successful mediators. *Negotiation Journal*, **21(3)**:365–376.

Hawley, A.H. 1950. Human Ecology: A Theory of Community Structure. Ronald Press.

Huang, L., L.-P. Morency and J. Gratch, 2010. Parasocial consensus sampling: Combining multiple perspectives to learn virtual human behavior. AAMAS.

Huang, Y. and M. Kallmann. 2010. Motion parameterization with inverse blending. Proceedings of the Third International Conference on Motion in Games.

Kallmann, M. and S. Marsella. 2005. Hierarchical motion controllers for real-time autonomous virtual humans. Proceedings of the 5th International Conference on Interactive Virtual Agents, pp. 12–14.

Kopp, S., B. Krenn, S. Marsella, A.N. Marshall, C. Pelachaud, H. Pirker, K.R. Thórisson and H. Vilhjalmsson. 2006. Towards a Common Framework for Multimodal Generation: The Behavior Markup Language. In 6th International Conference on Intelligent Virtual Agents, Marina del Rey, CA.

Kovar, L. and M. Gleicher. 2004. Automated Extraction and Parameterization of Motions in Large Data Sets, ACM Transactions on Graphics, Proceedings of SIGGRAPH, **23(3)**:559–568.

Lafferty, J., A. McCallum and F. Pereira. 2001. Conditional random fields: Probabilistic models for segmenting and labelling sequence data. In International Conference on Machine Learning.

Lee, J. and S. Marsella. 2010. Predicting speaker head nods and the effects of affective information. *IEEE Transactions on Multimedia*, **12(6)**:552–562.

Lee, J. and S. Marsella. 2006. Nonverbal behavior generator for embodied conversational agents. In Proc. of the 6th Int. Conf. on Intelligent Virtual Agents, pp. 243–255.

Morency, L.-P., A. Quattoni and T. Darrell. 2007a. Latent-dynamic discriminative models for continuous gesture recognition. Proceedings of the IEEE Conference on Computer Vision and Pattern Recognition, June.

Morency, L.-P., C. Sidner, C. Lee and T. Darrell. 2007b. Head gestures for perceptual interfaces: The role of context in improving recognition. *Artificial Intelligence*. **171(8–9)**:568–585.

Morency, L.-P., I. de Kok and J. Gratch. 2008. Predicting listener backchannels: A probabilistic multimodal approach. Conference on Intelligent Virutal Agents.

Ozkan, D., K. Sagae and L.-P. Morency. 2010. Latent mixture of discriminative experts for multimodal prediction modeling. In International Conference on Computational Linguistics (COLING).

Pentland, A. 2004. Social dynamics: Signals and behavior. In IEEE Int. Conf. Developmental Learning, San Diego, CA, October.

Pettre, J. and J.P. Laumond. 2006. A motion capture-based control-space approach for walking mannequins. *Computer Animation and Virtual Worlds*, **17(2)**:109–126.

Rabiner, L.R. 1989. A tutorial on hidden markov models and selected applications in speech recognition. *Proceedings of the IEEE*, **77(2):**257–286.

Rose, C., B. Bodenheimer and M.F. Cohen. 1998. Verbs and adverbs: Multidimensional motion interpolation. *IEEE Computer Graphics and Applications*, **18:**32–40.

Shapiro, A. 2011. Building a character animation system. The Fourth International Conference on Motion in Games, pp. 98–108.

Stone, M., D. DeCarlo, I. Oh, C. Rodriguez, A. Stere, A. Lees and C. Bregler. 2004. Speaking with hands: Creating animated conversational characters from recordings of human performance, ACM transactions on graphics. Proceedings of SIGGRAPH, **23(3):**506–513.

Smith, A., T. Cohn and M. Osborne. 2005. Logarithmic opinion pools for conditional random fields. In Annual Meeting of the Association for Computational Linguistics (ACL), pp. 18–25.

Thiebaux, M., S. Marsella, A.N. Marshall and M. Kallmann. 2008. SmartBody: Behavior realization for embodied conversational agents. Proceedings of the 7th International Joint Conference on Autonomous Agents and Multiagent Systems, pp. 151–158.

Tsui, P. and G. Schultz. 1985. Failure of rapport: Why psychotherapeutic engagement fails in the treatment of Asian clients. *American Journal of Orthopsychiatry*, **55:**561–569.

Wang, Zhiyang, Lee Jina and Marsella Stacy. 2011. Towards More Comprehensive Listening Behavior Beyond the Bobble Head, in The 11th International Conference on Intelligent Virtual Agents (IVA). Hyperlink "http://www. ict.usc.edu/~marsella/publications/WangIVA11.pdf".

Ward, N. and W. Tsukahara. 2000. Prosodic features which cue back-channel responses in English and Japanese. *Journal of Pragmatics*, **23:**1177–1207.

Watzlawick, P., J.B. Bavelas and D.D. Jackson. 1967. Pragmatics of Human Communication: A Study of Interactional Patterns, Pathologies and Paradoxes. New York: Norton.

CHAPTER 15

Multimodal Fusion in Human-Agent Dialogue

Elisabeth André, Jean-Claude Martin,
Florian Lingenfelser and Johannes Wagner

1. Introduction

Sophisticated fusion techniques are an essential component of any multimodal system. Historically, systems aimed at analyzing the semantics of multimodal commands and typically investigated a combination of pointing and drawing gestures and speech. The most prominent example includes the "Put-that-there" system (Bolt, 1980) that analyzes speech in combination with 3D pointing gestures referring to objects on a graphical display. Since this groundbreaking work, numerous researchers have investigated mechanisms for multimodal input interpretation mainly working on speech, gestures and gaze while the trend is moving towards intuitive interactions in everyday environments. Since interaction occurs more and more in mobile and tangible environments, modern multimodal interfaces require a greater amount of context-awareness (Johnston et al., 2011).

At the same time, we can observe a shift from pure task-based dialogue to more human-like dialogues that aim to create social experiences. Usually, such dialogue systems rely on a personification of the user interface by means of embodied conversational agents or social robots. The driving force behind this work is the insight that a user interface is more likely to be accepted by the user if the machine is sensitive towards the user's feelings. For example, Martinovsky and Traum (2003) demonstrated by means of user dialogues with a training system and a telephone-based information system that many breakdowns in man-machine communication could be avoided if the machine was able to recognize the emotional state of the user and responded to it more sensitively. This observation shows that a system should not only analyze what the user said or gestured but also consider more subtle cues, such as psychological user states.

With the departure from pure task-based dialogue to more human-like dialogues that aim to create social experiences, the concept of multimodal fusion as originally known in the natural language community has to be extended. We do not only need fusion mechanisms that derive information on the user's intention from multiple modalities, such as speech, pointing gestures and eye gaze. In addition, fusion techniques are required that help a system assess how the user perceives the interaction with it. Accordingly, fusion mechanisms are required not only at the semantic level, but also at the level of social and emotional signals. With such systems, any interaction may indeed feature a task-based component mixed with a social interaction component. These different components may even be conveyed on different modalities and overlap in time. It is thus necessary to integrate a deeper semantic analysis in social signal processing on the one side and to consider social and emotional cues in semantic fusion mechanisms on the other side. Both streams of information need to be closely coupled during fusion since they can both include similar communication channels. For example, a system may fuse verbal and nonverbal signals to come up with a semantic interpretation, but the same means of expression may also be integrated by a fusion mechanism as an indicator of cognitive load (Chen et al., 2012).

By providing a comparative analysis of semantic fusion of multimodal utterance and fusion of social signals, this chapter aims to give a comprehensive overview of fusion techniques as components of dialogue system that aim to emulate qualities of human-like communication. In the next section, we first present taxonomies for categorizing fusion techniques focusing on the relationship between the single modalities and the level of integration. Section 3 addresses the fusion of semantic information, whereas Section 4 is devoted to the fusion of social signals. To enable a better comparison of issues handled in the two areas, both sections follow a similar structure. We first introduce techniques for fusing information at different levels of abstraction and discuss attempts to come up with standards to represent information to be exchanged in fusion engines. After that we discuss challenges that arise when moving from controlled laboratory environments to less controlled everyday scenarios. Particular attention is given to the use of fusion mechanisms in human-agent dialogue where the mechanisms for input analysis have to be tightly coordinated with appropriate feedback to be given by the agent. Section 5 presents approaches that combine semantic interpretation with social cue analysis either to increase the robustness of the analysis components or to improve the quality of interaction. Section 6 concludes the chapter and gives an outline for future research.

2. Dimensions of Description

Most systems rely on different components for the low-level analysis of the single modalities, such as eye trackers, speech and gesture recognizers, and make use of one or several modality integrators to come up with a comprehensive interpretation of the multimodal input. In this context, two fundamental questions arise: How are the single modalities related to each other and at which level should they be integrated?

2.1 Relationships between modalities

Several combinations of modalities may cooperate in different manners. Martin et al. (1998) mention the following cases: equivalence, redundancy, complementarity, specialization and transfer.

When several modalities cooperate by *equivalence*, this means that a command or a chunk of information may be produced as an alternative, by either of them. For example, to consider the needs of a variety of users, a multimodal interface might allow them to specify a command via speech or as an alternative by pressing a button.

Modalities that cooperate by *redundancy* produce the same information. Redundancy allows a system to ignore one of the two redundant modalities. For example, when a user uniquely specifies a referent via speech and uses at the same time an unambiguous pointing gesture, only one modality needs to be considered to uniquely identify the referent.

When modalities cooperate by *complementarity*, different chunks of information are produced by each modality and have to be integrated during the interpretation process. A classical example includes a multimodal command consisting of a spoken utterance and a pointing gesture both of which contribute to the interpretation.

When modalities cooperate by *specialization*, this means that one modality provides the frame of interpretation for another. Specialization occurs, for example, when a user points to a group of objects and specifies the intended referent by verbally providing a category that distinguishes it from alternatives.

Cooperation by *transfer* means that a chunk of information produced by one modality is used by another modality. Transfer is typically used in hypermedia interfaces when a mouse click triggers the display of an image.

Different modalities may also be used *concurrently*, i.e. produce independent chunks of information, i.e. chunks without any semantic overlap, at the same time. For example, a user may say "Hello" and at the same time point to an object. Here the chunks of information should not be merged. Earlier systems usually did not allow for a concurrent use of modalities, but required an *exclusive* use of modalities. For example, the user may utter a greeting and point to an object, but not at the same time.

While the relationship between modalities has mainly been discussed for multimodal user commands, little attempts have been made to specify the relationship between social signals. However, modalities that convey social signals may cooperate in a similar manner as modalities that convey semantic information. For example, different dimensions of emotions, such as valence and arousal, may be expressed by different channels of communication, such as the face or the voice. It is important to note, however, that it is hard to deliberately employ given channels of communication for the expression of social signals.

Different modalities do not always convey *congruent* pieces of information. In the case of semantic information, little robust input processing components typically lead to incongruent pieces of information. In the case of social signals, incongruent pieces of information often result from the fact that users are not equally expressive in all modalities. In particular, the attempt to conceal social signals may result into an inconsistent behavior.

Another classification concerns the timing of modalities. Here, we may basically distinguish between the *sequential* use of modalities and the *parallel* use of modalities which overlap in time. Semantically related modalities may overlap in time or may be used in sequence. If they are merged, the temporal distance should, however, not be too large. Algorithms for the fusion of social signals usually start from the assumption that social signals that refer to particular user state, such as frustration, emerge at exactly the same time interval. We will later see that such an assumption may be problematic.

Other researchers use similar terms to describe relationships between modalities. See Lalanne et al. (2009) for an overview.

2.2 Levels of integration

Basically, two main fusion architectures have been proposed in the literature depending on at which level sensor data are fused.

In the case of *low-level fusion*, the input from different sensors is integrated at an early stage of processing. Low-level fusion is therefore often also called *early fusion*. The fusion input may consist of either raw data or low-level features, such as pitch. The advantage of low-level fusion is that it enables a tight integration of modalities. There is, however, no declarative representation of the relationship between various sensor data which aggravates the interpretation of recognition results.

In the case of *high-level fusion*, low-level input has to pass modality-specific analyzers before it is integrated, e.g. by summing recognition probabilities to derive a final decision. High-level fusion occurs at a later stage of processing and is therefore often also called *late fusion*. The advantage of high-level fusion is that it allows for the definition of declarative rules to combine the interpreted results of various sensors. There is, however, the danger that information goes lost because of a too early abstraction process.

3. Multimodal Interfaces Featuring Semantic Fusion

In this section, we focus on semantic fusion that combines the meaning of the single modalities into a uniform representation.

3.1 Techniques for semantic fusion

Systems aiming at a semantic interpretation of multimodal input typically use a late fusion approach at a decision level and process each modality individually before fusion (see Figure 1a). Usually, they rely on mechanisms that have been originally introduced for the analysis of natural language.

Johnston (1998) proposed an approach to modality integration for the QuickSet system that was based on unification over typed feature structures. The basic idea was to build up a common semantic representation of the multimodal input by unifying feature structures which represented the semantic contributions of the single modalities. For instance, the system was able to derive a partial interpretation for a spoken natural language reference which indicated that the location of the referent was of type "point'". In this case, only unification with gestures of type "point'" would succeed.

Kaiser et al. (2003) applied unification over typed feature structures to analyze multimodal input consisting of speech, 3D gestures and head direction in augmented and virtual reality. Noteworthy is the fact that the system went beyond gestures referring to objects, but also considered gestures describing how actions should be performed. Among others, the system was able to interpret multimodal rotation commands, such as "Turn the table <rotation gesture> clockwise." where the gesture specified both the object to be manipulated and the direction of rotation.

Another popular approach that was inspired by work on natural language analysis used finite-state machines consisting of $n + 1$ tapes which represent the n input modalities to be analyzed and their combined meaning (Bangalore and Johnston, 2009). When analyzing a multimodal utterance, lattices that correspond to possible interpretations of the single input streams are created by writing symbols on the corresponding tapes. Multiple input streams are then aligned by transforming their lattices into a lattice that represents the combined semantic interpretation. Temporal constraints are not explicitly encoded as in the unification-based approaches described above, but implicitly given by the order of the symbols written on the single tapes. Approaches to represent temporal constraints within state chart mechanisms have been presented by Latoschik (2002) and more recently by Mehlmann and André (2012).

3.2 Semantic representation of fusion input

A fundamental problem of the very early systems was that there was no declarative formalism for the formulation of integration constraints. A noteworthy exception was the approach used in QuickSet which clearly separated the statements of the multimedia grammar from the mechanisms of parsing (Johnston, 1998). This approach enabled not only the declarative formulation of type constraints, such as "the location of a flood zone should be an area", but also the specification of spatial and temporal constraints, such as "two regions should be a limited distance apart" and "the time of speech must either overlap with or start within four seconds of the time of the gesture".

Many recent multimodal input systems, such as SmartKom (Wahlster 2003), make use of an XML language for representing messages exchanged between software modules. An attempt to standardize such a representation language has been made by the World Wide Web Consortium (W3C) with EMMA (Extensible MultiModal Annotation markup language). It enables the representation of characteristic features of the fusion process: "composite" information (resulting from the fusion of several modalities), confidence scores, timestamps as well as incompatible interpretations ("one-of"). Johnston (2009) presents a variety of multimodal interfaces combining speech-, touch- and pen-based input that have been developed using the EMMA standard.

3.3 Choice of segments to be considered in the fusion process

Most systems start from the assumption that the complete input provided by the user can be integrated. Furthermore, they presume that the start and end points of input in each modality are given, for example, by requiring the user to explicitly mark it in the interaction. Under such conditions, the determination of processing units to be considered in the fusion process is rather straightforward. Typically, temporal constraints are considered to find the best candidates to be fused with each other. For example, a pointing gesture should occur approximately at the same time as the corresponding natural language expression while it is not necessary that the two modalities temporally overlap. However, there are cases when such an assumption is problematic and may present a system from deriving a semantic interpretation. For example, the input components may by mistake come up with an erroneous recognition result that cannot be integrated. Secondly, the user may unintentionally provide input, for example, by making a gesture that should not be taken as a gesture. In natural environments where users freely interact, the situation becomes even harder. Users permanently move their arms, but not every gesture is meant to be part of a system command. If eye gaze is employed as a means to indicate a referent, the determination of segments becomes even challenging. Users tend to fixate the objects with the eye they refer to. However, not every fixation is supposed to contribute to a referring expression. A first approach to solve this problem has been presented by Sun et al. (2009). They propose a multimodal input fusion approach to flexibly skip spare information in multimodal inputs that cannot be integrated.

3.4 Dealing with imperfect data in the fusion process

Multimodal interfaces often have to deal with uncertain data. Individual signals may be noisy and/or hard to interpret. Some modalities may be more problematic than others. A fusion mechanism should consider these uncertainties when integrating the modalities into a common semantic representation.

Usually, multimodal input systems combine several *n*-best hypotheses produced by multiple modality-specific generators. This leads to several possibilities of fusion, each with a score computed as a weighted sum of the recognition scores provided by individual modalities. In this vein, it may happen that a badly ranked hypothesis may still contribute to the overall semantic representation because it is compatible with other hypotheses. Thus, multimodality enables us to use the strength of one modality to compensate for weaknesses of others. For example, errors in speech recognition can be compensated by gesture recognition and vice versa. Oviatt (1999) reported that 12.5% of pen/voice interactions in QuickSet could be successfully analyzed due to multimodal disambiguation while Kaiser et al. (2003) even obtained a success rate of 46.4% for speech and 3D gestures that could be attributed to multimodal disambiguation.

3.5 Desktop vs. mobile environments

More recent work focuses on the challenge to support a speech-based multimodal interface on heterogeneous devices including not only desktop PCs, but also mobile devices, such as smart phones (Johnston, 2009).

In addition, there is a trend towards less traditional platforms, such as in-car interfaces (Gruenstein et al., 2009) or home controlling interfaces (Dimitriadis and Schroeter, 2011). Such environments raise particular challenges to multimodal analysis due to the increased noise level, the less controlled environment and multi-threaded conversations. In addition, we need to consider that users are continuously producing multimodal output and not only when interacting with a system. For example, a gesture performed by a user to greet another user should not be mixed up with a gesture to control a system. In order to relieve the users from the burden to explicitly indicate when they wish to interact, a system should be able to distinguish automatically between commands and non-commands.

Particular challenges arise in a situated environment because the information on the user's physical context is required to interpret a multimodal utterance. For example, a robot has to know its location and orientation as well as the location of objects in its physical environment, to execute commands, such as "Move to the table". In a mobile application, the GPS location of the device may be used to constrain search results for a natural language user query. When a user says "restaurants" without specifying an area on the map displayed on the phone, the system interprets this utterance as a request to provide only restaurants in the user's immediate vicinity. Such an approach is used, for instance, by Johnston et al. (2011) in the MTalk system, a multimodal browser for location-based services.

3.6 Semantic fusion in human-agent interaction

A number of multimodal dialogue systems make use of a virtual agent in order to allow for more natural interaction. Typically, these systems employ graphical displays to which a user may refer to using touch or mouse gestures in combination with spoken or written natural language input; for example, Martin et al. (2006), Wahlster (2003) or Hofs et al. (2010). Furthermore, the use of freehand arm gestures (Sowa et al., 2001) and eye gaze (Sun et al., 2008) to refer to objects in a 3D environment has been explored in interactions with virtual agents. Techniques for multimodal semantic fusion have also attracted interest in the area of human-robot interaction. In most systems, the user's hands are tracked to determine objects or locations the user is referring to via natural language; for example, Burger et al. (2011). In addition to the recognition of hand gestures, Stiefelhagen et al. (2004) make use of head tracking based on the consideration that users typically look at the objects they refer to.

While some of the agent-based dialogue systems employ unification-based grammars (Wahlster, 2003) or chart starts (Sowa et al., 2001) as presented in Section 3.1, others use a hybrid fusion mechanism combining declarative formalisms, such as frames, with procedural elements (Martin et al., 2006). Often the fusion of semantic information is triggered by natural language components which detect a need to integrate information from another modality (Stiefelhagen et al., 2004).

In addition, attempts have been made to consider how multimodal information is analyzed and produced by humans in the semantic fusion process. Usually what is being said becomes not immediately clear, but requires multiple turns between two interlocutors. Furthermore, people typically analyze speech in an incremental manner while it is spoken and provide feedback to the speaker before the utterance is completed. For example, a listener may signal by a frown that an utterance is not fully understood. To simulate such a behavior in human-agent interaction, a tight coupling of multimodal analysis, dialogue processing and multimodal generation is required. Stiefelhagen et al. (2007) propose to allow for clarification dialogues in order to improve the accuracy of the fusion process in human-robot dialogue. Visser et al. (2012) describe an incremental model of grounding that enables the simulation of several grounding acts, such as initiate, acknowledge, request and repair, in human-agent dialogue. If the virtual agent is not able to come up with a meaning for the user's input, it generates an appropriate feedback signal, such as a frown, to encourage more information from the user. As a consequence, the fusion process in this system may extend over a sequence of turns in a multimodal dialogue.

4. Multimodal Interfaces Featuring Fusion of Social Signals

Recently, the automatic recognition of social and emotional cues has shifted from a side issue to a major topic in human-computer interaction. The aim is to enable a very natural form of interaction by considering not only explicit instructions by human users, but also more subtle cues, such as psychological user states. A number of approaches to automated affect recognition have been developed exploiting a variety of modalities including speech (Vogt and André, 2005), facial expressions (Sandbach et al., 2012), body postures and gestures (Kleinsmith et al., 2011) as well as physiological measurements (Kim and André, 2008). Also, multimodal approaches to improve emotion recognition accuracy are reported, mostly by exploiting audiovisual combinations. Results suggest that integrated information from audio and video leads to improved classification reliability compared to a single modality—even with fairly simple fusion methods.

In this section, we will discuss applications with virtual humans and social robots that make use of mechanisms for fusing social and emotional signals. We will start off by discussing a number of design decisions that have to be made for the development of such systems.

4.1 Techniques for fusing social signals

Automatic sensing of emotional signals in real-time systems usually follows a machine learning approach and relies on an extensive set of labeled multimodal data. Typically, such data are recorded in separate sessions during which users are asked to show certain actions or interact with a system that has been manipulated to induce the desired behavior. Afterward, the collected data is manually labeled by human annotators with the assumed user emotions. Thus, a huge amount of labeled data is collected for which classifiers are trained and tested. An obvious approach to improve the robustness of the classifiers is the integration of data from multiple channels. Hence, an important decision to take concerns the level at which the single modalities should be fused.

A straightforward approach is to simply merge the features calculated from each modality into one cumulative structure, extract the most relevant features and train a classifier with the resulting feature set. Hence, fusion is based on the integration of low-level features at the feature level (see Figure 1b) and takes place at a rather early stage of the recognition process.

An alternative would be to fuse the recognition results at the decision level based on the outputs of separate unimodal classifiers (see Figure 1c). Here, multiple unimodal classifiers are trained for each modality individually and the resulting decisions are fused by using specific weighting rules. In the case of emotion recognition, the input for the fusion algorithm may consist of either discrete emotion categories, such as anger or joy, or continuous values of a dimensional emotion model (e.g. continuous representation of the valence or the arousal of the emotions). Hence, fusion is based on the integration of high-level concepts and takes place at a later stage of the recognition process.

Eyben et al. (2011) propose a mechanism that fuses audiovisual social behaviors at an intermediate level based on the consideration that behavioral events, such as smiles, head shakes and laughter, convey important information on a person's emotional state that might go lost if information is fused at the level of low-level features or at the level of emotional states.

Which level of modality integration yields the best results is usually hard to predict. Busso et al. (2004) report on an emotion-specific comparison of feature-level and decision-level fusion that was conducted for an audiovisual database containing four emotions, sadness, anger, happiness, and neutral state, deliberately posed by an actress. They observed for their corpus that feature-level fusion was most suitable for differentiating anger and neutral while decision-level fusion performed better for happiness and sadness. Caridakis et al. (2007) presented a multimodal approach for the recognition of eight emotions that integrated information from facial expressions, body gestures, and speech. They observed a recognition improvement of more than 10% compared to the most successful unimodal system and the superiority of feature-level fusion to decision-level fusion. Wagner et al. (2011a) tested a comprehensive repertoire of state-of-the-art fusion techniques including their own emotion-specific fusion scheme on the acted DaFEx corpus and the more natural CALLAS corpus. Results were either considerably improved (DaFEx) or at least in line with the dominating modality (CALLAS). Unlike Caridakis and colleagues, Wagner and colleagues found that decision-level fusion yielded more promising results than feature-level fusion.

W3C EmotionML (Emotion Markup Language) has been proposed as a technology to represent and process emotion-related data and to enable the interoperability of components dedicated to emotion-oriented computing. An attempt towards a language that is not limited to the representation of emotion-related data, but directed to the representation of nonverbal behavior in general has been recently made by Scherer et al. (2012) with PML (Perception Markup Language). As in the case of semantic fusion, the authors identified a specific need to represent uncertainties in the interpretation of data. For example, a gaze away from the interlocutor may signal a moment of high concentration, but also be an indicator of disengagement.

4.2 Acted versus spontaneous signals

Most emotion recognition systems still rely exclusively on acted data for which very promising results have been obtained. The way emotions are expressed by actors may be called prototypical, and independent observers would largely agree on the emotional state of these speakers. A common example includes voice data from actors for which developers of emotion recognition systems reported accuracy rates of over 80% for seven emotion classes. In realistic applications, there is, however, no guarantee that emotions are expressed in a prototypical manner. As a consequence, these applications still represent a great challenge for current emotion recognition systems, and it is obvious to investigate whether the recognition rates obtained for unimodal non-acted data can be improved by considering multiple modalities.

Unfortunately, the gain obtained by multimodal fusions seems to be lower for non-acted than for acted data. Based on a comprehensive analysis of state-of-the-art approaches to affect recognition, D'Mello and Kory (2012) report on an average improvement of 8.12% for multimodal affect recognition compared to unimodal affect recognition while the improvement was significantly higher for acted data (12.1%) than for spontaneous data (4.39%).

One explanation might be that experienced actors are usually able to express emotions consistently across various channels while natural speakers do not have this capacity. For example, Wagner et al. (2011a) found that natural speakers they recorded for the CALLAS corpus were more expressive in their speech than in their face or gestures—probably due to the fact that the method they used to elicit emotions in people mainly affected vocal emotions. As a consequence, they did not obtain a high gain for multimodal fusion compared to the unimodal speech-based emotion classifier. At least, they were able to handle disagreeing modalities in a way so that competitive results to the best channel could be achieved.

4.3 Offline versus online fusion

Characteristic of current research on the multimodal analysis of social and emotional signals is the strong concentration on posteriori analyses. Out of the many methods discussed in the recent analysis by D'Mello and Kory (2012), hardly any one of them was tested in an online scenario where a system responds to users' social and emotional signals while they are interacting with it. The move from offline to online analysis of social and affective cues raises a number of challenges for the multimodal recognition task. While in offline analysis the whole signal is available and analysis can fall back on global statistics, such a treatment is no longer possible for online analysis. In addition, offline analysis usually focuses on a small set of pre-defined emotion classes and neglects, for example, data that could not be uniquely assigned to a particular emotion class. Online analysis has, however, to take into account all emotion data. Finally, while there are usually no temporal restrictions for offline analysis, online analysis has to be very fast to enable a fluent human-robot dialogue. A fusion mechanism specifically adapted to the needs of online fusion has been used in the Callas Emotional Tree, an artistic Augmented Reality installation of a tree which responds to the spectators' spontaneous emotions reactions to it; see Gilroy et al. (2008). The basic idea of this approach is to derive emotional information from different modality-specific sensors and map it onto the 3D of the Pleasure-Arousal-Dominance model (PAD model) by Mehrabian (1980). In addition, to the input provided by a modality-specific sensor at a particular instance of time, the approach considers the temporal dynamics of modality-specific emotions by integrating the current value provided by a sensor with the previous value. The fusion vector then results from a combination of the vectors representing the single modality-specific contributions. Unlike traditional approaches to sensor fusion, PAD-based fusion integrates contributions from the single modalities in a frame-wise fashion and is thus able to respond immediately to a user's emotional state.

4.4 Choice of segments to be considered in the fusion process

Even though it is obvious that each modality has a different timing, most fusion mechanisms either use processing units of a fixed duration or linguistically motivated time intervals, such as sentences. Kim et al. (2005) suggested for a corpus consisting of speech and biosignals choosing the borders of the single segments in such a way that it lies in the middle between two spoken utterances. Lingenfelser et al. (2011) used the time interval covered by a spoken utterance for all considered modalities, i.e. audio and video. These strategies suffer from two major problems. First, significant hints for emotion recognition from different modalities are not guaranteed to emerge at exactly the same time interval. Second, they might occur in a shorter time period than a sentence only. Classification accuracy could be expected to improve, if modalities were segmented individually and the succession and corresponding delays between occurrences of emotional hints in different signals could be investigated more closely. A promising step into this direction is the event-based fusion mechanism developed for the Callas Emotional Tree (Gilroy et al., 2011). Rather than computing global statistics in a segmentation-based manner, the approach aims to identify changes in the modality-specific expression of an emotion and is thus able to continuously respond to emotions of users while they are interacting with the system.

4.5 Dealing with imperfect data in the fusion process

Most algorithms for social signal fusion start from the assumption that all data from the different modalities are available at all time. As long as a system is used offline, only this condition can be easily met by analyzing the data beforehand and omitting parts where input from one modality is corrupted or completely missing. However, in online mode, a manual pre-selection of data is not possible and we have to find adequate ways of handling missing information. Generally, various reasons for missing information can be identified. First of all, it is unrealistic to assume that a person continuously provides meaningful data for each modality. Second, there may be technical issues, such as noisy data due to unfortunate environmental conditions or missing data due to the failure of a sensor. As a consequence, a system needs to be able to dynamically decide which channels to exploit in the fusion process and to what extent the present signals can be trusted. For the case that data is partially missing a couple of treatments have been suggested in literature, such as the removal of noise or the interpolation of missing data from available data. Wagner et al. (2011a) present a comprehensive study that successfully applies adaptations of state-of-the-art fusion techniques to the missing data problem in multimodal emotion recognition.

While semantic fusion is driven by the need to exploit the complementarity of modalities, fusion techniques in social signal processing make less explicit use of modality-specific benefits. Nevertheless, such an approach might help improve the gain obtained by current fusion techniques. For example, there is evidence that arousal is recognized more reliably using acoustic information while facial expressions yield higher accuracy for valence. In addition, context information may be exploited to adapt the weights to be assigned to the single modalities. For example, in a noisy environment less weight might be given to the audio signal. A first attempt to make use of the complementarity of modalities has been by Wagner et al. (2011a). Based on evaluation of training data, experts for every class of the classification problem are chosen. Then the classes are rank ordered, beginning with the worst classified class across all classifiers and ending with the best one.

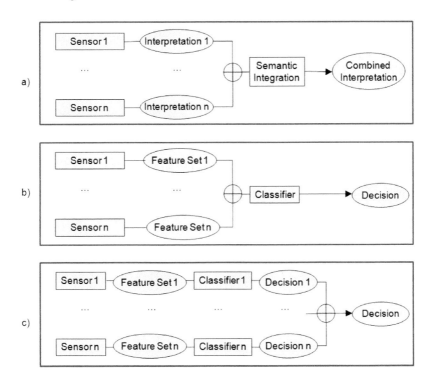

Figure 1. Different fusion mechanisms: (a) Semantic fusion, (b) feature-level fusion and (c) decision-level fusion.

4.6 Evaluation of schemes for social signal fusion

Since it is usually obvious which intention a user aims to convey to a system, the evaluation of schemes for semantic fusion is rather straightforward. On the opposite, the evaluation of fusion schemes for social and emotional signals raises a number of challenges. First of all, it is not obvious how to acquire ground truth data against which to evaluate the performance of automated emotion-recognition components. Users show a great deal of individuality in their emotional responses, and there is no clear mapping between behavioral cues and emotional states. To avoid a too-high degree of subjective interpretation, ground truth data are typically obtained by requesting multiple annotators label corpora. A voting scheme can be used if annotators disagree. One question to consider in this context is the amount of information to be made available to the annotators. Should the annotator just have access to one modality, such as speech or video, or to all available modalities? To provide a fair comparison between human and machine performance, it seems reasonable to make all the modalities that will be considered in the fusion process available to the human annotator. However, to acquire realistic ground truth values, one might consider giving the human annotator as much information as possible even if the fusion process will only employ part of it.

Another question concerns the processing units which should be taken into account for an evaluation. Since online fusion processes data frame by frame, an evaluation should consider time serious over a whole experimental session as opposed to computing global statistics over a longer period of time. One option might be to make use of an annotation tool, such as Feeltrace, that allows for the continuous annotation of data.

Instead of experimentally testing the robustness of the fusion process, a number of approaches have rather tested the effect of it. Gratch et al. (2007) presented an artificial listener that was able to recognize a large variety of verbal and nonverbal behaviours from a human user including acoustic features, such as hesitations or loudness, as well as body movements, such as head nods and posture shifts, and responded to it by providing nonverbal listening feedback. The system does not seem to employ a fusion engine, but rather responds to cues conveyed in a particular modality, such as a head nod, directly. An evaluation of the system revealed that the responsive agent was more successful in creating rapport with the human user than the non-responsive agent.

Typically, most experimental studies investigating the potential of social signal processing in human-agent dialogues have been performed offline, i.e. after the interaction between the human and the agent. Such an approach may, however, be problematic because the participants of an experiment might have forgotten what they experienced at a particular instance of time. As an alternative, Gilroy et al. (2011) present an evaluation approach which compares the results of the fusion process with the users' physiological response during the interaction with the system.

4.7 Social signal fusion in human-agent interaction

Starting the recent years, various attempts have been made to explore the potential of social signal processing in human interaction with embodied conversational agents and social robots.

Sanghvi et al. (2011) analyzed body postures and gestures as an indicator of the emotional engagement of children playing chess with the iCat robot. They came to the conclusion that the accuracy of the detection methods was high enough to integrate the approach into an affect recognition system for a game companion. Even though the approach above addressed an attractive scenario for multimodal social signal fusion, it was only tested in offline mode. An integration of the approach into an interactive human-robot system scenario did not take place.

Increasing effort has been made on the multimodal analysis of verbal and nonverbal backchannel behaviors during the interaction with a virtual agent. An example includes the previously mentioned artificial listener by Gratch et al. (2007) that aims to create rapport with a human interlocutor through simple contingent nonverbal behaviors. A more recent example is the virtual health care provider recently presented by Scherer et al. (2012). This agent is able to detect and respond multimodal behaviors related to stress and post-traumatic stress disorder. For example, when the patient pauses a lot in the conversation, the agent tries to encourage her to continue speaking. Even though both systems are able to analyze multiple modalities, they do not seem to employ a fusion engine, but rather directly respond to cues conveyed in a particular modality, such as a head nod.

An exemplary application that is based on a fusion algorithm adapted to the specific needs of online processing is the Affective Listener "Alfred" developed by Wagner et al. (2011b). Alfred is a butler-like virtual character that is aware of the user and reacts to his or her affective expressions. The user interacts with Alfred via acoustics of speech and facial expressions (see Figure 2). As a response, Alfred simply mirrors the user's emotional state by appropriate facial expressions. This behavior can be interpreted as a simple form of showing empathy.

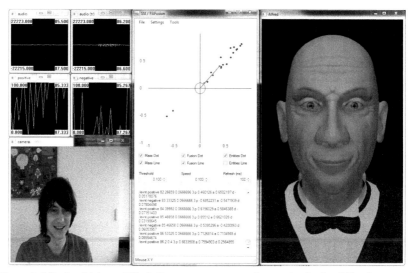

Figure 2. Affective Listener Alfred: the current user state is perceived using SSI, a framework for social signal interpretation (Wagner et al., 2011b) framework (upper left window); observed cues are mapped onto the valence and arousal dimensions of a 2D emotion model (upper middle window); values for arousal and valence are combined to a final decision and transformed to a set of Facial Animation Coding System (FACS) parameters, which are visualized by the virtual character Alfred (right window).

(Color image of this figure appears in the color plate section at the end of the book.)

The fusion approach is inspired by that developed for the Augmented Reality Tree. However, while Gilroy et al. (2011) generate one vector per modality, Wagner et al. (2011b) generate one vector for each detected event. This way they prevent sudden leaps in case of a false detection. Since the strength of a vector decreases with time, the influence of older events is lessened until the value falls under a certain threshold and is completely removed.

5. Exploiting Social Signals for Semantic Interpretation

Few systems combine semantic multimodal fusion for task-based command interpretation and multimodal fusion of social signals. A few studies nevertheless mention some interaction between the two communication streams. Such combinations occur in users' behaviors. For example, a user may say "Thanks" to a virtual agent and at the same time start a new command using gesture (Martin et al., 2006). In another study about multimodal behaviors of users when interacting with a virtual character embedded in a 3D graphical environment, such concurrent behaviors were also observed. In such cases, speech input was preferred for social communication with the virtual character ("how old are you?"), whereas 2D gesture input was used in parallel for task command (e.g. to get some information about one of the graphical object displayed in the environment) (Martin et al., 2006).

An example of a system managing these two streams includes the SmartKom system, which features adaptive confidence measures. While the user is speaking (possibly for task commands), the confidence value of the mouth area recognizer is decreased for the module that detected emotions expressed in user's facial expression (Wahlster, 2003). The SmartKom system thus uses a mixture of early fusion for analyzing emotions from facial expressions and speech and late fusion for analyzing the semantics of utterances.

Rich et al. (2010) presented a model of engagement for human-robot interaction that took into account direct gaze, mutual gaze, relevant next contribution and back channel behaviors as an indicator of engagement in a dialogue. Interestingly, the approach was used for modeling the behavior of both the robot and the human. As a consequence, it was able to explain failures in communication from the perspective of both interlocutors. Their model demonstrates the close interaction between the communication streams required for semantic processing and social signal processing because it integrates multimodal grounding with techniques for measuring experiential qualities of a dialogue. If communication partners fail to establish a common understanding of what a dialogue is about, it is very likely that they will lose interest in continuing the interaction.

Bosma and André (2004) presented an approach to the joint interpretation of emotional input and natural language utterances. Especially short utterances tend to be highly ambiguous when solely the linguistic data is considered. An utterance like "right" may be interpreted as a confirmation as well as a rejection, if intended cynically, and so may the absence of an utterance. To integrate the meanings of the users' spoken input and their emotional state, Bosma and André combined a Bayesian network to recognize the user's emotional state from physiological data, such as heart rate, with weighted finite-state machines to recognize dialogue acts from the user's speech. The finite-state machine approach was similar to that presented by Bangalore and Johnson (2009). However, while Bangalore and Johnston used finite-state machines to analyze the propositional content of dialogue acts, Bosma and André focused on the speaker's intentions. Their objective was to discriminate a proposal from a directive, an acceptance from a rejection, etc., as opposed to Bangalore and Johnston who aimed at parsing user commands that are distributed over multiple modalities, each of the modalities conveying partial information. That is, Bosma and André did not expect the physiological modality to contribute to the propositional interpretation of an utterance. Instead, the emotional input was used to estimate the probabilities of dialogue acts, which were represented by weights in the finite-state machines.

Another approach that fuses emotional states with natural language dialogue acts has been presented by Crook et al. (2012) who integrated a system to recognize emotions from speech developed by Vogt et al. (2008) into a natural language dialogue system order to improve the robustness of a speech recognizer. Their system fuses emotional states recognized from the acoustics of speech with sentiments extracted from the transcript of speech. For example, when the users employ words to express their emotional state that are not included in the dictionary, the system would still be able to recognize their emotions from the acoustics of speech.

6. Conclusion and Future Work

In this chapter, we discussed approaches to fuse semantic information in dialogue systems as well as approaches to fuse social and emotional cues. While the fusion of semantic information has been strongly influenced by research done in the natural language community, the fusion of social signals has heavily relied on techniques from the multimedia community. Recently, the use of virtual agents and robots in dialogue systems has led to stronger interactions between the two areas of research. The use of social signal processing in dialogue systems may not only improve the quality of interaction, but also increase their robustness. Vice versa research in social signal processing may profit from techniques developed for semantic fusion. Most systems that integrate mechanisms for the multimodal fusion of social signals in human-agent dialogue only consider a supplementary use of multiple signals. That is, the system responds to each cue individually, but does not attempt to resolve ambiguities by considering additional modalities. One difficulty lies in the fact that data have to be integrated in an incremental fashion while mechanisms for social signal fusion usually start from global statistics over longer segments. A promising avenue for future research might be to research to what extent techniques from semantic fusion might be included to exploit the complementary use of social signals. Among other things, this implies a departure from the fusion of low-level features in favor of higher level social cues, such as head nods or laughters.

Acknowledgement

The work described in this chapter is partially funded by the EU under research grants CEEDS (FP7-ICT2009-5), TARDIS (FP7-ICT-2011-7) and ILHAIRE (FP7-ICT-2009-C).

REFERENCES

Bangalore, S. and M. Johnston. 2009. Robust Understanding in Multimodal Interfaces. *Computational Linguistics*, **35(3)**:345–397.

Bolt, Richard A. 1980. Put-that-there: Voice and Gesture at the Graphics Interface. Proceedings of the 7'th Annual Conference on Computer Graphics and Interactive Techniques, SIGGRAPH '80, pp. 262–270. ACM, New York, NY.

Bosma, W. and E. André. 2004. Exploiting Emotions to Disambiguate Dialogue Acts. Proceedings of the 9th International Conference on Intelligent User Interfaces, IUI '04, pp. 85–92. ACM, New York, NY.

Burger, B., I. Ferrané, F. Lerasle and G. Infantes. 2011. Two-handed gesture recognition and fusion with speech to command a robot. *Autonomous Robots*, **32(2)**:129–147.

Busso, C., Z. Deng, S. Yildirim, M. Bulut, C.M. Lee, A. Kazemzadeh, S. Lee, U. Neumann and S. Narayanan. 2004. Analysis of emotion recognition using facial expressions, speech and multimodal information. International Conference on Multimodal Interfaces (ICMI 2004), pp. 205–211.

Caridakis, G., G. Castellano, L. Kessous, A. Raouzaiou, L. Malatesta, S. Asteriadis and K. Karpouzis. 2007. Multimodal emotion recognition from expressive faces, body gestures and speech. In Artificial Intelligence and Innovations (AIAI 2007), pp. 375–388.

Chen, F., N. Ruiz, E. Choi, J. Epps, A. Khawaja, R. Taib and Y. Wang. 2012. Multimodal Behaviour and Interaction as Indicators of Cognitive Load. ACM Transactions on Interactive Intelligent Systems, Volume 2, Issue 4, Article No. 22, pp. 1-36.

Crook, N., D. Field, C. Smith, S. Harding, S. Pulman, M. Cavazza, D. Charlton, R. Moore and J. Boye. 2012. Generating context-sensitive ECA responses to user barge-in interruptions. *Journal on Multimodal User Interfaces*, **6**:13–25.

Dimitriadis, D.B. and J. Schroeter. 2011. Living rooms getting smarter with multimodal and multichannel signal processing. IEEE SLTC newsletter. Summer 2011 edition, http://www.signalprocessingsociety.org/technical-committees/list/sl-tc/spl-nl/2011-07/living-room-of-the-future/

D'Mello, S.K. and J. Kory. 2012. Consistent but modest: A meta-analysis on unimodal and multimodal affect detection accuracies from 30 studies. International Conference on Multimodal Interaction (ICMI 2012), pp. 31–38.

Eyben, F., M. Wöllmer, M.F. Valstar, H. Gunes, B. Schuller and M. Pantic. 2011. String-based audiovisual fusion of behavioural events for the assessment of dimensional affect. Automatic Face and Gesture Recognition (FG 2011), pp. 322–329.

Gilroy, S.W., M. Cavazza and V. Vervondel. 2011. Evaluating multimodal affective fusion using physiological signals. Intelligent User Interfaces (IUI 2011), pp. 53–62.

Gilroy, S.W., M. Cavazza, R. Chaignon, S.-M. Mäkelä, M. Niranen, E. André, T. Vogt, J. Urbain, M. Billinghurst, H. Seichter and M. Benayoun. 2008. E-tree: Emotionally driven augmented reality art. *ACM Multimedia* pp. 945–948.

Gratch, G., N. Wang, J.Gerten, E. Fast and R. Duffy. 2007. Creating Rapport with Virtual Agents. Intelligent Virtual Agents (IVA 2007), pp. 125–138.

Gruenstein, A., J. Orszulak, S. Liu, S. Roberts, J. Zabel, B. Reimer, B. Mehler, S. Seneff, J.R. Glass and J.F. Coughlin. 2009. City browser: Developing a conversational automotive HMI. In Jr., Dan R. Olsen, Richard B. Arthur, Ken Hinckley, Meredith Ringel Morris, Scott E. Hudson and Saul Greenberg (eds.), Proceedings of the 27th International Conference on Human Factors in Computing Systems, CHI 2009, Extended Abstracts Volume, Boston, MA, April 4–9, pp. 4291–4296. ACM.

Hofs, D., M. Theune and R. den Akker. 2010. Natural interaction with a virtual guide in a virtual environment: A multimodal dialogue system. *Multimodal User Interfaces*, **3**:141–153.

Johnston, M. 1998. Unification-based Multimodal Parsing. In Proceedings of the International Conference on Computational Linguistics and the 36th Annual Meeting of the Association for Computational Linguistics (Coling-ACL), Montreal, Canada, pp. 624–630.

Johnston, M. 2009. Building multimodal applications with EMMA. In Crowley, James L., Yuri A. Ivanov, Christopher Richard Wren, Daniel Gatica-Perez, Michael Johnston, and Rainer Stiefelhagen (eds.), Proceedings of the 11th International Conference on Multimodal Interfaces, ICMI 2009, Cambridge, Massachusetts, USA, November 2–4, 2009, pp. 47–54. ACM.

Johnston, M., G. Di Fabbrizio and S. Urbanek. 2011. mTalk: A multimodal browser for mobile services. In INTERSPEECH 2011, 12th Annual Conference of the International Speech Communication Association, Florence, Italy, August 27–31, pp. 3261–3264. ISCA.

Kaiser, E., A. Olwal, D. McGee, H. Benko, A. Corradini, X. Li, P. Cohen and S. Feiner. 2003. Mutual Disambiguation of 3D multimodal interaction in augmented and virtual reality. In Proceedings of the 5th International Conference on Multimodal Interfaces, ICMI '03, pp. 12–19. ACM, New York, NY, USA.

Kim, J., E. André, M. Rehm, T. Vogt and J. Wagner. 2005. Integrating information from speech and physiological signals to achieve emotional sensitivity. INTERSPEECH 2005, pp. 809–812.

Kim, J. and E. André. 2008. Emotion recognition based on physiological changes in music listening. *IEEE Transactions on Pattern Analysis and Machine Intelligence*, **30(12):** 2067–2083.

Kleinsmith, A., N. Bianchi-Berthouze and A. Anthony Steed. 2011. Automatic Recognition of Non-Acted Affective Postures. *IEEE Transactions on Systems, Man and Cybernetics, Part B* **41(4):**1027–1038.

Lalanne, D., L. Nigay, P. Palanque, P. Robinson, J. Vanderdonckt and J.-F. Ladry. 2008. Fusion engines for multimodal input: A survey. In Proceedings of the 10th International Conference on Multimodal Interfaces ICMI 2008, pp. 153–160.

Latoschik, M.E. 2002. Designing transition networks for multimodal vr-interactions using a markup language. In Proceedings of ICMI' 02, pp. 411–416.

Lingenfelser, F., J. Wagner and E. André. 2011. A systematic discussion of fusion techniques for multi-modal affect recognition tasks. In Proceedings of the 13th International Conference on Multimodal Interfaces ICMI 2011, pp. 19–26.

Martin, J.C., R. Veldman and D. Béroule. 1998. Developing multimodal interfaces: A theoretical framework and guided propagation networks. In H. Burt, R.J. Beun and T. Borghuis (eds.), Multimodal Human-Computer Communication (Vol. 1374, pp. 158–187). Berlin: Springer Verlag.

Martin, J.-C., S. Buisine, G. Pitel and N.O. Bernsen. 2006. Fusion of Children's Speech and 2D Gestures when Conversing with 3D Characters. *Journal of Signal Processing. Special issue on Multimodal Human-computer Interfaces,* **86(12):**3596–3624.

Martinovsky, B. and D. Traum. 2003. Breakdown in Human-Machine Interaction: The Error is the Clue. Proceedings of the ISCA Tutorial and Research Workshop on Error Handling in Dialogue Systems, pp. 11–16.

Mehlmann, G. and E. André. 2012. Modeling multimodal integration with event logic charts. In Proceedings of the 14th ACM International Conference on Multimodal Interfaces, ICMI 2012, Santa Monica, USA, October 22–26, pp. 125–132.

Mehrabian, A. 1980. Basic Dimensions for a General Psychological Theory: Implications for Personality, Social, Environmental, and Developmental Studies. Cambridge, MA: Oelgeschlager, Gunn & Hain.

Oviatt, S.L. 1999. Mutual disambiguation of recognition errors in a multimodel architecture. In Williams, Marian G. and Mark W. Altom (eds.), Proceeding of the CHI '99 Conference on Human Factors in Computing Systems: The CHI is the Limit, Pittsburgh, PA, USA, May 15–20, pp. 576–583. ACM.

Rich, C., B. Ponsleur, A. Holroyd and C.L. Sidner. 2010. Recognizing engagement in human-robot interaction. Human-Robot Interaction, (HRI 2010), pp. 375–382.

Sandbach, G., S. Zafeiriou, M. Pantic and L. Yin. 2012. Static and dynamic 3D facial expression recognition: A comprehensive survey. *Image Vision Comput.*, **30(10):**683–697.

Sanghvi, J., G. Castellano, I. Leite, A. Pereira, P.W. McOwan and A. Paiva. 2011. Automatic analysis of affective postures and body motion to detect engagement with a game companion. Human Robot Interaction (HRI 2011), pp. 305–312. Scherer, S., Marsella, S., Stratou, G., Xu, Y., Morbini, F., Egan, A., Rizzo, A.A. and Morency, L.P. (2012). Perception markup language: Towards a standardized representation of perceived nonverbal behaviors. Intelligent Virtual Agents (IVA 2012), pp. 455–463.

Scherer, S., S. Marsella, G. Stratou, Y. Xu, F. Morbini, A. Egan, A. Rizzo and L.-P. Morency. 2012. Perception Markup Language: Towards a Standardized Representation of Perceived Nonverbal Behaviors. Intelligent Virtual Agents. Y. Nakano, M. Neff, A. Paiva and M. Walker, Springer-Verlag Berlin Heidelberg, **7502:**455–463

Sowa, T., M. Latoschik and S. Kopp. 2001. A communicative mediator in a virtual environment: Processing of multimodal input and output. Proc.

of the International Workshop on Multimodal Presentation and Natural Multimodal Dialogue—IPNMD 2001. Verona, Italy, ITC/IRST, pp. 71–74.

Stiefelhagen, R., H. Ekenel, C. Fügen, P. Gieselmann, H. Holzapfel, F. Kraft, and A. Waibel. 2007. Enabling multimodal human-robot interaction for the Karlsruhe humanoid robot. *IEEE Transactions on Robotics, Special Issue on Human-Robot Interaction*, **23(5)**:840–851.

Sun, Y., Y. Shi, F. Chen and V. Chung. 2009. Skipping spare information in multimodal inputs during multimodal input fusion. Proceedings of the 14th International Conference on Intelligent User Interfaces, pp. 451–456, Sanibel Island, USA.

Sun, Y., H. Prendinger, Y. Shi, F. Chen, V. Chung and M. Ishizuka. 2008. THE HINGE between Input and Output: Understanding the Multimodal Input Fusion Results in an Agent-Based Multimodal Presentation System. CHI '08 Extended Abstracts on Human Factors in Computing Systems, pp. 3483–3488, Florence, Italy.

Visser, T., D. Traum, D. DeVault and R. op den Akker. 2012. Toward a model for incremental grounding in spoken dialogue systems. In the 12th International Conference on Intelligent Virtual Agents.

Vogt, T. and E. André. 2005. Comparing feature sets for acted and spontaneous speech in view of automatic emotion recognition. Proceedings of the 2005 *IEEE International Conference on Multimedia and Expo*, ICME 2005, July 6–9, 2005, pp. 474–477. Amsterdam, The Netherlands.

Vogt, T., E. André and N. Bee. 2008. Emovoice—A framework for online recognition of emotions from voice. In André, Elisabeth, Laila Dybkjær, Wolfgang Minker, Heiko Neumann, Roberto Pieraccini, and Michael Weber (eds.), Perception in Multimodal Dialogue Systems, 4th IEEE Tutorial and Research Workshop on Perception and Interactive Technologies for Speech-Based Systems, PIT 2008, Kloster Irsee, Germany, June 16–18, 2008, Proceedings, volume 5078 of Lecture Notes in Computer Science, pp. 188–199. Springer.

Wagner, J., E. André, F. Lingenfelser and J. Kim. 2011a. Exploring fusion methods for multimodal emotion recognition with missing data. *T. Affective Computing*, **2(4)**:206–218.

Wagner, J., F. Lingenfelser, N. Bee and E. André. 2011b. Social signal interpretation (SSI)—A framework for real-time sensing of affective and social signals. *KI*, **25(3)**:251–256.

Wahlster, W. 2003. Towards symmetric multimodality: Fusion and fission of speech, gesture, and facial expression. In Günter, Andreas, Rudolf Kruse, and Bernd Neumann (eds.), KI 2003: Advances in Artificial Intelligence, 26th Annual German Conference on AI, KI 2003, Hamburg, Germany, September 15–18, 2003, Proceedings, volume 2821 of Lecture Notes in Computer Science, pp. 1–18. Springer.

Index

Color Plate Section

Chapter 1

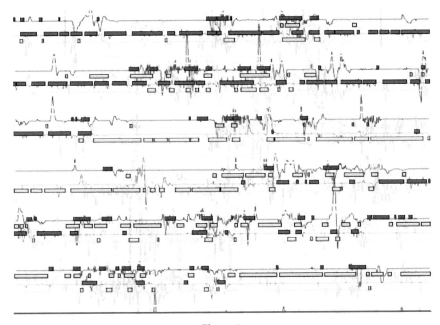

Figure 1.

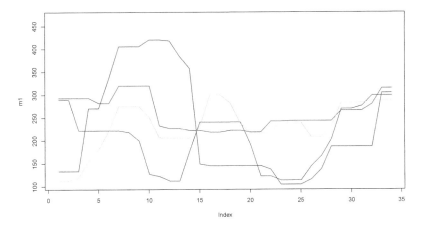

Figure 2.

Chapter 2

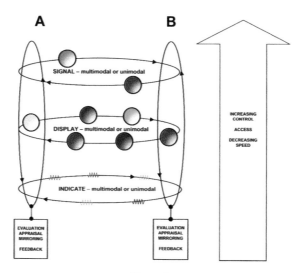

Figure 2.

Chapter 4

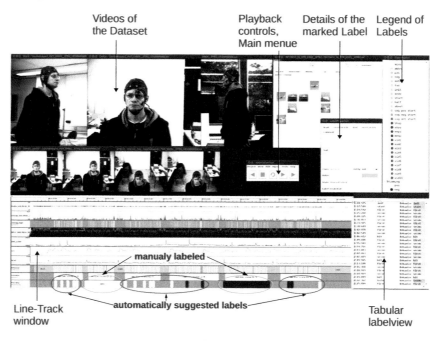

Figure 3.

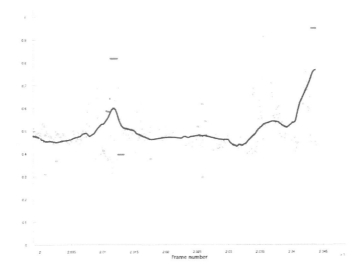

Figure 4.

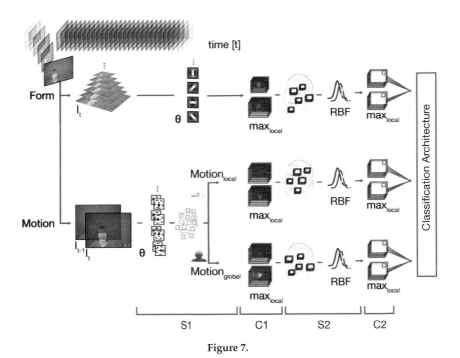

Figure 7.

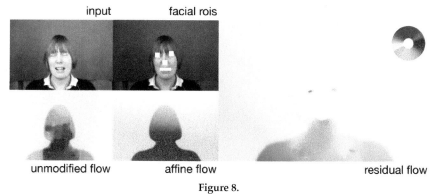

Figure 8.

Chapter 11

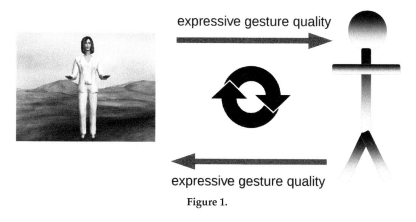

Figure 1.

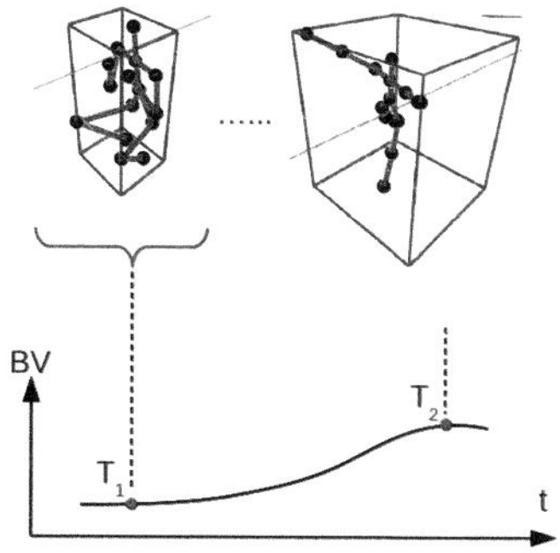

Figure 4.

Chapter 13

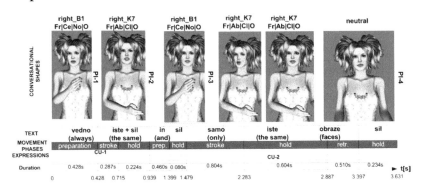

Figure 11.

Chapter 14

Figure 1.

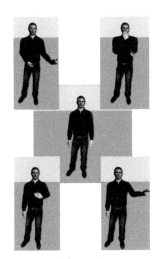

Figure 2.

Figure 3.

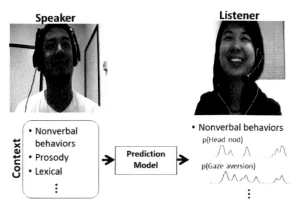

Figure 5.

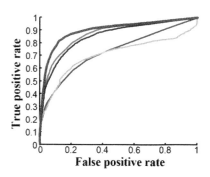

Figure 7.

Chapter 15

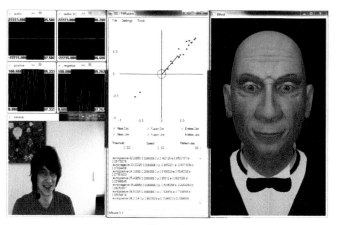

Figure 2.

T - #0195 - 251019 - C434 - 234/156/19 - PB - 9780367379292